Weed Biology

Weed Biology

Edited by **Jordan Smith**

SYRAWOOD
PUBLISHING HOUSE

New York

Published by Syrawood Publishing House,
750 Third Avenue, 9th Floor,
New York, NY 10017, USA
www.syrawoodpublishinghouse.com

Weed Biology
Edited by Jordan Smith

International Standard Book Number: 978-1-68286-138-7 (Hardback)

The publisher's policy is to use permanent paper from mills that operate a sustainable forestry policy. Furthermore, the publisher ensures that the text paper and cover boards used have met acceptable environmental accreditation standards.

Trademark Notice: Registered trademark of products or corporate names are used only for explanation and identification without intent to infringe.

Printed in the United States of America.

Contents

Permissions

List of Contributors

Preface

Weeds play a significant role in determining agricultural diversity. Weed biology as a specialised discipline aims to study the physiology, reproduction and life-cycle of weeds. This book provides a comprehensive overview of important topics of weed biology like weed taxonomy, weed management and control techniques, use of weeds, etc. Researches and case-studies by eminent experts and scientists are included in the book to highlight the present developments and responses to weed management. Students, researchers, agricultural scientists will find this book full of innovative insights and valuable knowledge.

This book is a result of research of several months to collate the most relevant data in the field.

When I was approached with the idea of this book and the proposal to edit it, I was overwhelmed. It gave me an opportunity to reach out to all those who share a common interest with me in this field. I had 3 main parameters for editing this text:

1. Accuracy – The data and information provided in this book should be up-to-date and valuable to the readers.
2. Structure – The data must be presented in a structured format for easy understanding and better grasping of the readers
3. Universal Approach – This book not only targets students but also experts and innovators in the field, thus my aim was to present topics which are of use to all

Thus, it took me a couple of months to finish the editing of this book.

I would like to make a special mention of my publisher who considered me worthy of this opportunity and also supported me throughout the editing process. I would also like to thank the editing team at the back-end who extended their help whenever required.

Editor

High sourgrass threshold interfere on chick-peas development in tropical conditions

Carita Liberato do Amaral[1]*, Marcelo Claro de Souza[4], Guilherme Bacarim Pavan[2], Marina Alves Gavassi[2] and Pedro Luis da Costa Aguiar Alves[3]

[1]Programa de Pós-Graduação em Agronomia (Produção Vegetal) – Departamento de Biologia Aplicada à Agropecuária, FCAV-UNESP, Jaboticabal-SP, Brazil.
[2]Bolsista de Iniciação Científica - Departamento de Biologia Aplicada à Agropecuária, FCAV-UNESP, Jaboticabal-SP, Brazil.
[3]FCAV-UNESP - Departamento de Biologia Aplicada à Agropecuária, Jaboticabal-SP, Brazil.
[4]Programa de Pós-Graduação em Biologia Vegetal – Departamento de Botânica, IB-UNESP, Rio Claro-SP, Brazil.

The grains of *Cicer arietinum* (chick-peas) is one of the richest sources of proteins among worldwide grain crops. However, 100% of chick-peas grains consumed in Brazil are imported. Aiming to contribute with agricultural technologies to yield this crop in Brazil, we studied the interference of sourgrass thresholds at the initial development of chick-peas in tropical conditions. We evaluated the competition effects of seven sourgrass densities on chick-peas crops from 15 to 45 days after sowing. These effects were evaluated according to interferences on chick-peas height, number of leaves and offshoots, stem diameter, leaf area and dry mass. We observed low interferences with up to 8 sourgrass plants m^{-2} and several interferences from 16 to 128 plants m^{-2} justifying the control of sourgrass at the initial development of chick-peas.

Key words: *Cicer arietinum*, *Digitaria insularis*, competition, interference, weed.

INTRODUCTION

The chick-peas (*Cicer arietinum* L.), belongs to the Fabaceae family, is an annual crop, diploid (2n=16), autogamous with complete pollination before opening (Maiti and Wesche-Ebeling, 2001; Biçer and Sakar,. 2010) rich in proteins, vitamins and minerals (El-Adawy, 2002), being considered the best source of proteins among legume crops around the world (Ferreira et al., 2006). Nowadays we import 100% of the grains of chick-peas consumed in Brazil increasing, the final market prices (FAO, 2011). If irrigated, this crop could be cultivated in São Paulo state, Brazil, as a potential autumn/winter crop (Valim and Batistuti, 2000), verifying the need of studies for the feasibility of its commercial production in Brazil.

This crop has low initial development which can be responsible for difficulties on its development in cohabitation with weeds (Maiti and Wesche-Ebeling, 2001), mainly at the initial stages (Chaudhary and Hussain, 2011). Weeds causes quantitative and qualitative losses to agricultural products, limiting the crop yield by direct competition for nutrients, light and water (Liu et al., 2009), affecting the development of crops, reducing their growth and production of grains (Pitelli, 1985). The degree of weed interference in crops is determined by crop strategies (variety, spacing and density), weed community

---

*Corresponding author. E-mail: caritaliberato@gmail.com.

(species, density and distribution), abiotic factors, edaphic conditions and period that weeds coexist with crop (Lemes et al., 2010).

The competitive threshold has been defined as the weed density above which crop yield is reduced beyond an acceptable amount (Oliver, 1988). Not only density, but also the physical position that one given species occupies in a plant community and its function determines its ecological niche (Lamego et al., 2005). In this way, the greater the overlap of species, more intense the competition for environmental resources can be (Radosevich and Holt, 1984), changing the morphology of the plants, affecting the weed-crop competition relationship (Lamego et al., 2005) mainly at the initial vegetative period.

The sourgrass [*Digitaria insularis* (L.) Mez ex Ekman] is a herbaceous perennial species that can reproduce by seed and/or rhizomes (Kissmann and Groth, 1997), and it can be considered as one of the most important weeds infesting perennial and annual crops in Brazil (Carvalho et al., 2011). In addition, the sourgrass seeds have high percentage of germination enabling this species to increase rapidly in number (Correia and Durigan, 2009). Actually there are observations of resistant-biotypes to the herbicide glyfosate in soybean and maize crops as well as in citrus and coffee orchards (Timossi, 2009). In this way, sourgrass may be a problem, in the future, in chick-peas crops mainly at the initial development of the culture.

In order to better understand the competition between chick-peas and sourgrass in Tropical conditions we evaluated the interference of sourgrass thresholds at the initial chick-peas development (height, diameter of stem, leaf number and area, dry mass and number of offshoots). Based on the assumption that sourgrass is one of the most aggressive weeds in Brazilian fields, we hypothesized this weed could be a potential problem in chick-peas crops reducing their development in different degrees depending on its threshold.

MATERIALS AND METHODS

Site description

The experiments were conducted in an open field at the São Paulo State University, Jaboticabal municipality, Brazil, using cement boxes (50 × 50 × 25 cm) filled with red latosol. According to Köppen (1948) climate classification system, the Jaboticabal-SP region can be described as Cwa with a dry season from April to September and a wet from October to May.

Chemical soil characteristics (Table 1) were determined according to Brazilian procedures (EMBRAPA, 1997). We determined pH, organic matter (OM), P, K, Ca, Mg, H + Al, base saturation (BS = K + Ca + Mg), cation exchange capacity (CEC = K + Ca + Mg + H + Al) and fertility rate [V% = 100(K + Ca + Mg)CEC^{-1}]. Based on the chemical soil analysis, we performed the correction of soil fertility before sowing the plants (Van Raij et al., 1997). Topdressing was done 20 days after sowing (DAS).

Experimental design

A randomized complete block design with four replicates was used for this experiment. We performed the evaluations at 15, 30 and 45 days after the beginning of the competition period between chick-peas and sourgrass. We considered the beginning of competition period immediately after planting the sourgrass seedlings together with the chick-peas crop.

Plant samples

The seeds of sourgrass were sown in trays with horticultural substrate Plantimax HT$^{®}$. Thirty days after the emergence of the sourgrass seedlings, two seeds of chick-peas were sown in the center of each box along with the transplanting of seedlings of sourgrass, when they were approximately 3 to 4 cm shoot long. The seedlings of sourgrass were planted into seven densities (threshold) per box, being the treatment C used as a control group (C, T1, T2, T3, T4, T5, T6 corresponding to 0, 4, 8, 16, 32, 64 and 128 sourgrass plants m^{-2} respectively). When the seedlings of chick-peas reached about 7 cm shoot long was performed, the thinning to one plant of chick-peas per box. Daily, the plants were wet according to necessity.

At 15 and 30 DAS after the beginning of competition period, we evaluated the chick-peas height, stem diameter, number of leaves and offshoots. Forty-five days after the competition period started, we evaluated the same parameters at 15 and 30 DAS plus chick-peas leaf area and dry mass. The biometric evaluations were done by a graduate scale and the leaf area was measured with a Li-cor LI-3000A leaf area mater. The chick-peas plants were dried in an oven (60°C) up to constant mass for dry mass determination.

Statistical analysis

For all parameters evaluated among 15 and 45 DAS, we performed the one-way ANOVA comparing the averages by Tukey test at 5%.

RESULTS

The competition between chick-peas and sourgrass interfered negatively at the development of chick-peas. At 15 DAS, we observed significant reduction on chick-peas height, 16.8% for T5 and 15.2% for T6 treatments in relation to C, which grew up weed free. This interference was clearer from 30 to 45 DAS, when we observed several reductions on chick-peas height from T3 to T6 (Figure 1). At 30 DAS, we observed reductions between 21.6 and 25.8% and at 45 DAS the reductions were between 21.2 to 22.7%.

We also observed several interferences at the development of chick-peas stem diameter. At 15 DAS, all treatments (from T1 to T6) differed significantly in relation to control group C, reducing the diameter of stems from 17.1 to 22.5%. Indeed, we observed this interference reduced from 30 to 45 DAS, being more expressive in T2 for 30 DAS and T3 for 45 DAS (Figure 2). For 30 DAS, we observed reductions in the order of 11.7 and 25.9% and at 45 DAS we observed reductions between 25.2 and 28.0%.

Table 1. Macronutrient contents and fertility parameters, Jaboticabal – SP, Brazil.

pH	OM	P	K	Ca	Mg	H+AL	BS	CEC	V
CaCl$_2$	g dm^{-3}	mg dm^{-3}			mMol$_c$ dm^{-3}				%
5.2	12.0	35.0	1.7	19.0	8.0	25.0	28.7	53.7	53.0

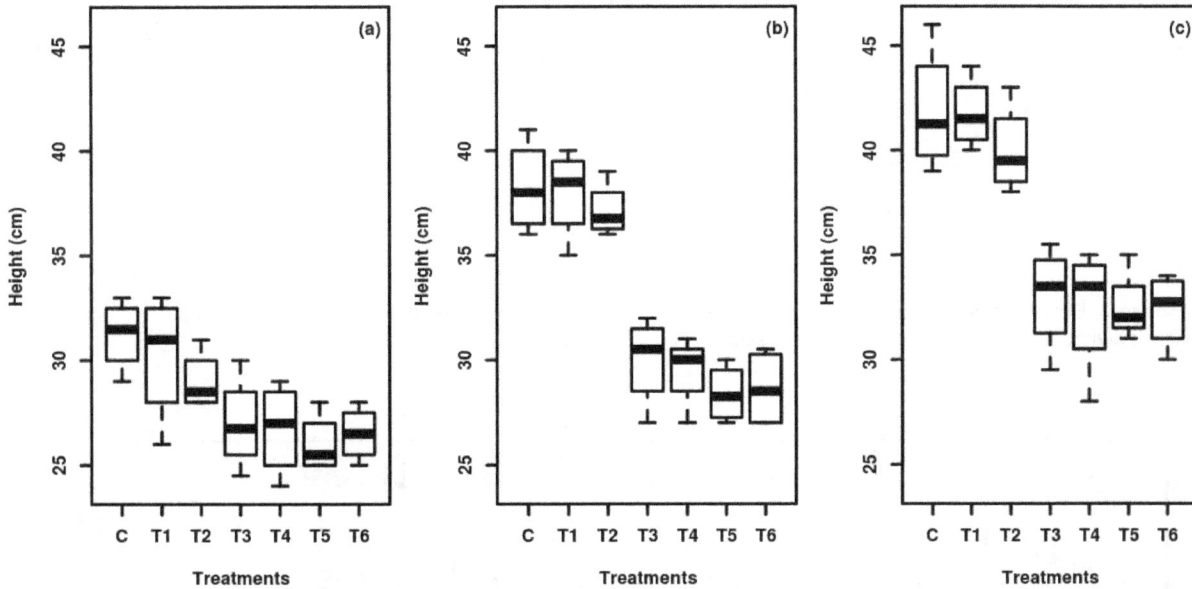

Figure 1. Boxplot of sourgrass plant densities effects on chickpeas height at 15 (a), 30 (b) and 45 (c) days after sowing. The line in the middle of each box indicates the 50th percentile of the observed distribution data; the pot and bottom parts of each box represent the 25th and 75th percentiles of the observed distribution data, respectively; the bottom and top error bars of each box are the 5th and the 95th percentiles of the observed distribution data.

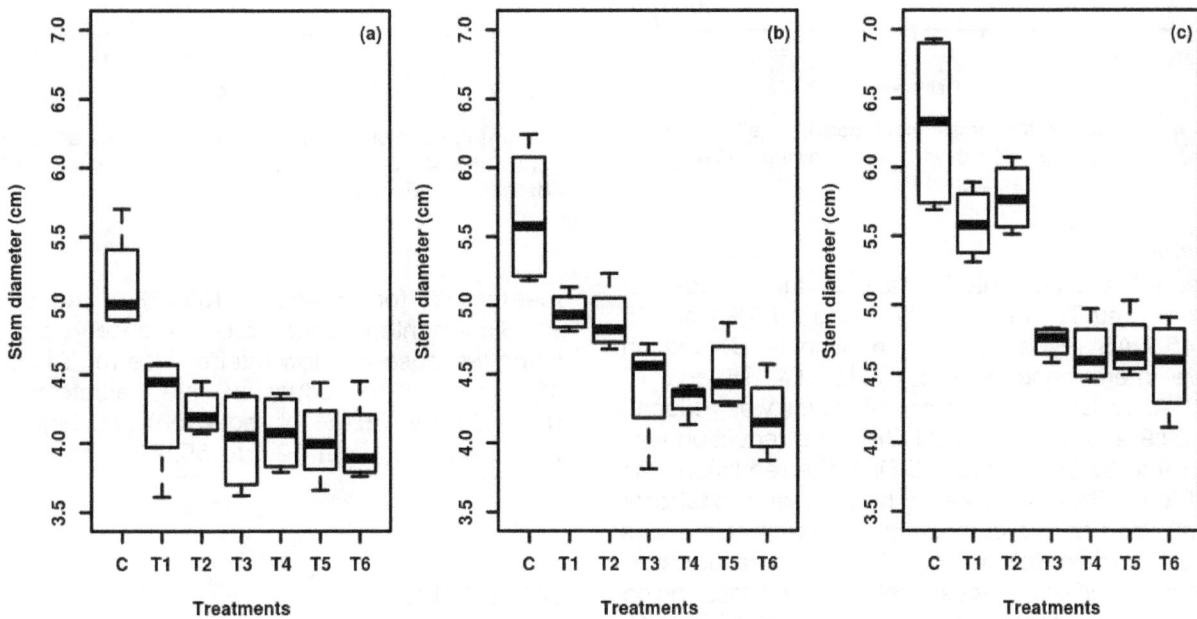

Figure 2. Boxplot of sourgrass plant densities effects on chickpeas stem diameter at 15 (a), 30 (b) and 45 (c) days after sowing. Boxplot characteristics are as described in Figure 1.

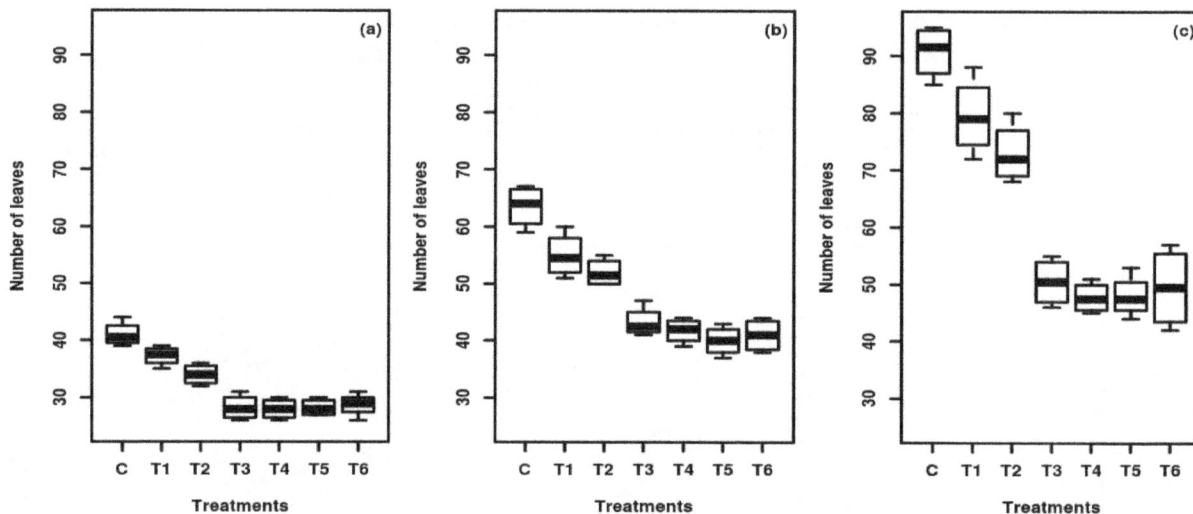

Figure 3. Boxplot of sourgrass plant densities effects on chickpeas number of leaves at 15 (a), 30 (b) and 45 (c) days after sowing. Boxplot characteristics are as described in Figure 1.

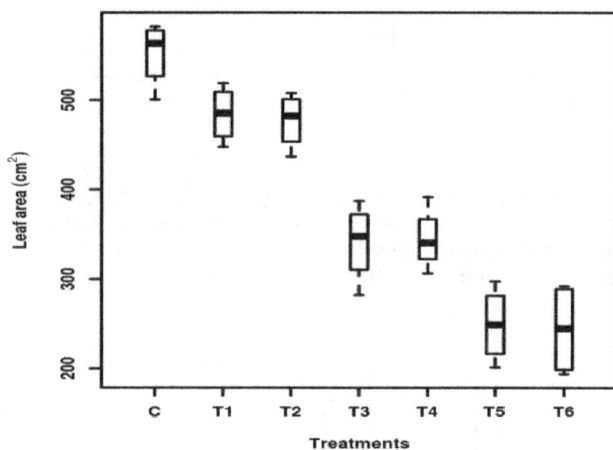

Figure 4. Boxplot of sourgrass plant densities effects on chickpeas leaf area 45 days after sowing. Boxplot characteristics are as described in Figure 1.

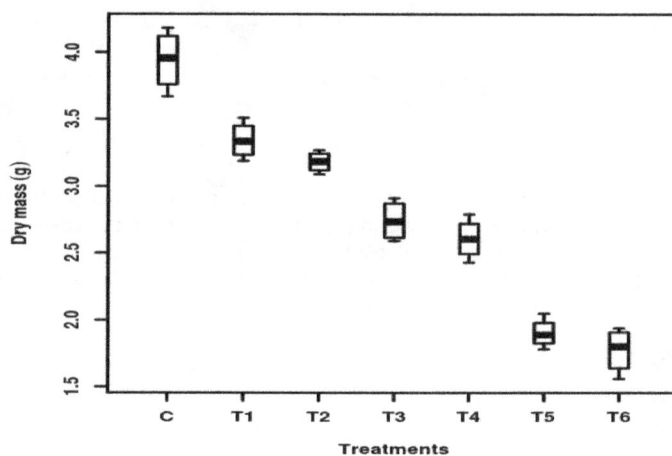

Figure 5. Boxplot of sourgrass plant densities effects on chickpeas dry mass 45 days after sowing. Boxplot characteristics are as described in Figure 1.

We observed a small interference at the number of leaves at T2 and T3 (between 17.0 and 19.6% from 15 DAS to 45 DAS respectively) in relation to C, and an aggressive interference caused by T3, T4, T5 and T6 (Figure 3). At 15 DAS, the number of leaves was reduced between 29.9 and 31.7%; for 30 DAS, the reduction was among 31.9 to 37.0% and for 45 DAS the reduction was from 44.4 to 47.7% respectively. The number of offshoots was not affected by the competition between chick-peas and sourgrass (data not shown). The leaf area of chick-peas suffered different levels of interference, being possible to observe three different levels in relation to C: no interference for the treatments T1 and T2, intermediate for T3 and T4 (38.0% of reduction) and high

interference for T5 and T6 (55.0% of reduction) (Figure 4). Similar interference could be observed for dry mass when we observed low interference for T1 and T2 (from 16.0 to 19.0% of reduction), intermediate for T3 and T4 (from 30.0 to 39.0% of reduction) and high interference for T5 and T6 (from 52.0 to 55.0% of reduction) (Figure 5).

DISCUSSION

Agricultural producers cannot tolerate excessive yield losses from weeds, being weed control necessary when the costs of control are smaller than the losses caused by competition (Oliver, 1988). To determine the relation costs/losses it does necessary a have a better understand about the thresholds of weeds on the culture. The increase in weeds density in an agricultural system can affect the quantity and quality of available resources for crops, affecting their development (Ballare and Casal, 2000).

It is easy to find several papers talking about sourgrass chemical control (Melo et al., 2012; Correia et al., 2010) and sourgrass glyphosate resistance (Carvalho et al., 2012, 2011), but we lack information about the interferences caused by this weed on crop development and crop losses caused by the competition between sourgrass and crops. In our study we clearly observed that high densities of sourgrass interfered negatively at the initial development of chick-peas reducing the height, stem diameter, leaf area and dry mass. Reductions caused by weeds on crop leaf area represent the competitive ability of the weeds (Procópio et al., 2004) and the interferences caused competition between weeds and crops reflect the weed aggressiveness (Silva et al., 2009). This interference acts on the CO_2 balance reducing photosynthesis, development and dry mass accumulation. In this job we observed several reductions on chick-peas dry mass accumulation and leaf area mainly for the treatments with high thresholds, suggesting that chick-peas plants can compete, without suffering severe damages, just in small infestations of sourgrass.

These observations make sense because, as observed for other crops, sourgrass plants have low capacity of interference in low densities due to its slow initial growth (Machado et al., 2006). Not only high densities of sourgrass can interfere on chick-peas development. Whish et al. (2002) observed that increasing densities of *Avena sterilis* and *Rapistrum rugosum* reduced the chick-peas production. So as observed for *A. sterilis* and *R. rugosun*, high infestations of sourgrass can be responsible for losses in chick-peas production, affecting its dry mass accumulation and leaf area, reducing the grains production.

The interference of weeds at the initial crop development could reduce the grain yield, stressing the culture and causing morpho-physiological changes (Lamego et al., 2005). Considering we are interested at the development of agricultural technologies to produce chick-peas in Brazil, the knowledge of weed threshold levels is essential for efficient herbicide use (Van Heemst, 1985) reducing environmental pollution and yield loses.

Conclusions

We conclude that chick-peas crop is sensitive to sourgrass competition threshold. Based on our results we observed tolerable interference up to 8 plants m^{-2} and

severe interference since 16 plants m^{-2}, justifying a weed control to avoid yield losses. Nonetheless, sourgrass is not the only potential weed that can interfere on chick-peas crops in São Paulo state, being necessary to enlarge this study for more potential weeds.

ACKNOWLEDGEMENTS

Carita L. do Amaral, Guilherme B. Pavan and Marcelo C. de Souza acknowledges São Paulo Reasearch Foundation (FAPESP) for each fellowship granted. Pedro L. C. A. Alves acknowledges National Council of Scientific and Technological Development (CNPq) for granting researcher scholarship. All the authors extend their acknowledgement to the Faculdade de Ciências Agrárias e Veterinárias (UNESP) for the facilities that were provided for this research and for Henrique Tozzi for the English review.

REFERENCES

El-Adawy AT (2002). Nutritional composition and antinutritional factors of chickapeas (*Cicer arietinum* L.) under going different cookin methods and germination. Plant Food Hum. Nutr. 57(1):83-97.

Ballare CL, Casal JJ (2000). Light signals perceived by crop and weed plants. Field Crops Res. 67(2):149-160.

Biçer BT, Sakar D (2010). Inheritance of pod and seed traits in chickpea. J. Environ. Biol. 31(5):667-669.

Carvalho LB, Alves PLCA, Gonzalez-Torralva F, Cruz-Hipolito HE, Rojano-Delgado AM, De Prado R, Gil-Humanes J, Barro F, Luque de Castro MD (2012). Pool of resistance mechanisms to glyphosate in *Digitaria isularis*. J. Agric. Food Chem. 60(2):615-622.

Carvalho LB, Cruz-Hipolito HE, Gonzalez-Torralva F, Alves PLCA, Christoffoleti PJ, De Prado R (2011). Detection of sourgrass (*Digitaria insularis*) biotypes resistant to glyphosate in Brazil. Weed Sci. 59(2):171-176.

Chaudhary SU, Hussain M (2011). Weed management in chickpea grown underrice based cropping system of Punjab. Crop Environ. 2(1):28-31.

Correia NM, Durigan JC (2009). Manejo químico de plantas adultas de Digitaria insularis com glyphosate isolado e em mistura com chlorimuronethyl ou quizalofop-p-tefuril em área de plantio direto. Bragantia. 68(3):689-697.

Correia NM, Leite GJ, Garcia LD (2010). Resposta de diferentes populações de *Digitaria insularis* ao herbicida glyphosate. Planta Daninha. 28(4):769-776.

EMBRAPA (1997). Manual de méodos de análises do solo. Serviço Nacional de Levantamento e Conservação do Solo, Rio de Janeiro.

FAO. Food and Agriculture Organization of The United Nations (2011). Base de dados FAOSTAT. Disponível em: <http://faostat.fao.org>. Acesso em: 12 jun. 2011.

Ferreira ACP, Brazaca SGC, Arthur V (2006). Alterações químicas e nutricionais do grão-de-bico (*Cicer arietinum* L.) cru irradiado e submetido à cocção. Ciênc. Tecnol. Aliment. 26(1):80-88.

Köppen W (1948). Climatología con un estudio de los climas de la Tierra. Mexico: Ed. Fondo de Cultura Económica-Pánuco.

Kissmann KG, Groth D (1997). Plantas infestantes e nocivas. 2.ed. São Paulo: BASF. Tomo. I:825.

Lamego FP, Fleck NG, Bianchi MA, Vidal RA (2005). Tolerância a interferência de plantas competidoras e habilidade de supressão por cultivares de soja - I. Resposta de variáveis de crescimento. Planta Daninha 23(3):405-414.

Lemes N, Carvalho LB, Souza MC, Alves PLCA (2010). Weed interference on coffee fruit production during a four-year investigation after planting. Afr. J. Agric. Res. 5(10):1138-1143.

Liu JG, Mahoney KJ, Sikkema PH, Swanton CJ (2009). The importance of light quality in cropweed competition. Weed Res. 49(2):217-224.

Machado AFL, Ferreira LR, Ferreira FA, Fialho CMT, Tuffi Santos LD, Machado MS (2006). Análise de crescimento de Digitaria insularis. Planta Daninha. 24(4):641-647.

Maiti R, Wesche-Ebeling P (2001). Advances in chickpea science. Enfield: Science Publishers Inc. p. 360.

Melo MSC, Rosa LE, Brunharo CACG, Nicolai M, Christoffoleti PJ (2012). Alternativas para o controle químico de capim-amargoso (Digitaria insularis) resistente ao glyphosate. Rev. Bras. Herb. 11(2):195-203.

Oliver LR (1988). Principles of Weed Threshold Research. Weed Technol. 2(4):398-403.

Pitelli RA (1985). Interferências de plantas daninhas em culturas agrícolas. Inf. Agropec. 11(129):16-27.

Procópio SO, Santos JB, Silva AA, Martinez CA, Werlang RC (2004). Características fisiológicas das culturas de soja e feijão e de três espécies de plantas daninhas. Planta daninha 22(2):211-216.

Radosevich SR, Holt JS (1984). Weed Ecology: Implications for vegetation management. John Wiley & Sons (Eds), New York, p. 263.

Silva AF, Concenço G, Aspiazú I, Ferreira EA, Galon L, Coelho ATCP, Silva AA, Ferreira FA (2009). Interferência de plantas daninhas em diferentes densidades no crescimento da soja. Planta Daninha 27(1):75-84.

Timossi PC (2009). Management of Digitaria insularis sprouts under no-till corn cultivation. Planta Daninha 27(1):175-179.

Valim MFCFA, Batistuti JP (2000). Efeito da extrusão termoplástica no teor de lisina disponível da farinha desengordurada de grão-de-bico (Cicer arietinum L.). ALAN 50:270-273.

Van Heemst HDJ (1985). The influence of weed competition on crop yield. Agric. Syst. 18:81-93.

Van Raij B, Cantarella H, Quaggio JÁ, Furlani AMC (1997). Recomendações de adubação e calagem para o Estado de São Paulo. 2. ed. Campinas : Instituto Agronômico, (IAC. Boletim Técnico, 100).p.285.

Whish JPM, Sindel BM, Jessop RS, Felton WL (2002). The effect of row spacing and weed density on yield loss of chickpea. Aust. J. Agric. Res. 53(12):1335-1340.

Competitive capacity of cassava with weeds: Implications on accumulation of dry matter

Daniel Valadão Silva[1]*, José Barbosa Santos[2], João Pedro Cury[1], Felipe Paolinelli Carvalho[1], Enilson Barros Silva[2], José Sebastião Cunha Fernandes[2], Evander Alves Ferreira[2] and Germani Concenço[4]

[1]Universidade Federal dos Vales do Jequitinhonha e Mucuri – UFVJM, Bahia, Brazil
[2]Programa em Pós Graduação em Produção Vegetal – UFVJM, Bahia, Brazil
[4]Empresa Brasileira de Pesquisa Agropecuária- Agropecuária Oeste, Brazil.

This work aimed to determine the competitive ability of two varieties of cassava against six weed species at initial growth stages, in relation to the allocation of dry mass in plants. The trial was conducted as a factorial experiment, with two cassava genotypes (IAC - 12 and Periquita) under competition with six weed species (*Euphorbia heterophylla*, *Bidens pilosa*, *Cenchrus echinatus*, *Amaranthus spinosus*, *Commelina benghalensis* and *Brachiaria plantaginea*), plus eight treatments corresponding to cassava varieties and weed species free from competition. The period of competition between cassava varieties and weeds was 75 days after crop emergence, when shoot and root mass were collected for evaluation of leaf area as well as, dry mass accumulation and distribution along plant organs (roots, leaves and stem). Cassava varieties presented smaller dry mass accumulation when under competition with weed species. Roots were the most affected organ by the competition. On the other hand, partition of dry mass in weeds was barely affected. In general terms, cassava variety Periquita was the most tolerant genotype to the competition and *B. plantaginea* was the weed species with higher competitive ability.

Key words: Weed species, cassava genotypes, competition, dry matter accumulation.

INTRODUCTION

Brazil currently ranks second as the world largest producer of cassava (*Manihot esculenta* Crantz) after Nigeria (FAO, 2010). However, current national average yields are relatively low (around 14 t ha^{-1}) if considering the fact that yield potential can be as high as 150 t ha^{-1} of tuberous roots (IITA, 2005). Cassava is a crop adapted to adverse environmental conditions, reaching satisfactory yields even in poor soils and without use of improved practices or technology. However, according to Alburquerque et al. (2008), this crop is highly susceptible to competition with weeds, and yield losses may be as high as to 90%, depending on the duration of competition and weed management practices adopted.

Cassava growers usually do not pay attention to weed control especially, at the early stage of crop and weed growth, as they were supposed to (Albuquerque et al., 2008). However, weed competition at initial stages of cassava development are among major factors contributing to the low income usually obtained with this crop (Pacheco et al., 1974).

In order to plan an appropriate and economical weed management program for cassava, the knowledge of the critical period of weed competition as well as, intensity of competition must be established (Carvalho et al., 2004). However, most of the studies about weed-crop competition are focused mainly in their occurrence and impact on yields without examining characteristics of both crop and weed species, and the mechanisms of their

*Corresponding author. E-mail: danielvaladaos@yahoo.com.br.

competition (Radosevich et al., 1996).

Many authors relate that morpho-physiological characteristics can be directly related to higher competitive ability of crops, as germination and emergence (Carvalho and Christoffoleti, 2008), height (Mcdonald, 2003), leaf area index (Haefele et al., 2004), solar radiation interception (Carvalho and Christoffoleti, 2008) and leaf density of top of plant (Caton et al., 2001). However, the accumulation and allocation capacity of dry matter depends on each species, and not always do the researchers pay attention on the necessary relevance of this.

Researches of plant competition can be used to predict losses of crop production due to co-existence with weeds and to determine optical levels or control periods of weed community. The context of the present work is to determine biological characteristics associated with the competitive ability of cassava varieties against weeds in terms of growth effect and dry mass partitioning within the plant.

MATERIALS AND METHODS

The trial was conducted under greenhouse environment. The soil used for this study was typical Red-Yellow Dystrophic Latosoil with medium texture. The soil after air drying was passed through a 5 mm mesh sieve before use. Soil chemical analysis showed the following results: pH (water) 5.4; organic matter 1mg kg^{-1}; P, K and Ca of 1.4; 10 and 0.5 mg dm^{-3}, respectively; Mg, Al, H + Al and CTC effective 0.2; 0.4; 4.4 and 1.7 cmolc dm^{-3}, respectively. In order to make the soil suitable as substract, 3.0 g dm^{-3} of dolomitic limestone, 2.2 g dm^{-3} of super simple phosphate or single super phosphate (P$_2$O$_5$) and 0.4 g dm^{-3} of potassium chloride or potassium oxide (K$_2$O) were applied. Nitrogen was applied 30 days after crop emergence, at dose of 0.4 g dm^{-3} of urea (45% N), previously dissolved in water. Irrigations were done by an automatic sprayer system. A factorial scheme 2 × 6 + 8 was adopted, constituted by combination of two cassava genotypes [IAC - 12 and Periquita] under competition with six weed species: *Euphorbia heterophylla* (EPHHL), *Bidens pilosa* (BIDPI), *Cenchrus echinatus* (CCHEC), *Amaranthus spinosus* (AMASP), *Commelina benghalensis* (COMBE) and *Brachiaria plantaginea* (BRAPL), plus eight additional treatments corresponding to the cassava varieties and weed species planted without other plant in competition. Treatments were arranged as a completely randomized design with four replications, and each bucket with capacity of 5 L (25 × 21 cm of diameter and height, respectively), represented one experimental plot.

Seedlings of *C. benghalensis* were transplanted while the other weed species were planted directly in the experimental plots at the same time cassava was planted, allowing weed and crop emergence to occur at the same time. For planted species, desired densities were established through thinning. The trial consisted of the same density of weed and cassava plants (one plant of each species per vase) except for *E. heterophylla*, which were two plants per vase after thinning. Weed and crop densities were predetermined by phytosociological studies in areas where cassava is cultivated on the same soil type (data not presented).

The period of coexistence between cassava varieties and weed species was 75 days after crop emergence. This intermission was established with the intention of quantifying damages by coexistence during the critical period of interference (CPIC) of weed species, which can be extended up to 100 days after crop emergence (Carvalho et al., 2004).

At the end of this period, cassava plants and weeds were sampled and dry mass was determined individually for roots, stems and leaves. For the weed species *C. echinatus* and *B. plantaginea*, leaf sheats were added to leaves and culms were considered as stems. For determination of leaf area, all leaves of each cassava plant were scanned and analyzed by the software Digital Areas Determiner (DAA) (Ferreira et al., 2008). Later, plant material was washed in distilled water and dried in oven under forced air circulation, at 70°C, until constant weight. Dry mass was determined in electronic balance with precision of 0.0001 g. Based on those data, leaf area ratio [LAR = (leaf area/total dry mass)], specific leaf area [SLA = (leaf area/dry mass of leaves)] and crop growth rate [GRC = (final dry mass/number of days between planting and harvest)] were determined. Also, partitioning of dry mass between plant parts (roots, stems, leaves) was determined for both crop and weed species.

As it was a greenhouse trial, all treatments were repeated twice in order to increase precision and accuracy of the results. All data was subjected to analysis of variance and treatment means and when significant, were compared by Scheffé test at 10% probability. In addition, contrasts estimates were used for comparison between plants of the same species under different situations; free from interference or under competition. Means comparison by Scheffé test was established due to the loss of some plots and to the determination of contrasts a posteriori. The significance level of 10% was used as recommended by Carraher and Rego (1981), due to Scheffé's high rigor in indicating differences between treatments. Pearson correlation between variables was also established for cassava plants.

RESULTS

Cassava varieties showed smaller overall dry mass accumulation as a function of weed interference (Table 1). Averages over genotype, dry mass of leaves, stems and roots were 30, 35 and 22%, respectively, when compared with the control affecting directly the accumulation of total crop dry mass (reduction of around 72%). The level of interference changed as a function of weed species and cassava variety. In general terms, it was observed that variety Periquita was the most tolerant to competition imposed by weeds. Under interference of *B. pilosa*, this variety had a superior dry mass accumulation in leaves and roots than IAC - 12 (Table 1). Under competition with *B. pilosa* and *C. benghalensis*, the average dry mass accumulation by crop plants as a whole and in roots was around 31 and 60%, respectively, when compared to the control without weed competition (Table 1). On the average, total dry mass and roots dry mass accumulation of the cassava varieties was reduced by 21 and 11% respectively, due to competition from *A. spinosus* and *E. heterophylla* when compared to the check (Table 1). When cassava was under competition with *C. echinatus*, total and root dry mass accumulation was reduced by 27 and 17%, respectively, compared to the control treatment without weed competition. *B. plantaginea* had a higher competitive ability against cassava causing reduction in dry mass accumulation for all parts of the crop plants, which translates into a loss of about 10, 13 and 18%, respectively in roots, leaves and stems, when compared to the check free of interference (Table 1).

Table 1. Effect of weed species interference on cassava cultivar dry mass accumulation at 75 days after emergency.

Treatment	TDM IAC – 12	TDM Periquita	TDM $\bar{x}$	RDM IAC – 12	RDM Periquita	RDM $\bar{x}$	LDM IAC – 12	LDM Periquita	LDM $\bar{x}$	SDM IAC – 12	SDM Periquita	SDM $\bar{x}$
Control[1]	29.97^Aa	27.45^Aa	28.71^a	11.28^Aa	11.58^Aa	11.43^a	13.99^Aa	11.50^Aa	12.74^a	4.71^Aa	4.37^Aa	4.54^a
EPHHL	6.13^Ac	6.57^Acd	6.31^cd	2.16^Ac	0.36^Bd	1.44^de	3.04^Ab	5.16^Aab	3.89^bc	0.94^Ac	1.05^Ab	0.98^c
BIDPI	5.09^Bc	13.01^Abc	9.05^c	1.41^Bc	5.54^Ac	3.48^c	2.63^Bb	5.44^Aab	4.04^bc	1.05^Abc	2.03^Aab	1.54^bc
CCHEC	6.17^Bc	9.16^Abcd	7.67^cd	1.65^Ac	2.14^Ad	1.89^cd	2.92^Ab	4.49^Ab	3.70^bc	1.61^Abc	2.54^Aab	2.07^bc
AMASP	4.05^Bc	7.88^Abcd	5.32^cd	1.09^Ac	1.37^Ad	1.18^de	2.16^Ab	4.71^Ab	3.01^c	0.80^Ac	1.81^Aab	1.13^c
COMBE	17.76^Ab	15.90^Ab	16.83^b	5.20^Bb	8.69^Ab	6.95^b	9.32^Aa	4.54^Bb	6.93^b	3.24^Aab	2.67^Aab	2.95^ab
BRAPL	2.47^Ac	3.10^Ad	2.83^d	0.25^Ac	0.34^Ad	0.30^e	1.52^Ab	1.85^Ab	1.71^c	0.70^Ac	0.91^Ab	0.8^c
CV (%)[2]	29.26			24.5			41.89			43.90		

[1]TDM, Total dry matter; RDM, root dry matter; LDM, leaves dry matter; SDM, stem dry matter. Means followed by the same letter in line (capital letter) and in column (lower case) for each variable (plant organ) did not differ by Scheffé test at 10% error probability; [2] variation coefficient; $\bar{x}$, average of weed interference; [1], control absentee of weed interference; EPHHL, *Euphorbia heterophylla*; BIDPI, *Bidens pilosa*; CCHEC, *Cenchrus echinatus*; AMASP, *Amaranthus spinosus*; COMBE, *Commelina benghalensis*; BRAPL, *Brachiaria plantaginea*.

Dry mass partitioning between plant parts of cassava varieties as a function of the weed species caused the interference (Figure 1). Under interference of *C. echinatus*, *A. spinosus* and *B. plantaginea* a larger relative allocation of dry mass in stems of the crop was noticed, however, with a consequent reduction of dry mass accumulation in roots (dry mass was 12, 5 and 15% higher in stems and about 14, 17 and 29% lower in roots, respectively to the weed species (Figure 1).

Under interference of *C. benghalensis*, a smaller relative dry mass allocation was observed in cassava leaves (39%) when compared to the check (44%) (Figure 1). On the other hand, under competition with *E. heterophylla*, a smaller relative dry mass allocation in crop roots (22%) was observed in relation to the check (40%).

Changes in pattern of dry mass allocation by cassava plants also influenced dry mass distribution in weed species (Figure 2). A relatively higher dry mass accumulation was observed in leaves of *B. pilosa* and *C. echinatus* (34 and 43% respectively), when compared to the check (30 and 40%, respectively) (Figure 2).

Unlike *C. benghalensis*, the weed species *B. plantaginea* in co-existence with cassava accumulated larger amounts of dry mass in roots, in detriment of leaves (dry mass 4% superior in roots and 3% inferior in leaves, in relation to the respective check competition (Table 2).

The leaf area index (LAI) and crop growth rate (CGT), were the only growth analysis parameters for cassava that were significant (Table 3).

Competition with weeds caused reductions in leaf area of cassava (Table 3). Under interference of *E. heterophylla*, *B. pilosa*, *C. echinatus*, *A. spinosus* and *C. benghalensis*, cassava leaf area was 31, 36, 22, 24 and 66%, respectively, of the value observed at checks free from these species. *B. plantaginea* presented again a superior competitive ability than the other weeds, because it caused crop leaf area reduction of around 89%

(only 11% of the observed at the respective check) (Table 3).

In general terms, the average growth rate of cassava shifted according to the weed species present, but was significant only when varieties were under competition with *A. spinosus*, *C. benghalensis* and *B. plantaginea* (Table 3).

At the correlation analysis between variables, it is possible to observe that a given variable does not influence all the others at the same degree, once significant interaction was not observed between all pairs of variables analyzed (Table 4).

Crop leaf area and growth rate are related in a negative and positive way to the total dry mass of leaves, stems and roots of cassava.

In this case, plants with larger leaf area are usually more capable of capturing sun radiation; however, it does not mean a linear relation between amounts of light intercepted and amount of dry mass accumulated by these species. Specific leaf area and leaf area ratio were not significantly

Table 2. Effect of cassava cultivar competition on weed dry matter at 75 days after emergency.

Species	TDM				RDM				LDM				SDM			
	IAC – 12	Periquita	$\bar{x}$	Test[1]	IAC – 12	Periquita	$\bar{x}$	Test[1]	IAC – 12	Periquita	$\bar{x}$	Test[1]	IAC – 12	Periquita	$\bar{x}$	Test[1]
EPHHL	18.97b	16.35b	15.43Ad	12.31Ac	3.67b	4.64b	3.97Ad	3.87Ac	7.13c	5.62c	6.70Ad	6.91Ac	8.17b	6.62b	6.70Ab	5.64Ab
BIDPI	35.82b	29.80b	33.72Acd	35.09Abc	7.92b	5.02b	7.02Acd	7.85Abc	12.10b	9.95bc	10.77Ad	10.38Ac	15.80b	14.82b	15.93Ab	16.86Ab
CCHEC	67.60b	58.35b	59.92Abc	55.33Abc	21.10b	21.47b	23.42Ab	26.62Abc	30.02a	24.09ab	28.64Aab	31.01Aa	16.49b	12.79b	16.40Ab	19.04Ab
AMASP	72.18b	64.07ab	69.76Ab	70.19Ab	24.77b	30.51ab	28.56Ab	31.37Aab	26.06a	21.82ab	25.60Ab	27.05Aa	16.17b	11.74b	15.34Ab	16.32Ab
COMBE	23.26b	32.12b	33.88Acd	43.15Abc	5.19b	3.05b	6.36Ad	9.72Abc	10.69c	14.13bc	14.24Acd	16.98Ab	8.97b	11.90b	12.85Ab	16.46Ab
BRAPL	149.23a	130.80a	134.16Aa	126.21Aa	66.95a	55.85a	57.78Aa	52.85Aa	36.34a	34.29a	34.87Aa	34.34Aa	45.95a	40.66a	41.51Aa	39.02Aa
CV (%)[2]			42.13				52.24				30.03				37.33	

TDM, Total dry matter; RDM, root dry matter; LDM, leaves dry matter; SDM, stem dry matter. Means followed by the same letter in line (capital letter) and in column (lower case) for each variable (plant organ) did not differ by Scheffé test at 10% error probability; [1] control absentee of weed interference; [2] variation coefficient; $\bar{x}$, average of cassava cultivars; EPHHL, *Euphorbia heterophylla*; BIDPI, *Bidens pilosa*; CCHEC, *Cenchrus echinatus*; AMASP, *Amaranthus spinosus*; COMBE, *Commelina benghalensis*; BRAPL, *Brachiaria plantaginea*.

Table 3. Growth parameter estimates of cassava cultivars under interference of different weed species, at 75 days after emergency.

Treatment	LA (cm² pl⁻¹)			GRC (g pl⁻¹)			SLA (cm² g⁻¹)			LAR (cm² g⁻¹)		
	IAC – 12	Periquita	$\bar{x}$	IAC – 12	Periquita	$\bar{x}$	IAC – 12	Periquita	$\bar{x}$	IAC – 12	Periquita	$\bar{x}$
Control[1]	20.77Aa	19.89Aa	20.42a	0.33Aa	0.31Aa	0.32a	0.67ns	0.73ns	0.70ns	1.45ns	1.69ns	1.55ns
EPHHL	5.32Ac	7.28Abc	6.30cd	0.07Ac	0.07Acd	0.07cd	0.94	1.12	1.03	2.22	1.31	1.77
BIDPI	5.57Bbc	9.26Abc	7.42c	0.06Ac	0.14Abc	0.10c	1.11	0.71	0.91	2.26	1.71	1.99
CCHEC	4.19Ac	4.80Ac	4.40cd	0.07Ac	0.10Abcd	0.09cd	0.76	0.65	0.72	1.87	1.34	1.69
AMASP	3.39Bc	6.55Abc	4.97cd	0.04Ac	0.09Abcd	0.06cd	1.06	0.89	0.97	2.34	1.41	1.88
COMBE	15.47Aa	12.07Bb	13.52b	0.20Ab	0.18Ab	0.19b	0.80	0.75	0.77	1.44	3.11	2.40
BRAPL	1.30Ac	3.37Ac	2.34d	0.03Ac	0.03Ad	0.03d	0.61	1.30	0.95	0.96	3.08	2.02
CV (%)[2]		25.61			29.26			31.13			56.62	

LA, Leaf area; GRC, crop growth rate; SLA, specific leaf area; LAR, leaf area ratio. Means followed by the same letter in line (capital letter) and in column (lower case) for each variable (plant organ) did not differ by Scheffé test at 10% error probability; [ns] no significant; [1] Control absent of weed interference; [2] Variation coefficient; $\bar{x}$ - Average of cassava cultivars; EPHHL, *Euphorbia heterophylla*; BIDPI, *Bidens pilosa*; CCHEC, *Cenchrus echinatus*; AMASP, *Amaranthus spinosus*; COMBE, *Commelina benghalensis*; BRAPL, *Brachiaria plantaginea*.

correlated with any other variable, except for the interaction between them (Table 4).

DISCUSSION

Both architecture and growth characteristics of the crop can be directly related to its ability to compete favorably with weed species at the initial stages of their growth and development. Periquita's ability to form more uniform and dense canopy earlier, due its branching habit, gave it advantage over the weed species. These characteristics reduce weeds access to sunlight as well as, its photosynthesis rate. However, according to Moura (2000), branching of cassava plants does not confer advantages to the crop in relation to roots yield. *Bidens pilosa* and *Commelina benghalensis* are among the major weeds in areas where cassava is grown, pre-senting potential infestation problem than many

Table 4. Pearson lineal correlation matrix between the analyzed variables of cassava cultivars (IAC - 12 and Periquita) after 75 days of emergency.

Variable	Interaction	Variable	Interaction
TDM × LDM	+ 0.94**	SDM × LA	- 0.49**
TDM × SDM	+ 0.92**	SDM × SLA	- 0.03^{ns}
TDM × RDM	+ 0.93**	SDM × LAR	+ 0.21^{ns}
TDM × LA	- 0.49**	SDM × GTC	+ 0.92**
TDM × SLA	- 0.05^{ns}	RDM × LA	- 0.47**
TDM × LAR	+ 0.19^{ns}	RDM × SLA	- 0.07^{ns}
TDM × GTC	+ 1.00**	RDM × LAR	+ 0.24^{ns}
LDM × SDM	+ 0.88**	RDM × GTC	+ 0.93**
LDM × RDM	+ 0.76**	LA × SLA	+ 0.01^{ns}
LDM × LA	- 0.43**	LA × LAR	- 0.18^{ns}
LDM × SLA	- 0.07^{ns}	LA × GTC	- 0.49**
LDM × LAR	+ 0.10^{ns}	AFE × LAR	+ 0.67**
LDM × GTC	+ 0.94**	AFE × GTC	- 0.05^{ns}
SDM × RDM	+ 0.79**	RAF × GTC	+ 0.19^{ns}

Ns, Non-significant interaction, **, significant interaction at 1% probability by Pearson matrix; TDM, total dry matter; LDM, leaf dry matter; RDM, roots dry matter; SDM, stem dry matter; LA, leaf area; SLA, specific leaf area; LAR, leaf area ratio; GTC, growth tax of culture.

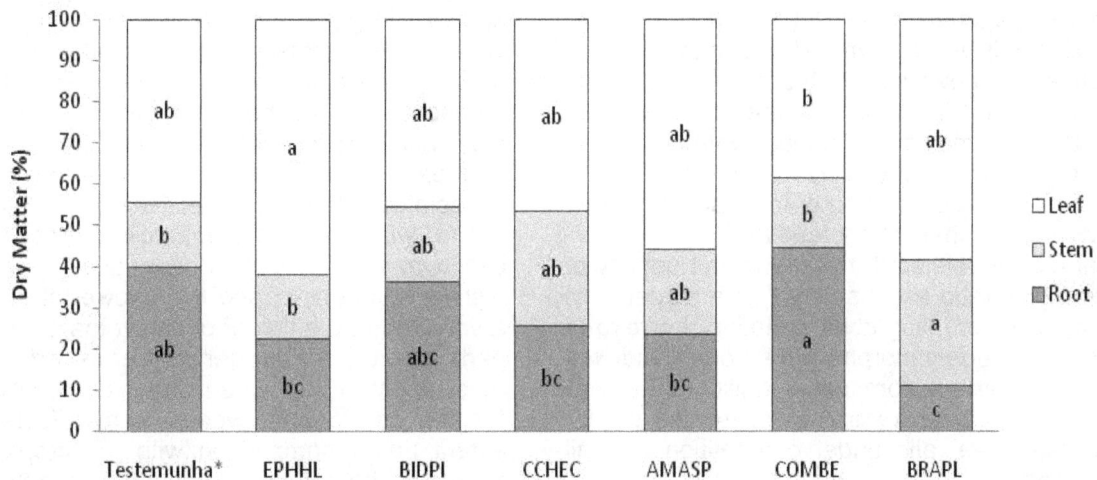

Figure 1. Average percentage distribution of dry matter between vegetative components of cassava cultivars (IAC - 12 and Periquita) under interference of different weeds. Means followed by same letter in each variable (plant organ) did not differ by Scheffé test at 10% error probability *Cassava cultivars absent of interference from weed; EPHHL, *Euphorbia heterophylla;* BIDPI, *Bidens pilosa;* CCHEC, *Cenchrus echinatus;* AMASP, *Amaranthus spinosus;* COMBE, *Commelina benghalensis;* BRAPL, *Brachiaria plantaginea.*

other weed species (Albuquerque et al., 2008). They are also predominant in potato (Ossom and Rhykerd, 2007), bean (Cury et al., 2011) and maize (Carvalho et al., 2011) fields. At varietal level, cassava under interference from different weed species showed some levels of competitive ability and tolerance to competition, with the consequences of lower dry mass accumulation mainly in roots in relation to more competitive weed species. However, it was observed that in relation to the weed

community, weed species behaved from a neutral way in relation to the effects of competition imposed by cassava plants.

According to Carvalho et al. (2011), not only biomass accumulation but also biomass allocation is a fundamental aspect in the competition between plant species. The pattern of dry mass allocation between plant organs denotes it to be a variable more important than the total amount of biomass accumulated as regarding

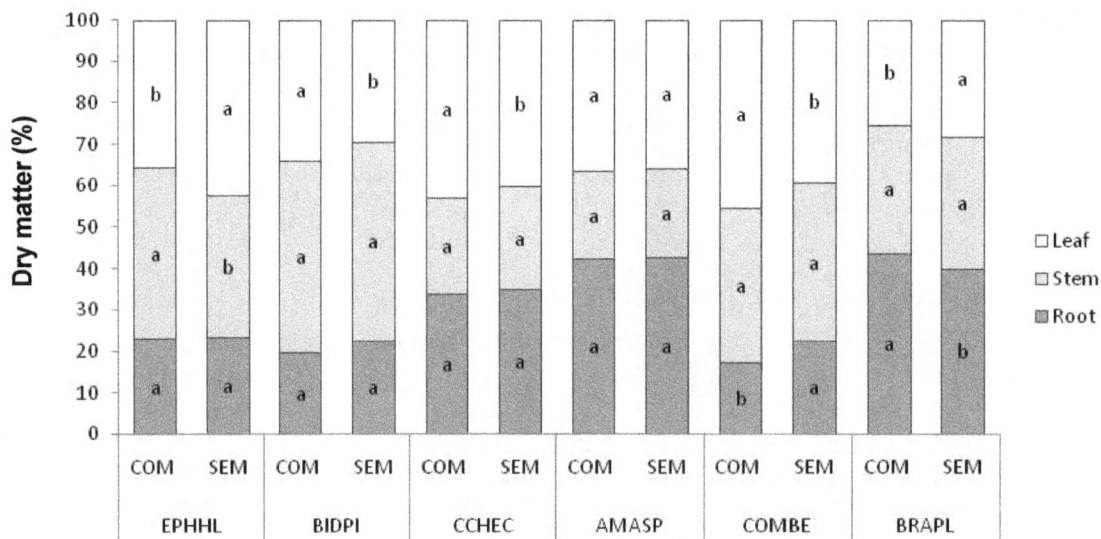

Figure 2. Average percentage distribution of dry matter between vegetative components of different weeds in competition with cassava cultivars (IAC - 12 and Periquita). Means followed by same letter for each variable (plant organ) did not differ by Scheffé test at 10% error probability; COM/Average of weeds in competition with cassava cultivars; SEM/Weeds absent of competition with cassava cultivars.

the tolerance mechanisms of crops to the competition with weed species (Ngouajio et al., 2001).

Dry mass allocation in *C. benghalensis* also followed the same tendency; however, that species shifted its ability in forming roots. It is possible that those weeds, when subjected to the co-existence with cassava, induced the allocation of dry mass in shoots to the detriment of roots, in order to attempt to close its canopy faster and allow maximum light interception.

Specific leaf area is related to thickness and density of a leaf, and leaf area ratio is considered as a measure of plant's capacity in intercepting solar radiation (Ferreira et al., 2008). Probably certain morpho-physiological indexes are not associated with the competitive ability of the crop against weeds once cassava varieties presented similar leaf area and leaf area ratio under competition with all weed species (Table 3).

In general terms, results showed that the leaf area of variety Periquita was superior to IAC - 12, when subjected to interference by *B. pilosa* and *A. spinosus*. However, under competition with *C. benghalensis*, IAC - 12 presented larger leaf area. This characteristic, associated to superior shoot dry mass accumulation (leaf and stem) (Table 1), would provide to that variety faster formation of a dense canopy which would reduce the effect of initial interference imposed by weed species under field conditions. In cassava, positive correlation between leaf area or leaf area duration and increase of root yields was observed, indicating that leaf area is crucial to determining both crop growth rate and roots development rate (Cock et al., 1979).

All weed species under competition with cassava, had similar dry mass to the respective check. On the other

hand, the crop was practically suppressed by those species. In general terms, variety Periquita was the one with most tolerated weed competition. *B. plantaginea* demonstrated to be the weed species with higher competitive ability, because it affected both leaf area and dry mass partitioning by the crop. Both specific leaf area and leaf area ratio were not morpho-physiological indices associated with the competitive ability of cassava plants.

The wide set of competitive relationships to what cassava was submitted was due to the presence of distinct competitors and this showed that there is always a variation in the flux of photosynthates between different parts of the plant. In general terms, crop roots were the most affected part when under competition with weeds. On the other hand, weeds were affected at a smaller extent under competition with the crop with almost no hazard to the dry mass accumulation in these species.

ACKNOWLEDGEMENTS

We are grateful to Conselho Nacional de Desenvolvimento Científico e Tecnológico (CNPq), Fundação de Amparo à Pesquisa do Estado de Minas Gerais (FAPEMIG) and the Coordenação de Aperfeiçoamento de Pessoal de Nível Superior (CAPES) for financial support in execution of this work.

REFERENCES

Alburquerque JAA, Sediyama T, Silva AA, Carneiro JES, Cecon PR, Alves JMA (2008). Interferência de plantas daninhas sobre a produtividade da mandioca (*Manihot esculenta*). Planta Daninha

26:279-289.

Carraher TN and Rego LLB (1981). O realismo nominal como obstáculo na aprendizagem da leitura. Caderno de Pesquisa 39:3-10.

Carvalho JEB, Araújo AMA, Azevedo CLL (2004). Período de controle de plantas infestantes na cultura da mandioca no Estado da Bahia. Cruz das Almas: EMBRAPA-CNPMF. Technical Communication. p. 109.

Carvalho SJP, Christoffoleti PJ (2008). Competition of *Amaranthus* species with dry bean plants. Scientia Agricola 65:239-245.

Carvalho FP, Santos JB, Cury JP, Valadão Silva D, Braga RR, Byrro ECM (2011). Alocação de matéria seca e capacidade competitiva de cultivares de milho com plantas daninhas. Planta Daninha 29:373-382.

Caton BP, Mortimer AM, Foin TC, Hill JE, Gibson KD, Fischer AJ (2001). Weed shoot morphology effects on competitiveness for light in direct-seeded rice. Weed Res. 41:155-163.

Cock JH, Franklin D, Sandoval D, Juri P (1979). The ideal cassava plant for maximum yield. Crop Sci. 19:271-279.

Cury JP, Santos JB, Valadão Silva D, Carvalho FP, Braga RR, Byrro ECM, Ferreira EA (2011). Produção e partição de matéria seca de cultivares de feijão em competição com plantas daninhas. Planta Daninha 29:149-158.

FAO (2010). Organização das nações unidas para agricultura e alimentação. Available at: https://www.fao.org.br (Accessed: 20 February, 2010).

Ferreira EA, Concenço G, Silva AA, Reis MR, Vargas L, Viana RG, Guimarães AA, Galon L (2008). Potencial competitivo de biótipos de azévem (*Lolium multiflorum*). Planta Daninha 26:261-269.

Ferreira OGL, Rossi FD, Andrighetto C (2008). DDA: Software for determination of leaf area, leaf area index and loin eye area - Version 1.2. Santo Augusto.

Haefele SM, Johnson DE, Bodj DM, Wopereis MCS, Miezan KM (2004). Field screening of diverse rice genotypes for weed competitiveness in irrigated lowland ecosystems. Field Crops Res. 88:39-56.

IITA (2010). Cassava productivity in the lowland and midaltitude agro-ecologies of Sub-Saharan Africa. Available at: www.iita.org/cms/articlefiles/92 IITA%20MTP%202001_2003.pdf (last accessed: 30 March 2010).

Mcdonald GK (2003). Competitiveness against grass weeds in field pea genotypes. Weed Res. 43:48-58.

Moura GM (2009) Interferência de plantas daninhas na cultura de mandioca (*Manihot esculenta*) no estado do Acre. Planta daninha 18:235-240.

Ngouajio M, McGiffen Junior ME, Hembre KJ (2001). Tolerance of tomato cultivars to velvetleaf interference. Weed Sci. 49:91-98.

Pacheco C, Chavarria PL, Mata RH (1974). Herbicidas em pré-emergência en el cultivo de la yuca (*Manihot esculenta* Crantz.). Costa Rica: Estación Experimental Fábio Banchit. Technical Commun. 1:12.

Ossom EM, Rhykerd RL (2007). Lime effects on weeds and sweet potato yield, Int. J. Agric. Biol. 9:755-758.

Radosevich SR, Holt J, Ghersa C (1996). Physiological aspects of competition. In: Holt J and Ghersa C (eds.), Weed ecology: Implication for Managements, Vol. 2. John Wiley & Sons, New York, USA. pp. 217-301.

3

Evaluation of brown seaweed (*Padina pavonica*) as biostimulant of plant growth and development

Asma Chbani, Hiba Mawlawi and Laurence Zaouk

Doctoral School for Sciences and Technology, Azm Centre for Research in Biotechnology and its application, Lebanese University, El Miten Street, Tripoli, Lebanon.

An innovative horticulture nutrient and biodegradable support is described in this paper for replacing plastic culture pots. This support is prepared with *Luffa aegyptica*, plant having a water holding capacity higher than that of the regular soil and that is also biodegradable. Brown seaweed *Padina pavonica* was incorporated as an organic fertilizer of plant growth. Chemical analysis of the aqueous extract of this alga showed the presence of macronutrients such as nitrogen (N), phosphorus (P) and potassium (K) necessary for development and growth of plants. Agar-agar was added as a solidifying agent. A medium containing only soil and another containing soil with chemical fertilizer served as controls. Sunflower seeds grown in medium supplemented with brown seaweed; (*P. pavonica* + agar (4% or 6%) + *L. aegyptica* have a growth rate (length and diameter of the stem, number of leaves) that is slower than the plants grown in a medium with a comparable amount of the soil with chemical fertilizer. However, the plants in the soil and others in the soil with chemical fertilizer and the media (seaweed + *L. aegyptica* + agar 4%) have not completed their growth while the plants grown in the media (seaweed + *L. aegyptica* + agar 6%) continued to grow. A biodegradability test showed that a piece of support (seaweed + agar 1.5% + *L. aegyptica*) presented a degradation rate higher than the support with only Luffa and agar 1.5%, while a piece of plastic had not degraded. The results of our study have shown that this support has helped to extend the duration of growth and enhanced the quality of the plants. Ultimately, the fabricated support presented fertilizer properties, water retention and biodegradability and could serve in horticulture as an alternative to plastic pots and chemical fertilizer.

Key words: Brown seaweed, *Padina pavonica*, Biostimulant, Luffa, water retention, biodegradability.

INTRODUCTION

Compared to organic fertilizers, chemical fertilizers are not sufficient to procure alone all the minerals and nutrients required by the plant: nitrogen, phosphorus, potassium, magnesium and other trace elements. Indeed, chemical fertilizers are not compatible with organic farming and have adverse effects on health and the environment. In addition to their high cost, they alter the quality, fertility, structure and humus of the soil. Modern agriculture is looking for new biotechnological advances that would allow a reduction in the use of chemical inputs without affecting crop yield or the farmer's income. In recent years, the use of natural seaweed as fertilizer has allowed for substitution in place of conventional synthetic fertilizer (Hong et al., 2007).

The agricultural sector is by far the largest user of water in the world. Referring to the consumption of water by evapo-transpiration, for 80 to 90% of the water is used in agriculture. Unfortunately, the efficiency of water use is very low and does not exceed 45% with more than 50% of losses, so it is more in agriculture than in other sectors that one can achieve substantial water savings. To meet these challenges, we must dramatically improve the use of water resources and technology and irrigation management. Moreover, the culture pots commonly used in horticulture are composed of non-biodegradable plastic. As a result, the plant with its clod of earth must be removed from the plastic pot before being planted in the ground or into a larger pot. The pot containing the plant initially is thrown by the consumer and thus contributes to environmental pollution.

Today, there are pots for growing plants made from biodegradable materials. The advantage of using such materials is that the plant can be buried with its support as it degrades naturally. Among the materials used in the manufacture of biodegradable pots are cellulose fibers, peat, starch or mixtures thereof. One of the major drawbacks of this type of materials is the non-existence of nutrients for the plant. On the contrary, the degradation of a device for packaging made of this material involves microorganisms in the soil whose activity requires nitrogen consumption. The plant is then deprived of part of the nitrogen present in its environment. The issue of chemical fertilizers, water shortage and non-biodegradable plastic pots is then at the heart of the debate and the cause of the ecological impacts threatening the health sector, agriculture and environment.

The main objective of this study is to producing a horticultural biodegradable nutrient support that does not require a large consumption of water, with a relatively low cost of production. The materials for the manufacture of this support are: The brown alga *Padina pavonica* as biofertilizer, lignocellulosic fiber (*Luffa aegyptica*) as water retention agents and the Agar-agar as a plasticizer. This support according to this composition will be fully biodegradable. The composition of this material has not been documented in any scientific publications to date. In Lebanon, the coastal marine environment that contains at least 243 algal species (Lakkis and Novel-Lakkis, 2000) is sadly untapped. In the literature, *P. pavonica* has already been studied for cytotoxic activity (Ktari and Guyot, 1999) for antibacterial activity (Chbani et al., 2011) and for antioxidant properties (Mawlawi et al., 2012) but has never been an object of research up to date as fertilizer. This has guided the choice of this seaweed in our experiments.

MATERIALS AND METHODS

Collection of seaweed: Biofertilizer agent

The brown alga *P. pavonica* belonging to the family Dictyotaceae was freshly collected manually from the coastal area of the Mediterranean, El Mina (34° 26'N-35° 50'E) in Tripoli, Lebanon during March 2011. It was washed with seawater to remove all the unwanted impurities such as adhering sand particles and epiphytes. Then, the seaweed was transported in polythene bags just moistened in the laboratory (Cabioc'h et al., 2006).

Preparation of seaweed liquid extracts (SLE) for physico-chemical analyses

The seaweed were thoroughly washed using tap water and ultrapure water to remove sea salts on the surface of samples. After that, fresh algae were cut into small pieces and boiled with ultrapure water under stirring for one hour on a hot plate. Then, the extract was filtered (Kumar and Sahoo, 2011).

To study and compare the composition of minerals and nutrients of *P. pavonica* liquid extract to other algae liquid extract used in other research, a dosage of the primary nutrients (N, P, K), secondary nutrients (Ca, Mg, S), trace elements (Cu, Zn, Fe, Co, Mn, Cl, Na), alkalinity (HCO_3^-) and undesirable elements (Ni, Pb, Cr Al, Cd) were analyzed by the method described by Normalization French Association (AFNOR). The pH of the SLE was directly measured using a pH meter ORION type. Each assay was performed twice.

Preparation of the water retention agent

The mature fruit of the species *L. aegyptica* (belonging to the family Cucurbitaceae) used in small pieces as a water retention agent in this study.

Agar-agar: Solidifier agent

The Agar-agar solidifier agent is extracted from red algae (Rhodophyceae) belonging to the families of Gelidiaceae. The Agar used was Fluka ® type (Cat. No.05040, Sigma-Aldrich Chemie GmbH) was prepared according to two different dilutions (4% and 6%) in water to compare the growth of sunflower plants.

Manufacture of horticultural biofertilizer and biodegradable support

Brown seaweed *P. pavonica* previously were dried at room temperature away from light and then ground in the form of fragments of 1 to 2 mm^2. The support was composed with algae *P. pavonica* (5%), crushed Luffa (3%) in pieces and Agar (92%) diluted with water. The hot mixture was then poured into molds and allowed to cool and solidify at room temperature. Finally, after cooling, the support was removed from the molds (Figure 1).

Germination and growth of sunflower seeds were conducted on the two supports, one containing only the soil and the other containing soil with chemical fertilizer were used as controls.

Experiment

Eight white sunflower seeds, Vilmorin type, with uniform shape, size, color and weight were incubated in water at 30°C for 48 h before inoculation. Every two sunflower seeds treated were sown at a depth of 1 cm in each support: (seaweed + *L. aegyptica* + Agar 4%), (seaweed + *L. aegyptica* + Agar 6%), (soil 100%) and (soil + chemical fertilizer). Then the 4 supports were placed in mini-greenhouse at 22 to 28°C, 70 to 85% relative humidity, 600 to 1,000 µmol photons m^{-2} s^{-1} light intensity and 12 h photoperiod during the period of observation. The seeds were watered regularly. The root system, the length of the stem, the stem diameter and number of

Figure 1. Biofertilizer support: a: Support with seaweed (*Padina pavonica*) + Lupha + Agar (4%). b: Support with seaweed (*Padina pavonica*) + Lupha + Agar (6%).

Figure 2. Biodegradability test of 3 fragments: a: Plastic fragment, b: Fragment with Luffa aegyptica + Agar and c: fragment of biofertilizer support with Luffa aegyptica + seaweed + Agar.

leaves were monitored by foot slide with an interval of 2 days within 18 days and after transfer in the soil. All samples were realized in triplicate.

Biodegradability test of nutritional support

A fragment (50 mm long, 20 mm wide and 2 mm thick) of each supports; plastic pots (Figure 2a), control support (*L. aegyptica* + Agar) (Figure 2b) and nutritional support (seaweed + *L. aegyptica* + Agar) (Figure 2c) was buried in 15 mm underground for 4 weeks. The soil was watered twice a week.

Testing of water holding capacity of nutritional support

To determine the volume of water retained in the nutritional support compared to that of the soil, a test of water-holding capacity of these materials has done achieved. Two supports (seaweed + *L. aegyptica* + Agar 6%) and (100% soil), were dried at 30 °C and were weighed. Each support was then immersed in a water bath at room temperature until saturation and was weighed. The water holding capacity of each support is the measure of the volume of water retained per 100 g of support.

RESULTS

Chemical constituents

The chemical constituents of *P. pavonica* seaweed extract were analyzed and presented in Table 1. The

Table 1. Chemical constituents of *Padina pavonica* seaweed liquid extract.

Chemical constituents	(mg/L)
Nitrogen	10.90
Phosphorus	9.26
Potassium	160.13
Calcium	110.22
Magnésium	1.20
Sulphur	235.00
Zinc	-
Copper	-
Sodium	73.40
Chloride	85.09
Iron	-
Manganese	0.22
Cobalt	-
Carbonate HCO$_3^-$	207.40
Nickel	-
Plomb	-
Chrome	-
Aluminium	-
Cadmium	-

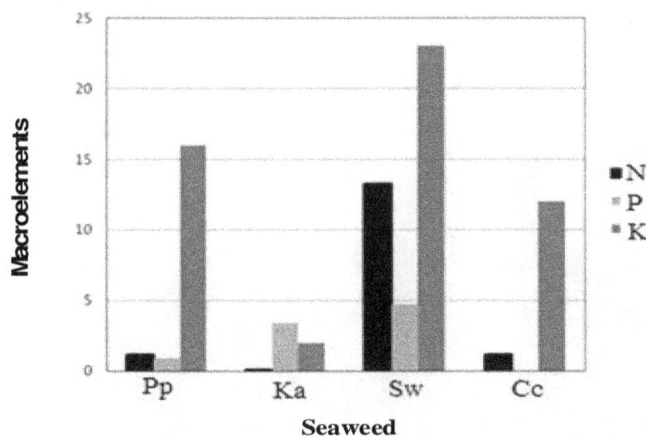

Figure 4. The micronutrients (Ca, Mg and SO$_4^{2-}$) measured in seaweeds: *Padina pavonica* (Pp), *Kappaphycus alvarezzii* (Ka), *Sargassum wightii* (Sw) and *Caulerpa chemnitzia* (C.c).

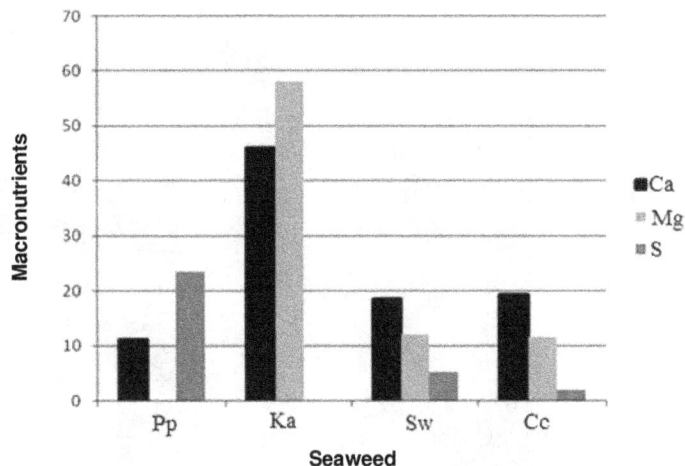

Figure 3. The macronutrients (N, P and K) measured in seaweeds: *Padina pavonica* (Pp), *Kappaphycus alvarezzii* (Ka), *Sargassum wightii* (Sw) and *Caulerpa chemnitzia* (C.c).

color of the SLE of *P. pavonica* is brown and the pH was 7.22. Figures 3 and 4 represent the comparison of macronutrients (N, P,K) and micronutrients (Ca, Mg, SO$_4^{2-}$) of the brown seaweed *Sargassum wightii* (Sw), the green seaweed *Caulerpa chemnitzia* (Cc) (Sivasankar et al., 2006), the red seaweed *Kappaphycus alvarezzii* (Ka) (Rathore et al., 2009) and the brown alga *P. pavonica* (Pp) liquid extract. For the macronutrients N, P and K, the liquid extract of Pp is rich in potassium and has nitrogen content lowest after Ka. P$_2$O$_5$ is the lowest (this element is not detected in the liquid extract of the

green alga (Cc). The amount of potassium K is lower than that of the brown seaweed (Sw) but higher than that of the red seaweed Ka and the green seaweed (Cc). For micronutrients elements Ca, Mg, SO$_4^{2-}$, the liquid extract of Pp is rich in sulfate S$_2$O4$^-$ has the lowest levels of calcium.

Plant growth

The growth of the stem of the sunflower plant is lower in the support with seaweed + *L. aegyptica* + agar (no significant difference between the agar at 4 and 6%) and is higher in the soil and in the soil with chemical fertilizer (Figures 5 and 6). The diameter of the stem is lower (flexible stem) into the soil and into support with seaweed + *L. aegyptica* + agar 4% and higher (rigid stem) in support with seaweed + *L. aegyptica* + agar 6% and in soil with chemical fertilizer (Figure 7). The growth of leaves is highest in the soil with chemical fertilizer followed by the support with seaweed + *L. aegyptica* + agar (6%) (Figure 8). Figure 9 represents the roots of plants grown in the support with seaweed+ *L. aegyptica* + agar 6% who are visible, thick and crossed. The roots also have white color characteristic of a good development. No root system is observed in the support with seaweed + Luuffa aegyptica + agar 4%. The roots of the plants grown in soil and in soil with chemical fertilizer are also visible, thin, long and pale. After 20th day growth, the sunflower plants began to yellow in the media with chemical fertilizer (Figure 10a), in support with seaweed + *L. aegyptica* + agar (4%) (Figure 10b) and in the soil (Figure 10c). Only the plant grown in the medium with seaweed; *P. pavonica* + *L. aegyptica* + agar (6%) has a good growth (Figure 10d). 66 days after transfer in the soil (Figure 11a) all plants have faded. Only the sunflower plant grown in the support with seaweed *P. pavonica*

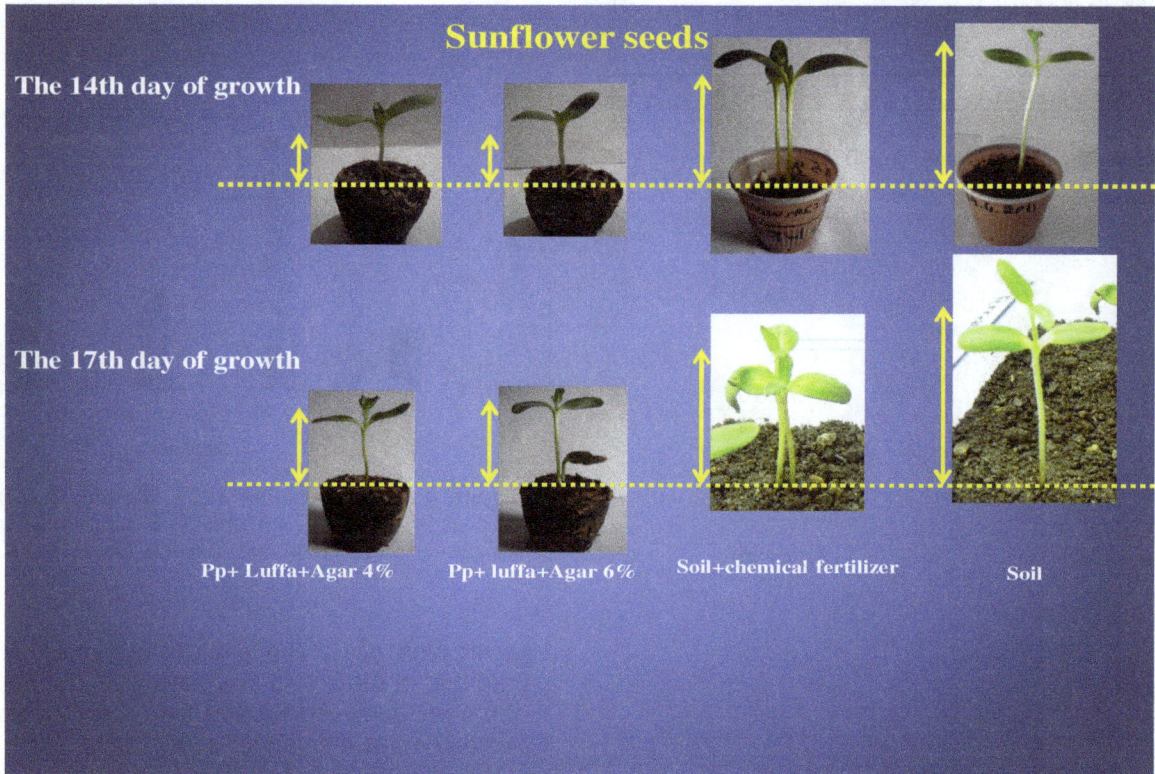

Figure 5. The sunflower plants in the 14th and 17th days of growth.

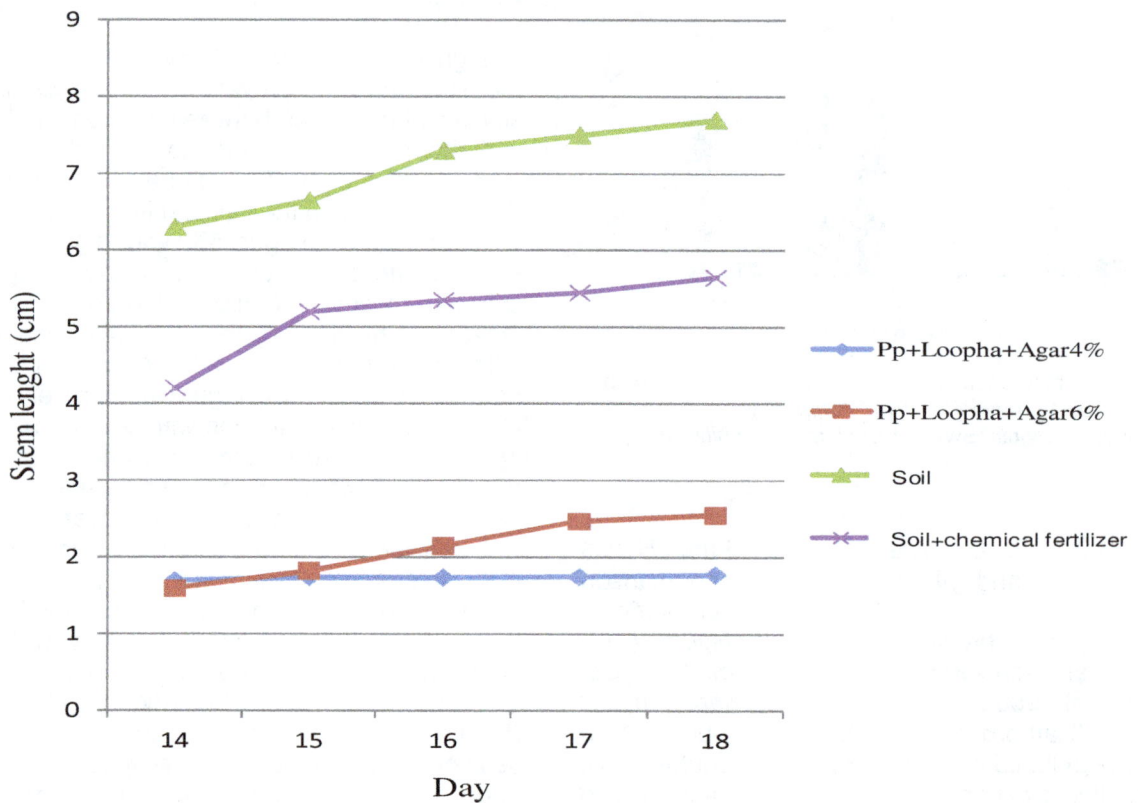

Figure 6. Variation in the length of the stem of sunflower plants over time.

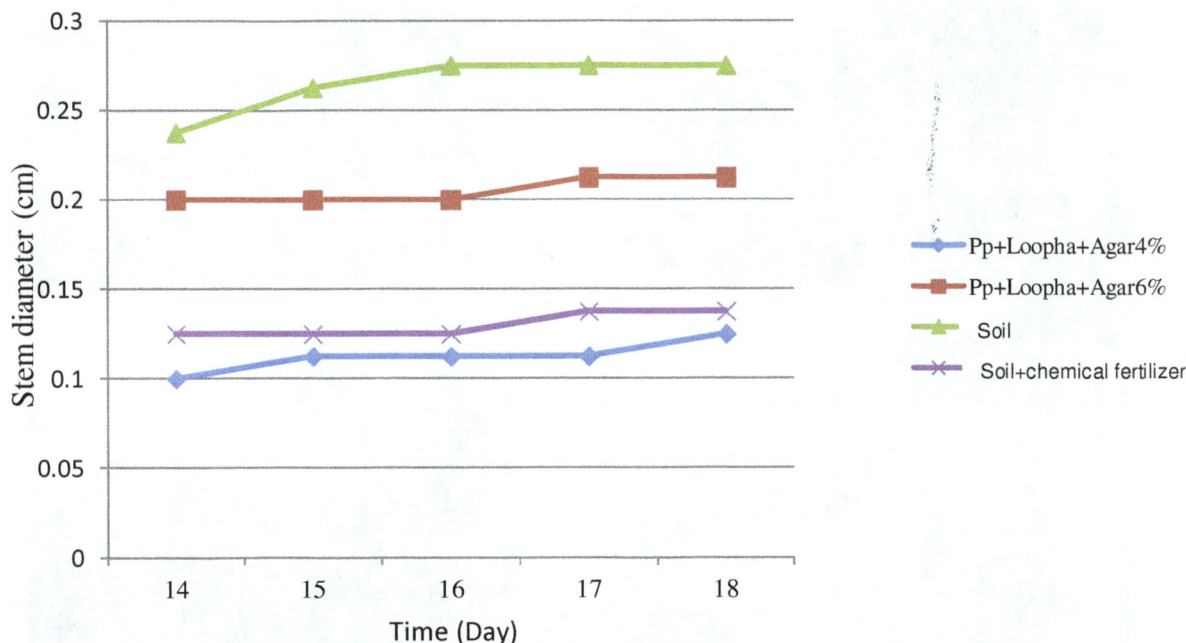

Figure 7. Variation in the diameter of the stem of sunflower plants over time.

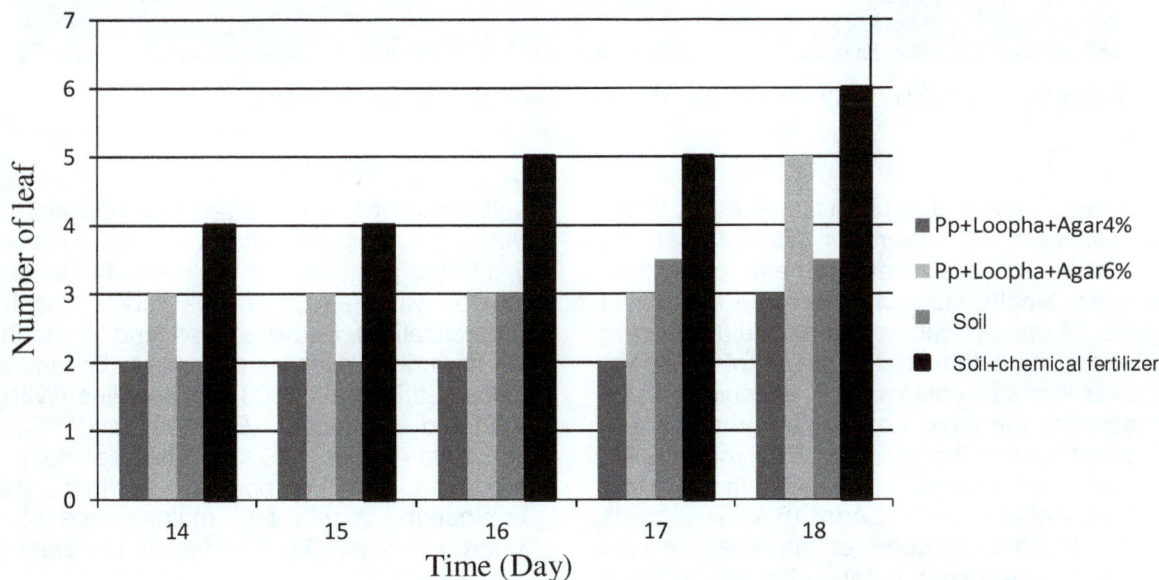

Figure 8. Variation in number of leaves of sunflower plants during 18 days.

+ *L. aegyptica* + agar 6% continued to grow in the soil (Figure 11b).

Degradation and water holding capacity

After 4 weeks, degradation fragments of the support containing the seaweed + *L. aegyptica* + Agar (Figure 12c) is higher than that of fragments containing only fiber *L. aegyptica* and agar (Figure 12b). As expected, no degradation of the plastic fragment is observed (Figure 12a). The support (Seaweed + Luffa + Agar) has a water holding capacity which is three times that of the support with the soil (Table 2).

DISCUSSION

P. pavonica nutrients (N, P, K, Ca, S) composition is not negligible compared to other brown seaweed; *S. wightii*,

Figure 9. The root system in the four media at the end of the eighteenth day of growth.

a green seaweed; *C. chemnitzia* (Sivasankar et al., 2006) and a red seaweed; *K. alvarezzii* (Rathore et al., 2009). The results of the measurement of growth parameters (stem length, stem diameter and number of leaf) during the 18 days of culture did not yield expected results cited in literature. Indeed, the growth of sunflower plants was lower in media containing *P. pavonica* + Luffa + Agar compared to the growth of plants in the soil with chemical fertilizer and in the soil only. The root system after 18 days of growth was more developed in the media containing *P. pavonica* + Luffa + Agar (6%). The plants grown in soil with chemical fertilizer and medium with seaweed and Agar (4%) began to fade after the 20th day of growth up until they started to wilt. The plant in the medium with seaweed + Luffa + Agar (6%) continued to grow.

This implies that, since the algae was added to the medium in mashed form without being submitted to a decomposition process, such as transformation into a powder form, and because of its ability to retain water and mineral elements and slowly release them to the plant (capacity of water storage) on a longer period of time. The algae can retain nutrients and release them slowly to the plants. It contains important nutrients such as nitrogen (N), phosphorus (P) and potassium (K), calcium (Ca) and sulfate (SO_4^{2-}) that will become available to the plants after decomposition (Khan et al., 2009). The decomposition of the algae is taken in charge by micro-organisms that degrade the algae into carbon dioxide, water and nutritious substances for the algae (mineralization). Seaweed also tend to enrich the medium with hormones such as cytokinins (Durand et al., 2003), auxins (Stirk et al., 2003), gibberellins (Wildgoose et al., 1978) and betaines (Blunden and Gordon, 1986; Whapham et al., 1993). They contain in addition, precursors of compounds elicitors that promote germination, growth and maintenance of plant health (Kloareg et al., 1996). Our results correlate with those in the literature.

Indeed, several scientific studies (Blunden, 1991; Crouch and Van Staden, 1994; Moore, 2004; Norrie and Keathley, 2006; Hong et al., 2007; Khan et al., 2009; Kumar and Sahoo, 2011) have demonstrated the action of seaweed extracts in stimulating plant growth. These extracts are known to enhance the growth of vegetables, fruits and other crops (Blunden, 1991; Crouch and van Staden, 1994; Washington et al., 1999).

In addition, when applied to seeds, added to the soil or sprayed on crops and vegetative stages of flowering, seaweed extracts can stimulate seed germination (Hong et al., 2007), growth (Moore, 2004; Khan et al., 2009) and yield of various crops (Arthur et al., 2003; Norrie and

Figure 10. Aspect of plant grown in different medias and transferred in the soil at the 20th day a: Plant grown in soil with chemical fertilizer, b: Plant grown in support with seaweed + Luffa + Agar (4%), c: Plant grown in soil, d: Plant grown in support with seaweed + Luffa + Agar (6%).

Keathley, 2006; Masny et al., 2004; Ei-Zeiny, 2007; Kumar and Sahoo, 2011). The observation of good color and good root and good leaf development in support with seaweed and Agar (6%), compared to those observed in the soil and the soil with chemical fertilizer, could be explained by an efficient uptake of nutrients at these levels. This is consistent with Mugnai et al. (2007) who argues that bio-stimulants, even those containing varying

Figure 11. (a): Sunflower plant after 66th day growth in the soil, (b): in support with *Padina pavonica* + Luffa + agar (6%) and transferred in soil after 18 days.

Figure 12. Results of degradation test after 4 weeks. (a): Plastic fragment. (b): Fragment of support with Luffa + agar. (c) Fragment of support with seaweed (*Padina pavonica*) + Luffa + Agar.

levels of mineral fertilizers, cannot provide all the essential nutrients in amounts that plants need (Schmidt et al., 2003), but one of their main functions is to increase the absorption of minerals by the plant, thus improving nutrient use efficiency at the roots (Vernierie et al., 2005) and leaves (Mancuso et al., 2006).

Table 2. Results of water holding capacity test of support with seaweed+ Luffa+ Agar and of soil sample.

Parameter	Weight of the dry sample (g)	Weight of Saturated sample (g)	Amount of water (g) retained by 100 g of sample
Biofertilizer support	15	66.8	445.33
Soil	96	143.3	149.27

According to Metting et al. (1990), Jeannin et al. (1991) applications of seaweed extract reduced the shock of transplanting in souci seedlings, cabbage (Aldworth and van Staden, 1987) and tomato (Crouch and van Staden, 1992) by increasing the size and strength of the root (Khan et al., 2009). Products made from seaweed stimulate growth and root development, this stimulatory effect was more pronounced when the extracts were applied to an early growth stage of corn, and the response was similar to that of auxin, an important hormone promoter of root growth (Jeannin et al., 1991). Other authors (Ouhssine et al., 2007) have shown that the use of waste as a fertilizer for algae resulted in a major growth of maize root growth allowing an important and also that of the aerial part which is found by the number of leaves, diameter and stem length and yield.

Increased growth of plants, development of deep roots, improved training of lateral roots (Atzmon and van Staden, 1994) and the increase of the total root system were also reported by Slavik (2005). Similarly, treatment with algae extracts has improved both the ratio root/shoot and biomass accumulation in tomato seedlings by stimulating root growth (Crouch and van Staden, 1992). The biostimulants in general are able to affect root development by improving both the formation of lateral roots (Atzmon and van Staden, 1994; Vernieri et al., 2005) and the increase of the total root system (Thompson, 2004; Slavik, 2005; Mancuso et al., 2006). Improved root systems could be influenced by endogenous auxins and other compounds in the extracts (Crouch and Van Staden, 1992).

The seaweed extracts improve the absorption of nutrients through the roots causing additional and strong overall growth of the plant (Crouch et al., 1990). In addition, abiotic stresses such as drought, salinity and temperature extremes can affect the performance of most crops (Wang et al., 2003) and limit agricultural production worldwide. Collectively, studies suggest that the products generate algal tolerance to abiotic stress in plants and bioactive substances derived from algae confer tolerance to stress and improve plant performance (Khan et al., 2009).

The total absence of undesirable elements in the analysis of the aqueous extract of *P. pavonica* such as nickel, plumb, chromium, aluminum and cadmium means that the algae will not contribute to soil pollution. This reflects the quality of sea water in which they were collected. After 4 weeks, the biodegradability of the substrate produced is verified by the observation of a degradation fragment containing the seaweed

P. pavonica higher than that of fragments containing only fiber Luffa. No degradation of the plastic fragment is observed. These results confirm that the algae are effective in the degradation of the material in the ground. This contribution is particularly due to their nitrogen supply to soil microorganisms responsible for degradation.

The support achieved in this study presented a high capacity to retain water; this is due probably to the use of lignocellulosic fibers, contained in the mature fruit of the species *L. aegyptica*. The grinding of plant natural fibers gives advantageously calibrated granulation glitter that can be introduced into the composition of the carrier as an agent for water retention that is 100% natural and can thus be naturally degraded by microorganisms. The large capacity (345%) of the support to hold irrigation water would explain the first assumption that the water retained in the media could have a delayed effect on plant growth. This water allows the slow diffusion of the nutrients of the algae into the medium at a certain concentration.

In terms of costs, despite appearances, organic farming is as competitive as conventional farming. Biological agriculture almost certainly requires more employees and control because of the lack of use of pesticides. In addition, it is often limited in terms of area and is very costly in terms of certifications. But on the other hand, conventional farming has many negative externalities caused by the misuse of the land in excess pesticides, negative impacts on health, etc... Thus, several feasibility studies have shown that if these externalities were included in the calculations, organic production is at least as expensive as conventional production.

Conclusion

This study had as a main objective the production of a plant growing support which is both biodegradable and of low manufacturing cost. The constituent materials are seaweed (*P. pavonica*), vegetable fiber (Luffa) and Agar-agar. According to this composition, this material has been the subject of no scientific publications to date.

The results of our study have shown that on one hand, the support based on algae does not have a significant organic fertilizer short-term effect on growth of sunflower plants. But it does increase the duration of growth and increases the quality of the plants. Regarding the test of capacity to retain water produced, the media containing the Luffa has a holding capacity of water higher than the soil and equivalent three and a half times its dry weight. This contributes very much to lower water expenses.

Similarly, the biodegradability test conducted revealed that the degradation of the media containing algae is higher than that of the substrate containing only Luffa. These results confirm that algae contribute to the degradation of the soil by supplying nitrogen for the microorganisms responsible for degradation.

This embodiment is particularly advantageous since it provides a growing medium that can be planted with the plant; it allows the overcoming of the non-biodegradable plastics constituting the pots in which plants are proposed to consumers. On the other hand, the cost of support base made of seaweed (*P. pavonica*), vegetable fiber (Luffa) and Agar-agar is very affordable compared to the cost of the plastic pot when you consider also that it is biodegradable and does not contribute thereby to the pollution of soil and the environment.

REFERENCES

Aldworth SJ, van Staden J (1987). The effect of seaweed concentrates on seedling transplants. S. Afr. J. Bot. 53:187-189.

Arthur GD, Stirk WA, Van Staden J (2003). Effect of a seaweed concentrate on the growth and yield of three varieties of *Capsicum annuum*. S. Afr. J. Bot. 69:207-211.

Atzmon N, Van Staden J (1994). The effect of seaweed concentrates on the growth of *Pinus pinea* seedlings. New For. 8:279-288.

Blunden G, Gordon SM (1986). Betaines and their sulphono analogues in marine algae. Progress in phycological research. Biopress Ltd. Bristol. 4:39-80.

Blunden G (1991). Agricultural uses of seaweeds and seaweed extracts. Seaweed resources in Europe: uses and potential. Wiley, Chicester. pp. 65-81.

Cabioc'h J, Floc'h JY, Le Toquin A, Boudouresque CF, Meinesz A, Verlaque M (2006). Guide des algues des mers d'Europe. Delachaux et Niestlé S.A., Paris. P. 272.

Chbani A, Mawlawi H, Etahiri S (2011). Antibacterial activity of brown seaweed extracts of *Padina pavonica* from the Mediterranean coast of Lebanon. J. Phytothérapie 9:283-286.

Crouch IJ, Van Staden J (1992). Effect of seweed concentrate on the establishment and yield of greenhouse tomato plants. J. Appl. Phycol. 4:291-296.

Crouch IJ, Van Staden J (1994). Commercial seaweed products as biostimulants in horticulture. J. Home Consum. Hortic. 1:19-76.

Crouch IJ, Beckett RP, Van Staden J (1990). Effect of seaweed concentrate on the Growth and mineral nutrition of nutrient stressed lettuce. J. Appl. Phycol. 2:269-272.

Durand N, Briand X, Meyer C (2003). The effect of marine bioactive substances (NPRO) and exogenous cytokinins on nitrate reductase activity in *Arabidopsis thaliana*. Phys. Plant. 119:489-493.

Ei-Zeiny OAH (2007). Effect of bio-fertilizers and root exudates of two weeds as a source of natural growth regulators on growth and productivity of bean plants (*Phaseolus vulgaris* L.). Res. J. Agric. Biol. Sci. 3:440-446.

Hong DD, Hien HM, Son PN (2007). Seaweeds from Vietnam used for functional food, medicine and biofertilizer. J. Appl. Phycol. 19:817-826.

Jeannin I, Lescure JC, Morot-Gaudry JF (1991). The effects of aqueous seaweed sprays on the growth of maize. Bot. Mar. 34:469-473.

Khan W, Rayirath UP, Subramanian S, Jithesh MN, Rayorath P, Hodges DM, Critchley AT, Craigie JS, Norrie J, Prithiviraj B (2009). Seaweed Extracts as Biostimulants of Plant Growth and Development. J. Plant Growth Regul. 28(4):386-399.

Kloareg B, Broquedis M, Joubert JM (1996). Effets éliciteurs des biostimulants. L'Arboriculture Fruitière 498:39-42.

Ktari L, Guyot M (1999). A cytotoxic oxysterol from the marine alga *Padina pavonica* (L.) Thivy. J. Appl. Phycol. 11(6):511-513.

Kumar G, Sahoo D (2011). Effect of seaweed liquid extract on growth and yield of *Triticum aestivum* var. Pusa Gold. J. Appl. Phycol. 23(2):251-255.

Lakkis S, Novel-Lakkis V (2000). Distribution of pytobenthos along the coast of Lebanon, Mediterr. Marine Sci. 1(2):143-164.

Mancuso S, Azzarello E, Mugnai S, Briand X (2006). Marine bioactive substances (IPA extract) improve ion fluxes and water stress tolerance in potted *Vitis vinifera* plants. Adv. Hortic. Sci. 20:156-161.

Masny A, Basak A, Żurawicz E (2004). Effects of foliar applications of KELPAK SL and GOËMAR BM 86® preparation on yield and fruit quality in two strawberry. J. Fruit Ornam. Plant Res. 12:23-27.

Mawlawi H, Chbani A, Naja K (2012). Antioxidant and Antifungal activities of *Padina Pavonica* and *Sargassum Vulgare* from the Lebanese Mediterranean Coast. Adv. Environ. Biol. 6(1):42-48.

Metting B, Zimmerman WJ, Crouch IJ, Van Staden J (1990). Agronomic uses of seaweed and microalgae. In: *Introduction to Applied Phycology* (ed. AKATSUKA I.), SPB Academic publishing, Netherlands. pp. 589-627.

Moore KK (2004). Using seaweed compost to grow bedding plants. BioCycle 45:43-44.

Mugnai S, Azzarello E, Pandolfi C, Salamagne S, Briand X, Mancuso S (2007). Enhancement of ammonium and potassium root influxes by the application of marine bioactive substances positively affects *Vitis vinifera* plant growth. J. Appl. Phycol. 20:177-182.

Norrie J, Keathley JP (2006). Benefits of *Ascophyllum nodosum* marine-plant extract applications to 'Thompson seedless' grape production. (Proceedings of the Xth International Symposium on Plant Bioregulators in Fruit Production, 2005). Acta. Hortic. 727:243-247.

Ouhssine K, Ouhssine M, El Yachioui M (2007). L'application des déchets traités de l'algue *Gelidium sesquipédale* dans la culture du Maïs. Afrique Sci. 3(2):259-270.

Rathore SS, Chaudhary DR, Boricha GN, Ghosh A, Bhatt BP, Zodape ST, Patolia JS (2009). Effect of seaweed extract on the growth, yield and nutrient uptake of soybean (*Glycine max*) under rainfed conditions. South Afr. J. Bot. 75:351-355.

Schmidt RE, Ervin EH, Zhang X (2003). Questions and answers about biostimulants. Golf Course Manage. 71:91-94.

Sivasankar S, Venkatesalu V, Anantharaj M, Chandrasekaran M (2006). Effect of seaweed extracts on the growth and biochemical constituents of *Vigna sinensis*. Bioresour. Technol. 97(14):1745-1751.

Slavik M (2005). Production of Norway spruce (*Picea abies*) seedlings on substrate mixes using growth stimulants. J. For. Sci. 51:15-23.

Stirk WA, Novak O, Strnad M, Van Staden J (2003). Cytokinins in macroalgae. Plant Growth Regul. 41:13-24.

Thompson B (2004). Five years of Irish trials on biostimulants: the conversion of a skeptic. USDA For Serv. Proc. 33:72-79.

Vernieri P, Borghesi E, Ferrante A, Magnani G (2005). Application of biostimulants in floating system for improving rocket quality. J. Food Agric. Environ. 3:86-88.

Wang Z, Pote J, Huang B (2003). Responses of cytokinins, antioxidant enzymes, and lipid peroxidation in shoots of creeping bentgrass to high root-zone temperatures. J. Am. Soc. Hortic. Sci. 128:648-655.

Washington WS, Engleitner S, Boontjes G, Shanmuganathan N (1999). Effect of fungicides, seaweed extracts, tea tree oil and fungal agents on fruit rot and yield in strawberry. Aust. J. Exp. Agric. 39:487-494.

Whapham CA, Blunden G, Jenkins T, Hankins SD (1993). Significance of betaines in the increased chlorophyll content of plants treated with seaweed extract. J. Appl. Phycol. 5:231-234.

Wildgoose PB, Blunden G, Jeewers k (1978). Seasonal variation in gibberellins activity of some species of Fucaceae and Laminariaceae. Bot. Mar. 21:63-65.

4

Molecular detection of two cassava *Begomoviruses* in some parts of Southern Nigeria

Eni A. O. and Fasasi D. K.

Department of Biological Sciences, College of Science and Technology, Covenant University, KM 10 Idiroko Road, Canaanland, P. M. B. 1023 Ota, Ogun State, Nigeria.

Cassava mosaic disease (CMD), caused by an array of is the most economically important viral disease of cassava in sub-Saharan Africa. The most frequently reported in West Africa are African cassava mosaic virus (ACMV) and East African cassava mosaic Cameroon virus (EACMCV). In this study, 42 cassava leaves and 30 symptomatic weeds belonging to the Asteraceae, Cucurbitaceae and Leguminosae families were collected from backyard gardens in Edo, Ondo, Anambra, and Delta States in 2009. Deoxyribonucleic acid (DNA) extracts from these leaves were tested for ACMV and EACMCV in a multiplex polymerase chain reaction (PCR) assay. The PCR primers used were designed to amplify the replicase regions of DNA-A components of both viruses. Most of the cassava plants within the survey area were either symptomless or showed mild symptoms. ACMV was detected in 16% of cassava leaves from Edo State but not in any of the cassava leaves from the other three states. One weed sample each from Edo State (5.56%) and Ondo State (10%) were also positive for ACMV. EACMCV was not detected in any of the samples tested. The low virus occurrence observed from PCR results and the observed low incidence of the CMD characteristic mosaic symptoms on cassava leaves in the states sampled may be attributed to the use of CMD resistant or tolerant cassava varieties, and may be a result of the massive distribution of virus resistant cassava cuttings to these States by the International Institute of Tropical Agriculture (IITA).

Key words: African cassava mosaic virus (ACMV), East African cassava mosaic Cameroon virus (EACMCV), multiplex polymerase chain reaction (PCR).

INTRODUCTION

Cassava (*Manihot esculenta*) is one of the leading food and feed plants of the world, particularly in Africa (FAO, 2008; Nassar and Ortiz, 2007). It is grown mainly for its enlarged starch-filled root, which is used for human consumption and, commercially, for the production of animal feed and starch-based products (O'Hair, 1990). Young tender cassava leaves are consumed as vegetables in many regions in Africa as a source of protein, vitamins and mineral salts (Legg and Fauquet, 2004).

Cassava mosaic disease (CMD) is the major constraint to cassava production in Africa and is caused by eight distinct of the family *Geminiviridae*, commonly referred to as Cassava mosaic (CMGs) (Fauquet and Stanley, 2003; Thresh and Cooter, 2005; Fauquet et al., 2008). CMD results in stunting and severe reduction in the yield of the desired tuberous cassava root and is thus a production threat to cassava which feeds over 200 million people in sub-Saharan Africa. CMD has been reported as occurring at varying levels of incidence throughout the cassava belt of Africa (Ogbe et al., 2003b; Thresh et al., 1997). For several years after African cassava mosaic virus (ACMV) was confirmed as the causal agent of CMD, it remained the only known causal agent of CMD in Nigeria until the 1990s when East African cassava mosaic Cameroon virus (EACMCV) and several variants of the EACMV were diagnosed as additional causative agents (Ogbe et al., 2003a; 2003b; Ariyo et al., 2005; Ogbe et al., 2006; Alabi et al., 2008b).

Cassava plants infected with CMGs express a range of symptoms depending on factors like the virus strain or species, environmental conditions and the susceptibility of the cassava host (Legg and Thresh, 2003). CMD-affected plants are stunted with conspicuous foliar symptoms such as mosaic and chlorosis interfering with the photosynthetic ability of the plant to produce food for storage in the roots, thus, infected plants produces greatly diminished or no tuberous roots.

This study reports the testing of cassava and symptomatic weeds collected from backyard gardens in some states in Southern Nigeria, for ACMV and EACMCV.

MATERIALS AND METHODS

Sample leaves from cassava plants showing the CMD characteristic mosaic symptom were collected from farms in Edo, Delta, Ondo and Anambra States in July, 2009. A few non-symptomatic cassava leaves were also collected. Leave samples were also collected from symptomatic *Chromolaena odorata* (L.) King and Robinson, *Centrosema pubescens* Benth., *Senna alata* (Linn.) Roxb., and *Cucurbita* spp weeds within the vicinities of the sampled gardens, to ascertain their role as alternate/reservoir hosts for the viruses. Leaf samples were kept on ice while in the field and processed the same day by splicing and drying (rapidly) in anhydrous calcium chloride. Total deoxyribonucleic acid (DNA) was extracted from cassava and weed samples as previously described by Dellaporta et al. (1983).

The polymerase chain reaction (PCR) mixture consisted of 2.2 × PCR buffer, 0.25 mM of each dNTP, 0.533 μmol of each of the primers and 1 U of Taq DNA polymerase (Promega, Madison W1). The primers used for the multiplex PCR were CMBRep/F (5'-CRTCAATGACGTTGTACCA-3'), ACMVRep/R (5'-CAGCGGMAGTAAGTCMGA-3') and EACMVRep/R (5'-GGTTTGCAGAGAACTACATC-3') (Alabi et al., 2008b). Cycling conditions consisted of one cycle of denaturation at 94°C for 1 min, annealing at 48°C for 30 s and extension at 72°C for 1 min, followed by 36 cycles of denaturation at 94°C at 1 min, annealing at 48°C for 30 s and extension at 72°C for 1 min and a final extension at 72°C for 5 min.

The PCR amplified products were resolved by agarose gel electrophoresis and visualized under ultraviolet (UV) light using a gel documentation system. A 100 bp DNA molecular weight marker (Promega Corporation, Madison, W1) was run in each gel as a reference to estimate the size of the virus-specific DNA band in the PCR amplified products.

RESULTS

A total of 42 cassava leaves and 30 weeds were collected for virus testing. Seven (7) of the cassava leaves were non-symptomatic, while 35 were symptomatic. All the weeds tested showed leaf mosaic

symptoms, and they were tested to determine their role as alternative host for the viruses. The distribution of leaf mosaic both on cassava and weeds across the states is shown in Figure 1.

The expected amplicon sizes for ACMV and EACMCV are 368 and 650 bp, respectively. Only 8.33% of the 72 samples produced the expected amplicon size of 368 bp for ACMV. Four of the positive samples were cassava leaves from Edo state; three of these were symptomatic, while one was non-symptomatic (Table 1). One *C. pubescens* leaf sample from Edo State and one *S. alata* sample from Ondo State tested positive for ACMV. None of the samples from Delta and Anambra States tested positive to ACMV, and EACMCV was not detected in any of the samples tested in all the States.

DISCUSSION

This study describes our quest to determine the occurrence of ACMV and EACMCV in cassava and associated weeds in backyard gardens in some parts of Southern Nigeria. Most cassava plants in the gardens surveyed were non-symptomatic. The high incidence of uninfected cassava plants observed in this study differs from reports from previous surveys in Nigeria where high incidence of CMD symptoms were reported in cassava fields (Ogbe et al., 2003a, 2003b, 2006). This difference may be due to the massive distribution and use of healthy, resistant, and/or tolerant cassava varieties by the farmers in the sampled areas (Manyong et al., 2000). This clearly indicates that increasing farmers' awareness on the cause and nature of a disease and possible control measures ensures their participation in the eradication of the diseases. The control of CMD is achieved mainly by the use of resistant/tolerant varieties and by employing adequate phytosanitry techniques of destroying diseased plants and by the use of healthy stems.

The positive detection of ACMV in some of the weeds tested confirms previous reports that weeds serve as alternate host for (Monde et al., 2010; Alabi et al., 2008a; Ogbe et al., 2006) and further buttresses the role of phytosanitation in the control of ACMV. This information is also of epidemiological importance because infected weeds may serve as virus reservoirs and inoculum foci.

Although, most positive samples were symptomatic, the detection of ACMV in one of the non-symptomatic cassava leaves indicates the possibility of latent infection. This phenomenon had been previously reported during field surveys in 2002 and 2003 (Ogbe et al., 2006), and further highlights the need for laboratory certification of cassava planting materials.

The negative EACMCV result obtained for all the leaves tested may indicate either the absence of the virus in the fields sampled, or an extremely low incidence. The high incidence of negative symptomatic cassava leaves

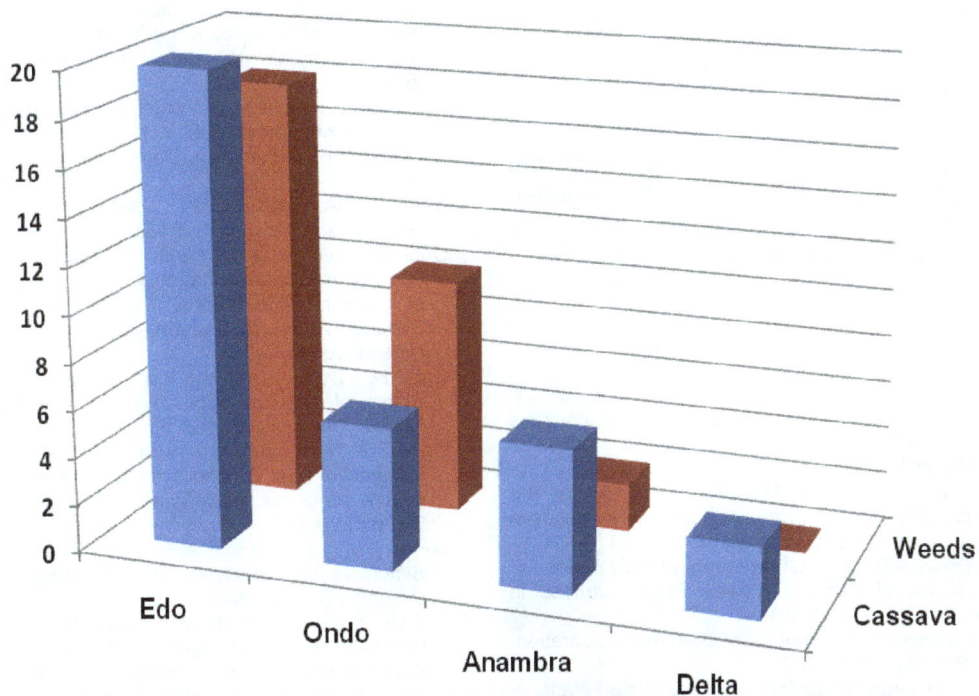

Figure 1. Distribution of leaf mosaic symptom on cassava leaves and weeds collected from Edo, Ondo, Anambra, and Delta States, Nigeria.

Table 1. Summary of cassava and weed samples collected and the result of multiplex PCR testing for ACMV and EACMCV in the surveyed samples from Edo, Delta, Ondo and Anambra states, Nigeria.

State	Plants	Number collected		ACMV		EACMCV	
		S	NS	S	NS	S	NS
Edo	Cassava	20	5	3	1	0	0
	Weed	18	0	1	0	0	0
Ondo	Cassava	6	0	0	0	0	0
	Weed	10	0	1	0	0	0
Anambra	Cassava	6	0	0	0	0	0
	Weed	2	0	0	0	0	0
Delta	Cassava	3	2	0	0	0	0
	Weed	0	0	0	0	0	0
Total		65	7	5	1	0	0

S, Symptomatic; NS, Non symptomatic.

may be indicative of the occurrence of other not tested for in this study and requires further investigation.

CMD remains a threat to cassava production in several African countries. The chances of emergence of new strains/variants of the causative are also very high due to the frequent reports of dual/multiple infections (Fondong et al., 2000; Harrison et al., 1997). Constant surveillance, phytosanitation, and improvement on existing diagnostics will continue to play major roles in ensuring that Nigeria retains its non-epidemic status and low epidemic index (Legg and Owor, 2003). It is also important to ensure that existing quarantine measures continue to be enforced.

ACKNOWLEDGEMENTS

We thank Dr. S. I. Smith and Mrs. Fowora of the Molecular and Biotechnology Department of the Nigerian Institute of Medical research for bench space, laboratory equipment and assistance to carry out this work.

REFERENCES

Alabi OJ, FO Ogbe, Bandyopadhyay R, Kumar PL, Dixon AGO, d'A Hughes J, Naidu RA (2008a). Alternate hosts of African cassava mosaic virus and East African cassava mosaic Cameroon virus in Nigeria. Arch. Virol. 153(9):1743-1747.

Alabi OJ, Kumar PL, Rayapati AN (2008b). Multiplex PCR for the detection of African cassava mosaic virus and East African cassava mosaic Cameroon virus in cassava. J. Virol. Methods 154:111-120.

Ariyo OA, Koerbler M, Dixon AGO, Atiri GI, Winter S (2005). Molecular Variability and Distribution of Cassava Mosaic Begomoviruses in Nigeria. J. Phytopathol. 153(4):226-231(6).

Dellaporta SL, Wood JJ, Hicks JB (1983). A plant DNA minipreparation: version II. Plant Mol. Biol. Rep. 1:19-21.

FAO (2008). Food and Agricultural Organization of the United Nations. Production Yearbook. FAO Statistics 2007. Rome, Italy: http://faostat.fao.org/.

Fauquet CM, Briddon RW, Brown JK, Moriones E, Stanley J, Zerbini M, Zhou X (2008). Geminivirus strain demarcation and nomenclature. Arch. Virol. 153:783-821.

Fauquet CM, Stanley J (2003). Geminivirus classification and nomenclature: progress and problems. Ann. Appl. Biol. 142:165-189.

Fondong V, Pita JS, Rey MEC, de Kochko A, Beachy RN, Fauquet CM (2000). Evidence of synergism between African cassava mosaic virus and the new double recombinant geminivirus infecting cassava in Cameroon. J. General Virol. 81:287-297.

Harrison BD, Zhou X, Otim-Nape GW, Liu Y, Robinson DJ (1997). Role of a novel type of double infection in the geminivirus-induced epidemic of severe cassava mosaic in Uganda. Ann. Appl. Biol. 131:437-448.

Legg JP, Fauquet CM (2004). Cassava mosaic geminiviruses in Africa. Plant Mol. Biol. 56:585-599.

Legg JP, Owor B (2003). Cassava mosaic disease in Africa: where are the epidemics? Afr. Crop Sci. Conf. Proc. 6:322-328.

Legg JP, Thresh JM (2003). Cassava virus diseases in Africa. In: Proceedings of the First International Conference on Plant Virology in Sub-Saharan Africa (4–8 June 2001, Ibadan, Nigeria), IITA, Ibadan, Nigeria. pp. 517-522.

Manyong VM, Dixon AGO, Makinde KO, Bokanga M, Whyte J (2000). Impact of IITA-improved germplasm on cassava production in sub-Saharan Africa. International Institute of Tropical Agriculture (IITA), Ibadan. Nigeria. P. 11.

Monde G, Walangululu J, Winter S, Bragard C (2010). Dual infection by cassava begomoviruses in two leguminous species (Fabaceae) in Yangambi, Northeastern Democratic Republic of Congo. Arch. Virol. 155(11):1865-1869.

Nassar NMA, Ortiz R (2007). Cassava improvement: challenges and impacts. J. Agric. Sci. 145:163-171.

Ogbe FO, Atiri GI, Dixon AGO, Thottappilly G (2003b). Cassava mosaic disease and its causal agents: the Nigerian situation In: Proceedings of the First International Conference on Plant Virology in Sub-Saharan Africa (4–8 June 2001, Ibadan, Nigeria), IITA, Ibadan, Nigeria. pp. 411-422.

Ogbe FO, Dixon AGO, d'A HughesJ, Alabi OJ, Okechukwu R (2006). Status of cassava begomoviruses and their new natural hosts in Nigeria. Plant Dis. 90:548-553.

Ogbe FO, Thottappilly G, Dixon AGO, Atiri GI, Mignouna HD (2003a). Variants of East African cassava mosaic virus and its distribution in double infections with African cassava mosaic virus in Nigeria. Plant Dis. 87:229-232.

O'Hair SK (1990). Tropical Root and Tuber Crops. In: Janick J, JE Simon (eds.) Advances in new crops. Timber Press, Portland. Oregon. pp. 424-428.

Thresh JM, Otim-Nape GW, Legg JP, Fargette D (1997). African cassava mosaic disease: The magnitude of the problem. In: Contributions of Biotechnology to Cassava for Africa: Proceedings of the Cassava Biotechnology Network, Third International Scientific Meeting (Kampala, Uganda, 26–31 August 1996). A.M. Thro and M.O. Akoroda (Eds.). Afr. J. Root Tuber Crops (Special Issue). pp. 13-19.

Thresh JM, Cooter RJ (2005). Strategies for controlling cassava mosaic virus disease in Africa. Plant Pathol. 54:587-614.

5

The effect of intercropping pattern of okra, maize, pepper on weeds infestations and okra yield

J. A. Orluchukwu[1] and Udensi E. Udensi[1,2]

[1]Department of Crop and Soil Science, University of Port Harcourt, P. M. B 5323 Port Harcourt, Nigeria.
[2]International Institute of Tropical Agriculture (IITA) Cassava Project, South-South and South Eastern Nigeria Zones, Nigeria.

Field study was conducted to evaluate the efficacy of intercropping pattern in reducing weed infestation in okra, maize and pepper intercrop; at the teaching and research farm of Rivers State University of Science and Technology Port Harcourt, Nigeria during 2009 and 2010 cropping season. Three intercropping pattern namely; alternate row intercropping, strip row intercropping and mixed intercropping were compared to sole cropping in a randomized complete block design replicated three times. The result reveal that weed biomass were significantly lower in both years in all forms of intercropping pattern compared to sole cropping or mono-cropping. Weed smothering efficiency in both years showed that mixed pattern (45.7%) >alternate row pattern (33.4%) > strip row pattern (11.5%). Crop yield were better in an intercrop system for maize and pepper in both years compared to sole crop. However, mean okra fruit yield was highest in sole cropping (3253 kg ha^{-1}) when compared to intercropping pattern. Maize yield was highest in mixed pattern (8,987 kg ha^{-1}) and lowest in sole cropping (6,955 kg ha^{-1}) while pepper fruit yield was highest in strip row pattern (5,435 kg ha^{-1}) and lowest in mixed pattern (1,562 kg ha^{-1}). The results from this study have shown that intercropping pattern has a great potential in reducing weed infestation in cropping systems especially in farming system with low external input. However, the effect of the intercrop pattern on yield may be variable, because viability may depend on the adaptation of planting pattern and selection of compatible crops.

Key words: Intercropping pattern, weed infestation, land equivalent ratio (LER).

INTRODUCTION

Okra (*Abelmoschus esculentus* (L.) Moench), is a member of the family Malvaceae widely cultivated mostly by peasant farmers in Nigeria as a fruit vegetable. It is found in almost every market in Nigeria (Akoroda et al., 1985) and Africa (Schippers, 2000). Okra is cultivated over a total area of about 1.5 million ha (Adejonwo et al., 1989). Smallholders in the tropics face the problem of maintaining productivity, due to a range of factors which factors which include weeds. Many crops grown in the the first 3 to 4 months after planting. For increased tropics

are susceptible to early weed competition during within the first 6 to 9 weeks of planting. Weed competition productivity of okra, weeds must be controlled timely reduces canopy development in most crops, and predisposes the crop to pest and disease infestation. Absence of weed control in crop farm may lead to crop losses of up to 100% (Nyam, 2005). Uncontrolled weeds cause okra yield losses ranging from 63 to 91% (Adejonwo et al., 1989). Weed control is by far the most labor-demanding field operation in okra production and the control is currently the cornerstone of increased production in Nigeria. The smallholders groups of farmers rely heavily on the traditional hand weeding for weed control. Hand weeding is probably the oldest method of weed control which has consistently proved inefficient and costly too (Agahiu et al., 2011). It is the popular weed control method used by more than 80% of the resource poor farmers who produces bulk of the food eaten in developing nation. Okra is popularly grown in mixtures with staple food crops such as yam, maize, cassava, cowpea and pepper or with various vegetable crops on small farm holdings (Muoneke and Asiegbu, 1996; Olasotan, 2001; Odeleye et al., 2005). The use of herbicides for weed control in such an intercropping system has not been found workable or popular, especially in small farm holdings, where various crops are planted simultaneously. The use of intercrop to smother weeds has been successful (Rao and Shetty, 1976). Recent studies have also addressed intercropping as an option for an integrated weed management, particularly in farming systems with low external inputs (Liebman and Davis 2000; Rana and Pal, 1999). It seems when used in conjunction with correct timing of hoe-weeding, the practice could prove satisfactory to smallholder farmers (Agahiu et al., 2011). Its appeal is enhanced by the added food value obtained from the component crops. The choice of the method must, however, be based on the optimum economic returns and resources available. Studies have indicated that weed population density and biomass production may be markedly reduced using crop intercropping (spatial diversification) strategies (Liebman and Elizabeth, 1993). Intercrop system, light interception and soil cover are usually increased compared with a monoculture, and yield loss due to weed competition is seen to be reduced. Therefore, intercropping can be seen as one option for reducing weed problems through non-chemical methods (Vandermeer, 1989).This study was designed to assess the efficacy of intercropping pattern in reducing weed infestation in okra, maize and pepper intercrop.

MATERIALS AND METHODS

Study area

The study was conducted at the teaching and research farm of the Rivers State University of Science and Technology, Port Harcourt, during the planting season of March, 2009 and April, 2010. Port Harcourt falls within latitude of 4° to 6°N and longitude of 7.010E with an elevation of 18 m above sea level (FAO, 1984). The rainfall pattern is essentially bimodal with peaks in June and September, while in April and August there are periods of lower precipitation. The annual rainfall averaged between 2000 mm and 4500 m (Ukpong, 1992; University of Uyo, 1997). The long rainy season is between April and October, while the dry season lasts from November to March with occasional interruption by sporadic down pours (Anderson, 1967). The mean monthly temperature ranges from 28 to 33°C. The highest temperature is experienced during the months of December through March and coincides with the overhead passage of sun (Enwezor et al., 1990). The experiment

was carried out on a Typic paleudult soil. The soil of the experimental site had the following characteristics: pH, 4.10, total nitrogen, 0.05%; available-P' 28.0 ppm; and K, 21.10 ppm (Allen et al., 1974). Soil analysis revealed the following texture: sand 85.6%, silt 9.0%, and clay 5.4%.

Planting

Three crops, maize, pepper and okra, were the component for the intercropping pattern. The maize cultivar used was Bende white, a local variety. The pepper cultivar NHV$_4$, a high yielding and early maturing variety and okra variety, NIHORT 47-4 were obtained from National Horticultural Research Institute (NIHORT) Ibadan. The experimental design was randomized complete block (RCB) design. Three types of intercropping pattern of okra with maize and pepper was studied to evaluate their influence on weed infestation. The cropping patterns were alternate rows, strip rows, mixed pattern and sole crop of okra as control. The alternate rows pattern was made up of two rows of maize followed by two rows of okra, followed by two rows of pepper, and this arrangement repeated three times to give a plot size of 9 × 3 m. The strip row pattern was made up of six rows of okra, six rows of pepper and six rows of maize. The mixed pattern was made up of a group containing six stands of each crop and randomly planted at six stands on the plot. There were replicated three times. However, the planting distance and number of stands were the same as in other plots. The sole cropping pattern (control) was made up of okra plants as six grown stands and contained a total of 108 stands per plot replicated three times. All crops were planted at 50 × 50 cm in both years, a nursery bed was prepared and pepper seeds planted a month before clearing the main field and were later transplanted. Okra, maize seeds and pepper seedlings were planted the same day. The plots were weeded at 6 weeks after planting (WAP) for all cropping pattern and at 9 WAP for sole cropping pattern only. It was not necessary again to weed the intercropping pattern due to ground cover, this is because the level of infestation will not have any effect on yield whether weeded or not.

Weed species abundance and cover estimate

The determination of weed infestation was made with a quadrat measuring 1 × 1 m, three random sample per plot were taken and the weed cover estimated by means of weed ground cover rate using a scale of 1-6 [where 1(0 to 5% weed cover), 2(5 to 25%), 3(25 to 50%), 4(50 to 75%), 5(75 to 95%) and 6(5 to 100%), Daubenmire, 1968; Ossom, 1986a]. In this scale, 1 represented the minimum weed density; 5 and 6 (all ground space completely covered by weeds) represented the maximum weed coverage. The weed species, and relative abundance were also recorded in each plot. In both years, above-ground weed biomass was determined by taking three quadrats samples of 1 × 1 m long a diagonal transect in each treatment plot at 6 WAP. The weeds were oven- dried at80°C for 48 h for biomass determination. Weed smothering efficiency of the different intercropping pattern was determined based on weed control efficiency according to Subramanian et al. (1991) as follows:

$$WSE\ (\%) = \frac{WDWT\ in\ monocrop - WDWT\ in\ intercrop\ pattern}{WWDT\ in\ monocrop} \times 100$$

Where, WSE = Weed smothering efficiency; WDWT = Weed dry weight.

Crop yield and land equivalent ratio (LER)

All crop yield and yield components were determined to evaluate

Table 1. Effect of intercropping pattern on relative abundance of weed at Port Harcourt in 2009 and 2010.

| Species | Relative abundance (%) | | | | | | | |
| | 2009 | | | | 2010 | | | |
	Sole	Alternate row	Strip	Mixed	Sole	Alternate row	Strip	Mixed
P. maximum	50	35	40	32	46	35	40	30
A. compresssus	30	10	20	5	15	35	20	25
Asphilia africana	16	4	2	8	7	0	5	8
Tridax procumbens	7	7	5	10	10	0	6	7
Sida acuta	6	5	4	2	8	0	7	4

performance from a net plot of 27 m^2. The LER was calculated as:

LER = (Yio/Yso) + (Yim/Ysm) + (Yip/Ysp) Where, Yio and Yso are the yields of okra in intercropped and monocrop, Yim and Ysm are the yields of maize in intercropped and monocrop, and Yip and Ysp are the yields of pepper in intercropped and monocrop, respectively. Where LER was more than 1.0, this indicates a positive intercropping advantage which shows that interspecific facilitation is higher than interspecific competition (Vandermeer, 1989).

Data analysis

Data from the trial were subjected to analysis of variance (ANOVA), and differences between means were separated using least significant difference (LSD) at 5% level of probability.

RESULTS

Weed species abundance and weed cover estimate

Guinea grass (*Panicum maximum)* was the dominant species at the experimental site followed by carpet grass (*Axonopus compressus*). The sole okra plot had the highest number of weed species cover with 3.5 score in 2009 and 3.0 in 2010 (Table 1). *P. maximum* was more abundant in the sole okra plot in both years compared to the various intercrop patterns with 50% (2009) and 46% (2010), respectively. The least weed cover in both years was found in mixed intercrop pattern plots with 2.0 (≤25%) and 1.0 (≤ 5%) weed ground cover, respectively (Figure 1).

Weed biomass and weed smothering efficiency

Weed biomass was significantly affected by cropping pattern (Figure 1). Weed biomass in sole cropping pattern of okra was significantly (P < 0.05) greater than in intercropping pattern with maize and pepper.

The sole okra cropping pattern (control) had the highest weed biomass (330.23 gm^{-2}) in 2009 and (310.85 gm^{-2}) in 2010. The mixed intercropping pattern had the least weed biomass, 185.2 and 163.57 gm^{-2} in 2009 and 2010, respectively (Figure 2). Weed smother efficiency (WSE)

was highest in mixed pattern in both years compared to the other forms of intercrop pattern (Table 2).

Crop yield

The okra fruit yield showed significant differences between cropping pattern in both years (Figure 3). The sole okra had the highest yield in both years 2009 (2857 kg ha^{-1}) and 2010 (3648 kg ha^{-1}) followed by strip row pattern 2453 and 2470 kg ha^{-1} in 2009 and 2010, respectively.

The mixed intercrop pattern was not different from alternate rows intercrop pattern in 2009. Averaged over the years okra fruit yield was as follows: sole okra crop (3253 kg ha^{-1}) > strip row (2462 kg ha^{-1}) > mixed intercrop pattern (2213 kg ha^{-1}) > alternate row pattern (1933 kg ha^{-1}) (Figure 3). Maize yield generally was higher in 2009 than in 2010, and was lower in sole pattern in both years (Figure 4). Maize yield when averaged over the years was as follows: mixed pattern (8,987 kg ha^{-1}) > alternate row pattern (8,220 kg ha^{-1}) > strip row pattern (7,853 kg ha^{-1}) and the least sole maize crop (6,955 kg ha^{-1}) (Figure 4). Averaged over the years pepper fruit yield on the other hand, was highest with the strip row pattern (5,435 kg ha^{-1}) and lowest with mixed pattern (1562 kg ha^{-1}). The year average for yields of alternate row and sole pepper patterns were 1693 and 1683 kg ha^{-1}, respectively (Figure 5).

Land equivalent ratio (LER)

The mean LER values were greater than 1.0 in all intercropping pattern. This means that intercropping pattern showed an advantage over sole cropping in reducing weed dry matter. In both years, strip rows pattern had the highest LER, 4.67 and 5.98, respectively (Table 3). From literature, the pepper yield in the present study is not abnormal.

DISCUSSION

Despite okra wide leaves and low growing canopies, in sole cropping it had the highest weed population and

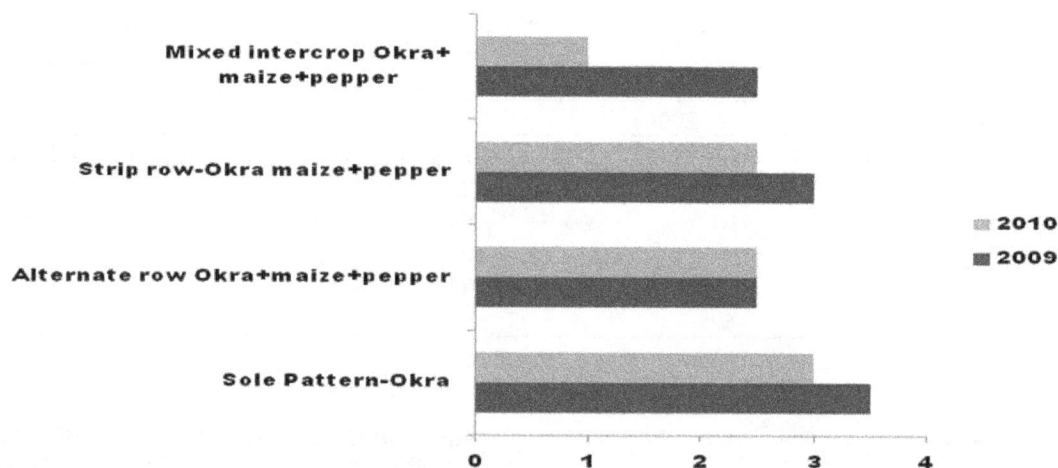

Figure 1. Weed cover estimate [(Cover scale (1-6): 1 = 0 to 5%, 2 = 5 to 25%, 3 = 25 to 50%, 4 = 50 to 75%, 5 = 75 to 95% and 6 = 95 to 100%) Daubenmire cover scale, 1968].

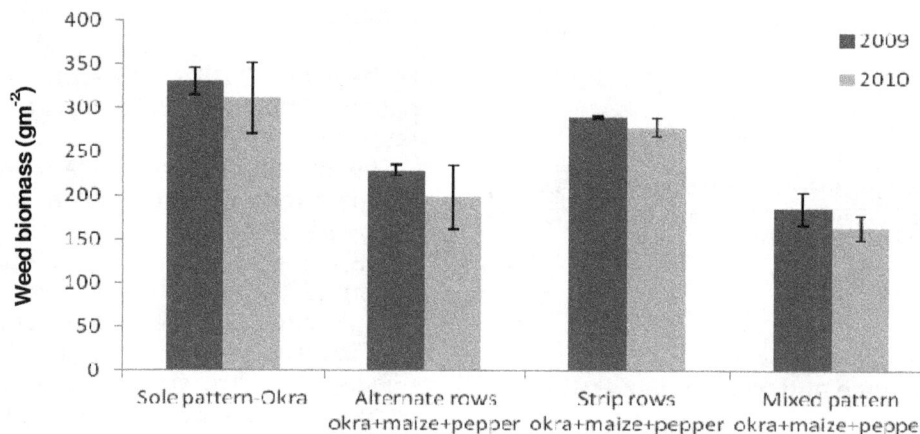

Figure 2. Effect of intercropping pattern of okra with maize and pepper on weed biomass [Error bars are standard error bars (±)].

Table 2. Weed smothering efficiency (WSE) of intercrop pattern.

Intercrop pattern	2009(%)	2010 (%)	Year average
Sole pattern okra	-	-	-
Alternate row - Okra + maize + pepper)	30.65	30.14	33.40
Strip row with- Okra + maize + pepper)	12.45	10.44	11.45
Mixed-pattern with Okra + maize + pepper	43.92	47.38	45.65

biomass in both years. This result corroborates the findings of McGill-Christ and Trenbath (1984) that sole cropping encourages weed growth and development, due mainly to sparse canopy. The low weed incidence in mixed intercropping pattern clearly showed the advantages of dense canopy and close covering of soil surface by crops of different leaf shapes and heights. The results of this work are also in conformity with those reported by Jones (1983), Hague et al. (2008).

The relatively low incidence of weeds in the intercrop plot irrespective of planting patterns in this trial could also be attributed to more photosynthetic active radiation (PAR) interception and possible interference from the component crops, in addition to ground cover effect. This finding corroborates the results of Eskandari and Ghanbari (2010), Eskandari and Kazemi (2011), Tripathi et al. (2008), Chikoye et al. (2006), Hugar and Palled (2008) and Agahiu et al. (2011) on the efficacy of intercrop

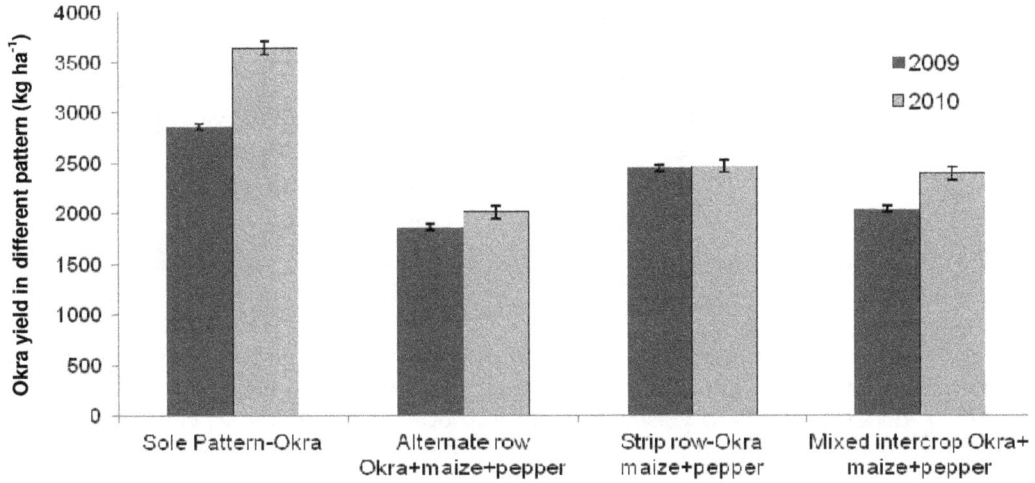

Figure 3. Okra yield in sole and in an intercrop pattern with maize and pepper.

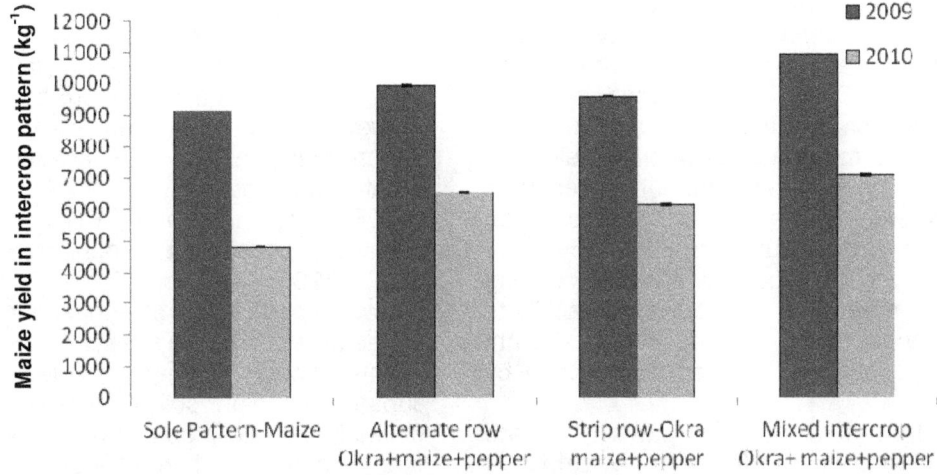

Figure 4. Maize yield in sole and in an intercrop pattern with okra and pepper.

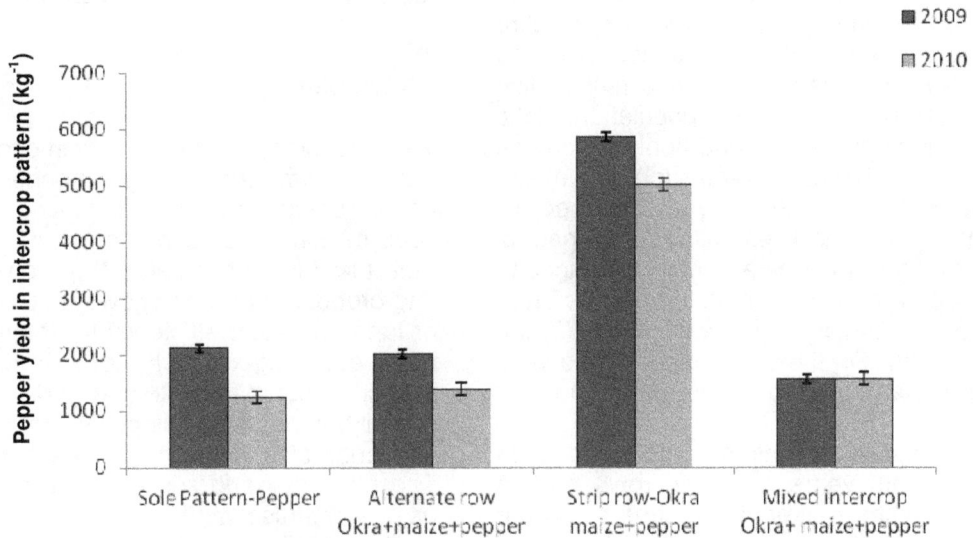

Figure 5. Pepper yield in sole and in an intercrop pattern with okra and maize.

Table 3. Intercropping pattern relative yield and LER.

Intercropping pattern	2009				2010			
	Relative yield of crops			LER	Relative yield of crops			LER
	Okra	Maize	Pepper		Okra	Maize	Pepper	
Alternate row	0.652	1.086	0.947	2.69	0.549	1.364	1.107	3.02
Strip row	0.858	1.049	2.762	4.67	0.677	1.280	4.027	5.98
Mixed	0.712	1.196	0.736	2.64	0,655	1.475	1.257	3.39

intercrop in reducing weed incidence. Similarly, the low weed biomass recorded in intercrop pattern agrees with earlier reports on reduced weed dry weight in intercropping systems (Eskandari and Ghanbari, 2010). Weed smothering efficiency calculated at 6 WAP clearly showed that all types of intercropping pattern had advantage or potentials of smothering weeds compared to sole cropping pattern. This result agrees with Singh et al. (2005), and Shah et al. (2011), on the weed smothering efficiency of intercropping. However, weed smothering efficiency was highest with the mixed intercropping pattern at 6 WAP. Low weed pressure experienced in this study can be attributed to intercropping effect. This result agrees with Maerek et al. (2009), who reported on reduced amount of resource consumption by weeds, in a productivity and weed suppression study of maize-pumpkin intercrops. The morphological and physiological differences among intercrop components may have resulted in their ability to occupy different niches, thus, causing more efficient utilization of natural resources by mixed stands than by pure stands.

In terms of okra fruit yield, the sole pattern had the highest fruit yield in both years followed by strip row intercrop pattern. This result agrees with the findings of Andrews (1972), that sole cropping may promote high productivity in some crops. However, it disagrees with Muoneka and Asiegbu (1996), who reported that in maize-okra intercropping yield, yield components of okra was increased. The high productivity due to sole cropping in okra is as a result of several agronomic factors like easier agronomic operations, plant population, little shading effect and non-competitive and non-interference effect from more aggressive crop competitors like maize (Rosenthal and Janzen, 1979). Shading effect especially from maize may have curtail efficient utilization of natural resources and restrict growth of okra from initial stages to harvest which resulted in yield competition in intercrop. Similar results were obtained by Hussain et al. (2003) and Haque et al. (2008). On the other hand, maize yield were noted higher under intercrop patterns than the sole crop.

This could be as a result of intraspecific competition from maize. During both years, the strip rows pattern clearly outperformed other treatments in terms of pepper fruit weight. This still agrees with the report of Okigbo and Green (1976) on the advantages of an intercropping

system in giving high yields through beneficial interactions from nearby intercrops. Andrews (1972) had consistently obtained yield increase from crops grown in mixture compared with crops grown sole.

Land utilization efficiency of intercrop patterns measured by LER values at all intercrops were higher than 1.0.

This means that land utilization efficiency for okra-maize-pepper intercropping pattern was more advantageous than sole cropping. Averaged over the years, the strip rows intercrop pattern had the best LER, while the alternate row intercrop pattern had the least. However, there was an overall advantage due to intercropping, as the LER in each intercrop pattern was greater than one (>1).

LER greater than one (LER > 1.0) have been reported with various maize intercropping (Saban et al., 2007; Carr et al., 1995). At about 70 DAP, the sole crop of okra showed some signs of senescence that depicted water stress or temperature stress.

This could suggest exposed soil surface and subsequent high soil temperature and moisture loss, unlike in the intercrop plots that had a better soil covering due to crops of different leaf canopies (Cobley, 1976). Therefore, it is proper to say that in addition to weed control advantage due smothering effect on weeds, the intercrop pattern also has the advantages of lowering soil temperature and conserving soil moisture.

Conclusion

Weed smothering efficiency with okra, in an intercropping pattern with maize and pepper suggest that the intercropping pattern achieved acceptable weed suppression benefits than do sole cropping pattern. Also, that the LER of greater than one recorded with the intercropping patterns shows that resource consumption or land utilization efficiency for intercropping pattern was more advantageous than for sole cropping. Choosing of the crop combinations and those crops best able to exploit soil nutrients will plays vital role in harnessing the efficiency of intercropping pattern in smothering weeds. The efficiency and sustainability of intercropping pattern as non-chemical method of weed management especially at the small farm level will depend on the choice of compatible crops and the optimum population to minimize interference. It will form a good component of integrated weed management at the low input farm level.

ACKNOWLEDGEMENT

The authors thank Mr Henry Okolie of the Department of Crop/Soil Science, Rivers State University of Science and Technology, P.M.B 5080, Port Harcourt, Nigeria for his technical support.

REFERENCES

Adejonwo KO, Ahmed, MK, Lagoke STO, KariKari SK (1989). Effects of variety and period of weed interference on growth and yield of okra (*Abelmoschus esculentus*). Nig. J. Weed Sci. 21(1&2):21-28.

Agahiu AE, Udensi UE Tarawali G, Okoye BC, Ogbuji RO, Baiyeri KP (2011). Assessment of weed management strategies and intercrop combinations on cassava yield in the middle belt of Nigeria. Afr. J. Agric. Res. 6(26):5729-5735

Akoroda MO, Ayin OA, Tijani IOA (1985). Edible fruit productivity and harvest duration of okra in southern Nigeria. Trop. Agric. (Trinidad) 63:(20)110-112.

Allen SE, Grinshaw HM. Parkinson JA, Quarmby C (1974). Chemical Analysis of Ecological materials, Oxford, Blackwell, Scientific Publications,

Andrews DJ (1972). Inter cropping with sorghum in Nigeria. Exp. Agric. p. 8. 139-150

Carr PM, Gardner JC, Schatz BJ, Zullnger SW, Guldan J (1995). Grain yield and biomass of a wheat-lentil intercrop. Agron. J. 87:547-579

Cobley LC (1976). An introduction Botany of Tropical Crops. Longman, New York. pp. 43-46.

Daubenmire RF (1968).Plant Communities: A textbook of Plant Synecology. Harper & Row, New York. p. 300.

Chikoye, David Ellis-Jones J, Patrick K, Udensi E.U, Simon EI, Ter-Rumun A (2006). Options for cogongrass (*Imperata cylindrica*) control in white Guinea yam (*Dioscorea rotundata*) and cassava (*Manihot esculenta*). Weed Technol. 20:784-792.

Eskandari H, Ghanbari A (2010). Effect of Different Planting Pattern of Wheat (*Triticum aestivum*) and Bean (*Vicia faba*) on Grain Yield, Dry Matte Production and Weed Biomass. Notulae Scientia Biol. 2(4):111-115.

Eskandari H. K Kazemi K (2011). Weed Control in Maize-Cowpea Intercropping System Related to Environmental Resources Consumption. Notulae Scientia Biologicae 3:57-60

Food and Agricultural Organization, FAO (1984) Agroclimatological Data in Africa, FAO publications, Rome.

Haque M, Sharma RP, Prasad S (2008). Weed control in maize based intercropping system under rainfed condition. *ISWS Biennial Conference on Weed Management in Modern Agriculture,* February 27-28, 2008, Pusa, Bihar. p. 118.

Hugar HY, Palled YB (2008) Effect of intercropping vegetables on maize and associated weeds in maize-vegetable intercropping systems. Karnataka. J. Agric. Sci. 21(2):159-161.

Hussain Nazim, Imran Haider Shamsi, Sherin Khan, Habib Akbar, Wajid Ali Shah (2003) Effect of legume intercrops and nitrogen levels on the yield performance of maize. Asian J. Plant Sci. 2(2):242-246.

Jones HG (1983). Plants and Micro Climate. A Quantitative Approach to Environmental Physiology. *Cambridge University Press.* Cambridge.

Liebman M, Davis AS (2000). Integration of soil, crop and weed management in low-external-input farming systems. Weed Res. 40: 27-47.

Liebman M, Elizabeth D (1993). Crop rotation and intercropping strategies for weed management. Ecol. Appl. 3(1):92-122.

Maereka EK, Madakadze RM, Ngakanda C (2009). Productivity and weed suppression in maize-pumpkin intercrops in small scale farming communities of Zimbabwe. Afr. J. Crop Sci. 9:93-102.

Mcgil-Christ CA, Trenbath BR (1984). A Revised Analysis of Plant Competition Experiments. Biometrics 27:659-671.

Muoneke CO, Asiegbu JE (1996). Evaluation of growth and yield advantage of okra and cowpea sown in mixture, pp. 100-105. In: Adedoyin, S. and A. Ayelaagbe (Eds.). Proceedings of 14th Hort. Soc. of Nig. Conference Held at Ago-Iwoye.

Nyam T (2005). Opening address: In Chikoye, D., J. Ellis-Jones, A. F. Lum, G. Tarawali and T. Avav (editors):Reducing Poverty through improved *Imperata* control. Proceedings of the second *Imperata* management stakeholders' conference, held in Makurdi, Benue State, Nigeria, 14-15 September 2004, IITA, Ibadan, Nigeria. p. 119

Odeleye FO, Odeleye OMO, Dada AO, Olaleye OA (2005). The response of okra to varying levels of poultry manure and plant population density under sole cropping. J. Food Agric. Environ. 3(3,4):68-74.

Olasotan FO (2001). Optimum population density for okra (*Abelmoschus esculentus* (L) Moench) in a mixture with cassava (*Manihot esculentus*) and its relevance to rainy season-based cropping system in south-western Nigeria. J. Agric. Sci. 136:207-214.

Okigbo BN Greenland DJ (1976). Intercropping Systems in Tropical Africa. In. R. L.Papendick, A.Sanchez and G.B. Triplet (editors). Multiple Cropping ASA. Special publication No. 27 Madison W.I. USA. pp. 63-101.

Ossom EM (1986a) Effect of Plant Population on yield and Weed infestation of Cassava maize intercropping. Ind. J. Agric. Sci. 56:732-734.

Rana KS, Pal M (1999). Effect of intercropping systems and weed control on crop-weed competition and grain yield of pigeon pea. Crop Res. 17:179-182.

Rao MR, Shetty SVR (1976). Some biological aspect of intercropping systems on crop-weed balance. Indian J. Weed Sci. 8:32-43.

Rosenthal GA, Janzen DH (1979). Herbivores, their interaction with secondary plant metabolites. Academic Press, New York, USA.

Saban Y, Mehmt A, Mustafa E (2007). Identification of Advantages of Maize-Legume intercropping over Solitary Cropping through Competion Indices in the East Mediterranean Region. Turk. J. Agric. 32:111-119

Schippers RR (2000). African indigenous vegetable. An overview of the cultivated species. Chatam UKS National Resources Institute (Technical Report).

Singh Mahender, Singh Pushpendra, Nepalia V (2005) Integrated weed management studies in maize based intercropping system. Indian J. Weed Sci. **37**(3 & 4):205-208.

Shah SN, Shroff, JC, Patel, RH, Usadadiya VP (2011). Influence of Intercropping and Weed Management practices on weed and yields of maize. Int. J. Sci. Nat. 2:47-50.

Subramanian S, Ali AM, Kumar RJ (1991). *All about weed control* Kalyani Publishers, New Delhi – 110002, India. p. 315.

Tripathi AK, Anand K Somendra N, Yadav RA (2008). Weed dynamics, productivity and net monetary returns as influenced by winter maize based intercropping systems in central U.P. *ISWS Biennial Conference on Weed Management in Modern Agriculture,* February 27-28, 2008, Pusa, Bihar. p. 120.

Ukpong IE (1992). The Structure and Soil Relations of Avicenia Mangrove Swamp in South Eastern Nigeria. Trop. Ecol. 33(3):5-16.

University of Uyo (1997). Agricultural Meteorological Station, University of Uyo Publications Uyo.

Vandermeer JH (1989). The ecology of intercropping. Cambridge.

6

Phelipanche aegyptiaca (Pers.) Pomel: A new record as a parasitic weed on apricot root in Turkey

Eda Aksoy, Z. Filiz Arslan and Naim Öztürk

Biological Control Research Station, Adana-Turkey.

This study is the first report of Egyptian broomrape (*Phelipanche aegyptiaca* (Pers.) Pomel) parasitizing apricot (*Prunus armeniaca* L., Rosaceae) trees in Turkey, Malatya Province, constitutes 52% of world apricot production. Most of the apricot production areas of Turkey are in Malatya, with 14 districts, 9 of them possess intensive apricot plantings that were surveyed for the frequency and density of broomrape in 2010. A total of 5 districts were found to be infested with *P. aegyptiaca* at frequencies ranging between 11 to 50%. Additionally, total 1415 quadrats (counting frame measuring 1 m^2) were used for the assessment of the broomrape and the percentage of quadrats infested with broomrape in Malatya was found to be 14%. The average frequency *P. aegyptiaca* was determined to be 14.8%, and its average density in all quadrats and in all infested quadrats (number/m^2) were 14.8 and 57.2, respectively. Unfortunately, broomrapes were counted high infestation rates (more than 200/m^2) in some of the infested apricot orchards. As a result of the study it is realised that, apricot production of Turkey and the world were under threat of broomrapes and urgent control measures must be taken in the region immediately.

Key words: Apricot, Egyptian broomrape, *Phelipanche aegyptiaca*, Turkey.

INTRODUCTION

Turkey ranks first in the world in fresh and dried apricot production and has an important potential due to its genetic apricot resources and ecological conditions. The annual world apricot production is approximately 2.5 to 3 million tons, with the 13.490.000 trees in Turkey producing approximately 661.000 tons of apricot per year. Malatya constitutes 52% of this production, 341.000 tons/year (Anonymous, 2009).

Approximately 90 to 95% of Malatya apricot orcards have been established for dried apricot varieties. Among all apricot varieties grown in the provinces; 73% is Hacihaliloglu, while 17% is Kabaaşi, the remaining are Soganci, Hasanbey and Cataloglu varieties (Unal, 2010).

Weeds are important in apricot orchards, and their injurious effects in Malatya have been acknowledged. According to the survey study conducted in Malatya, 109 weed species have been determined in apricot orchards (Koloren and Uygur, 2001).

Orobanche and *Phelipanche* species are root holoparasitic plants that cause severe damage to economically important dicotyledonous crops, depending entirely on their hosts for all of their nutritional requirements (Hershenhorn et al., 2009). These species pose a serious threat to global agricultural production, mainly because there are no practical methods to control them effectively (Gressel et al., 2004), and they are found in Southern and Eastern Europe, the Middle East and North Africa, and have recently been reported in the USA, Australia and some countries in South America (Rubiales et al., 2009). Species of *Orobanche* and *Striga* are among the most damaging parasitic weed species worldwide and a review of literature over the period since

1991 suggests that many million hectares are infested and that the losses amount to $ US billions annually (Parker, 2009).

In the last years *Phelipanche* and *Orobanche* species are continuously expanding into new areas which are considered as parasite free. Some new host crops for broomrapes were reported, like *Amygdalus communis* L., *Olea europaea* L. and *Quercus coccifera* L. which were parasitized by *Orobanche palaestina* Reut.; *Amygdalus communis* L., *Olea europaea* L., *Prunus armeniaca* L., and *Prunus persica* L., which were parasitized by *Orobanche cernua* L.; *Olea europaea* L. and *Amygdalus communis* L., which were parasitized by *Orobanche schultzii* Mutel. (Qasem 2009); *Olea europaea* L. which was parasitized *Orobanche aegptiaca* Pers. (Eizenberg et al., 2002). Some broomrapes that were considered as parasites of native plants are turning into pests of agricultural crops, like *Orobanche pubescens* d'Urv. It is known from the local flora, and was now found to cause damage in parsley fields and to *Tropaeolum majus* L. in ornamental gardens in Israel and *Orobanche amethystea* Thuill. and *Orobanche loricata* Reichb. are known as occasional weeds in Europe and were recently found for the first time in Israel (Joel and Eisenberg, 2002). Therefore monitoring *Orobanche* and *Phelipanche* species distribution and parasitism is important both for scientific and agricultural propose.

In Turkey, 36 species of *Orobanche* have been recorded (Gilli, 1982), but only four species cause significant damage to crops: *Orobanche ramosa* is harmful to tobacco (Demirkan and Nemli, 1993), *O. aegyptiaca* Pers. is parasiting red lentil (Uludag and Demir, 1997); *Orobanche cernua* Loefl. is harmful to sunflower (Petzoldt et al., 1994) and *O. crenata* Forsk. damages faba beans (Kıtıkı et al., 1993).

Controlling these weeds is difficult because broomrape species produce hundreds of thousands of minute seeds that are highly persistent in the soil and can easily be distributed to new areas. Moreover, owing to the intimate connection between these holoparasitic weeds and their hosts, effective and economically viable control system against the parasites have been developed for very few cultivated plants. The insufficiency of the countermeasures against broomrape contributes to the increasing importance of these weeds in agricultural areas (Bulbul et al., 2009).

Yield reduction caused by *Orobanche* is dependent on the timing and severity of the infestation, with yield losses generally ranging from 5 to 100%; the total loss, averaged across all broomrape species, is approximately 34% (Linke et al., 1989). According to some studies conducted in Turkey, yield losses in tomato caused by *O. ramosa* averaged 24% while losses of 82% have been estimated for faba bean under a high infestation of *O. crenata* (Aksoy and Uygur, 2008).

Specifically, *O. ramosa* and *O. aegyptiaca* is a problem that causes yield losses of some major annual crops in Turkey. It was firstly reported that Egyptian broomrape (*P. aegyptiaca*) also parasitizes a perennial plant, apricot, for Turkey and the world with this study.

MATERIALS AND METHODS

This survey study was conducted in the apricot orchards of Malatya during June to July 2010. The study was conducted in 9 of the total 14 districts that practice intensive apricot cultivation; 9 districts comprise 64390 ha, and 3.75% of the total area (231 ha) was surveyed.

The survey was conducted at 5 km intervals along the main road of the districts and sampling the closest apricot orchard (Uygur, 1985). For the assessment of the broomrape infestation, different number of quadrat (counting frame measuring one m^2) were used depending on the size of orchards: 10, 15, 20, 25 and 30 quadrats were used in apricot orchards that are 1 - 5, 6 - 10, 11 - 20, 21 - 50 and 50-more da in size, respectively. Each *Phelipanche* shoot was counted as one individual. The broomrape species was identified using *The Flora of Turkey* - Volume 7 (Gilli, 1982). Datas on diagnostic characters are given below.

Phelipanche aegyptiaca (Pers.) Pomel-(*O. aegyptiaca* Pers.). Usually branched, 25 to 40 cm. Calyx teeth entire, equaling or longer than calyx tube (Figure 1c). Corolla lavender-lilac, 23 to 27 mm (Figure 1a and b). Anthers long lanate-hairy (Figure 1d). These identified features show parallelism to the features given in *Flora of Turkey*.

The formulas for the broomrape indicence (%) in orchards, average densities of broomrapes in all quadrats counted and in all infested quadrats were calculated based on Odum (1971). The frequency of broomrape was calculated by dividing the number of orchards infested with broomrape by the number of total surveyed orchards. The general density of broomrape (plants/m^2) was calculated by dividing the total number of broomrape to the total number of quadrat. The special density of broomrape was calculated by dividing the total number of broomrape individuals by the total number of quadrat containing the weed.

RESULTS

During the survey study, broomrape plants were found densely around some apricot trees (Figure 2a, b, and c). Upon digging up the apricot roots, it was determined that the broomrape was parasitizing the young roots of the trees (Figure 2d). There isn't any broomrape infecting, annual weeds were seen in these areas. The broomrape in the apricot orchards of Malatya province was identified as Egyptian broomrape (*Phelipanche aegyptiaca* (Pers.) Pomel).

The average frequency of the weed was 14.8% throughout the surveyed districts of Malatya, with 5 of the 9 districts being infested with this weed, whereas the other 4 were not. The Yesilyurt district had the highest Egyptian broomrape frequency (50%), with the Darende, Yazıhan, Akcadag and Centre districts exhibiting frequencies of 33.3, 25, 13.3 and 11.1%, respectively (Table 1 and Figure 3).

According to the general densities provided in Table 1, the average density of broomrape in all quadrats was 14.8 (number/m^2) in the 9 districts of Malatya surveyed.

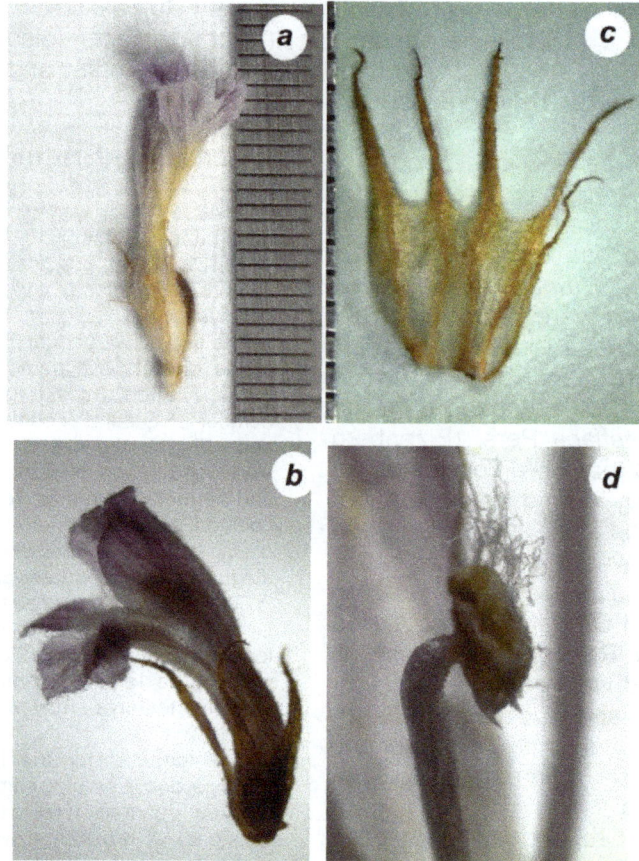

Figure 1. *Phelipanche aegyptiaca* (Pers.) Pomel. Flowers (a, b); calyx (c); anther (d).

Figure 2. Density of Egyptian broomrape in apricot orchards (a); *P. aegyptiaca* flowers (b); Mature Egyptian broomrapes (c); Apricot roots infested with broomrape (d).

Table 1. Frequency and density of *Orobanche ramosa* in the surveyed districts of Malatya.

Surveyed districts	Apricot planting area in provinces (ha)	Survey area (ha)	Rate of survey area to total area (‰)	Number of surveyed apricot orchard	Number of sample (quadrat)	Number of infested quadrat	Percentage of infested quadrat (%)	Frequency (%)	Number/m² (in all quadrats)	Number/m² (in all infested quadrats)
Akcadag	16200	55.5	3.37	15	350	45	12.9	13.3	25.9	201.4
Centre	11270	32.2	3.24	9	220	24	10.9	11.1	10.2	90
Darende	10850	47.6	4.26	12	255	72	28.2	33.3	18	54
Hekimhan	7680	16.7	2.15	5	105	0	0	0	0	0
Battalgazi	5550	19	4.14	6	130	0	0	0	0	0
Yazihan	4600	7.3	1.8	4	80	5	6.3	25	9.4	30.2
Dogansehir	4025	14	3.8	5	115	0	0	0	0	0
Yesilyurt	2665	26.5	10.6	4	110	52	47.3	50	69.5	139
Kale	1550	12	8.28	2	50	0	0	0	0	0
Total	64390	230.8		62	1415	198				
Average			3.75				14	14.8	14.8	57.2

The Yesilyurt District showed the highest value (69.5), whereas Akcadag (25.9), Darende (18), Centre (10.2) and Yazihan (9.4) Districts had lower general densities. The average density of *P. aegyptiaca* in all infested quadrats was 57.2 (number/m²) in Malatya. Although the density rate of Malatya, some of the apricot orchards in districts demonstrated a high infestation rate (more than 200). The density of Egyptian broomrape in infested quadrats was higher in Akcadag (201.4) than the other districts. Yesilyurt (139), Centre (90), Darende (54) and Yazihan (30.2) ranked below Akcadag (Table 1).

DISCUSSION

The host spectrum of broomrape species includes many annual crops that are members of the Solanaceae, Fabaceae, Apiaceae, Brassicaceae and Asteraceae families (Sauerborn, 1991; Parker and Riches, 1993; Riches and Parker, 1995). Similarly, some annual plants (tomato, red lentil, faba bean, sunflower, tobacco, hemp, eggplant, pepper, cabbage, radish, cucumber, carrot, squash and potato) have been reported to be host plants of broomrape in Turkey (Ekiz, 1964; Demirkan and Nemli, 1993; Kitiki et al., 1993; Aksoy-Orel and Uygur, 2003). However until now there has been no record of broomrape parasitism on any tree species in Turkey. *O. ramosa* had first been reported on apricot trees in Iraq in 1989 (Al-Khazarji et al., 1989), with the *O. ramosa* incidence increasing from 10% in 1986 to 16% in 1987. The density of *O. ramosa* was considered serious at 200 plants/per tree, and the infestation significantly reduced the total yield, fruit fresh and dry weights and fruit size in comparison to non-parasitized trees (Al-Khazarji et al., 1989). In the present study, we found that average frequency of Egyptian broomrape in the surveyed orchards was 14.8%.

Unfortunately, the cultivation of apricot trees in Malatya is at risk of infestation with broomrape. Growers have insufficient information about harmful effects, control and dispersal of broomrape. Consequently, the broomrape problem will expand in the area if sufficient control measures are not undertaken and the growers are not trained immediately.

Aim to solute the broomrape problem in apricot orchards, a project has been conducted since 2013. The project has been conducted in apricot orchards of Malatya provinces between the years of 2013 to 2015 and it was supported by "The Scientific and Technological Research Council of Turkey". As a result, harm effect of broomrape will be reduced and prevented dispersal of *P. aegyptiaca* more suitable space. In addition, effect of broomrapes to yield and quality of young trees and saplings will be investigated. The criteria on apricot quality are; number of flower buds, fruit size, fruit weight, water-soluble dry matter, pH, acidity, color analysis, yield per tree, and body diameter.

ACKNOWLEDGMENTS

The study was conducted under "National

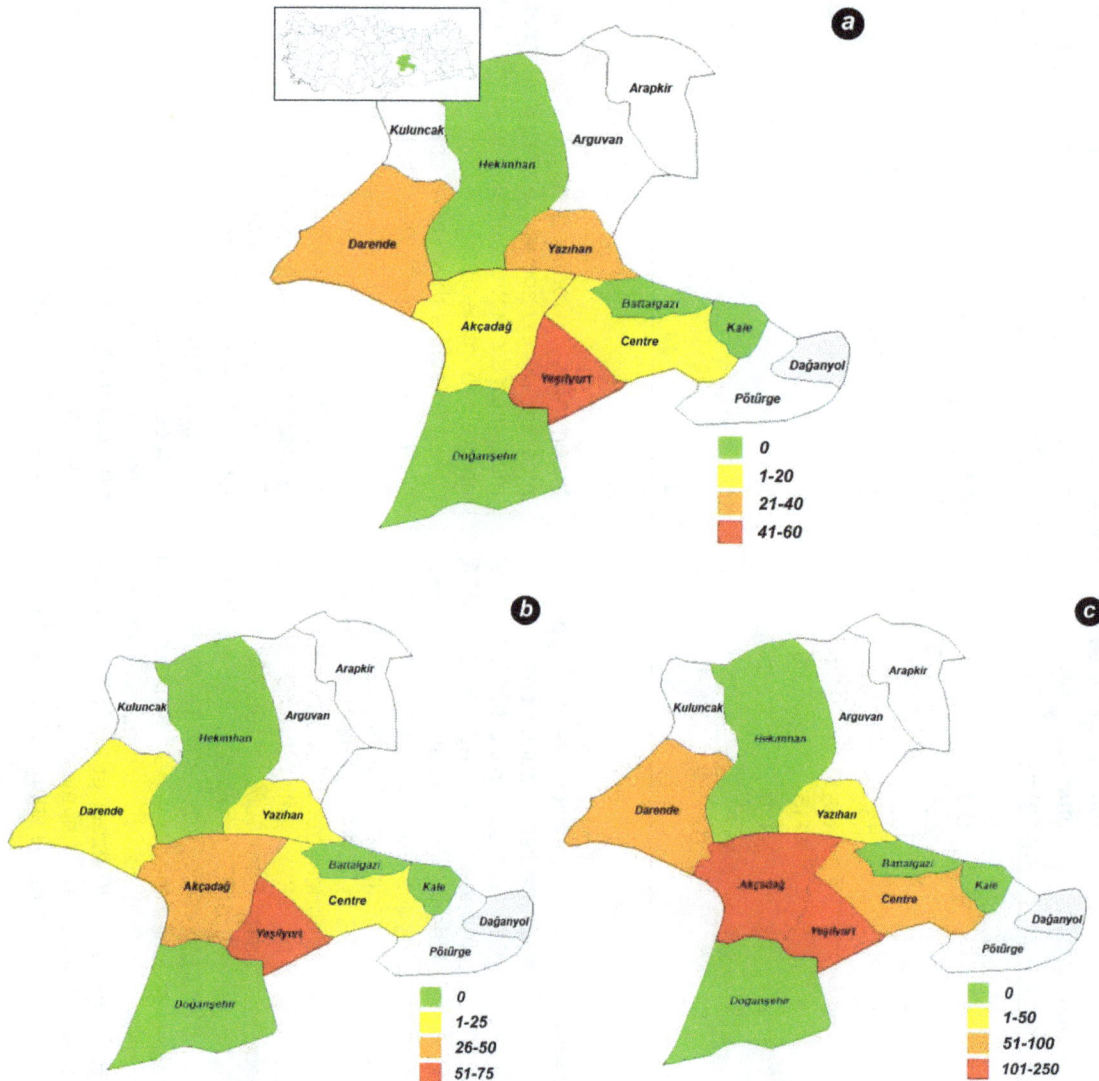

Figure 3. Frequency and density of *P. aegyptiaca* in the surveyed districts of Malatya. Frequency % (a), *Phelipanche* number/m² in all quadrats (b), *Phelipanche* number/m² in all infested quadrats (c).

Broomrape Project" (Project number: 105G080) supported by TUBITAK (The Scientific and Technological Research Council of Turkey).

REFERENCES

Al-Khazarji TO, Ishaak EY, Hameed KM (1989). Broomrape (*Orobanche ramosa*) on apricot (*Prunus armeniaca*) in Iraq. Arab J. Plant Protect. 7(2):147-152.

Aksoy-Orel E, Uygur FN (2003). Distrubition of *Orobanche* species in the East Mediterranean Region of Turkey. In: Proceedings of 7th EWRS Mediterranean Symposium. 6-9 May, 2003, Adana, Turkey. P. 131.

Aksoy E, Uygur FN (2008). Effect of broomrapes on tomato and faba bean crops. J. Turk. Weed Sci. 11(1):1-7.

Anonymous (2009). Turkish Statistical Institute, crop production statistics, Ankara, Turkey. Web page: http://www.turkstat.gov.tr

Bulbul F, Aksoy E, Uygur S, Uygur N (2009). Broomrape (*Orobanche* spp.) problem in the Eastern Mediterranean Region of Turkey. Helia. 32(51):141-152.

Demirkan H, Nemli Y (1993). Bazı domates çeşitlerinin *Orobanche ramosa* L.'ya duyarlılıklarının araştırılması. (Investigation on susceptibility of some tomato cultivars to *Orobanche ramosa* L.) Türkiye I. Herboloji Kongresi (Turkish Weed Science Congress), 3-5 Subat (February)1993-Adana. pp. 309-314.

Eizenberg H, Golan S, Joel DM (2002). First report of the parasitic plant *Orobanche aegyptiaca* infecting olive. Plant Dis. 86(7):814.

Ekiz E (1964). Türkiyede önemli bazı *Orobanche* türlerinin biyolojisi, morfolojisi ve konukçu bitkilere yaptığı zararlar uzerinde araştırmalar (Studies on biology, morphology and harm effect to host plants of some important *Orobanche* species in Turkey-Phd Thesis). Doktora Tezi. Ankara Universitesi Ziraat Fakultesi, Bilimsel Arastırma ve Incelemeler: 238:129.

Gilli A (1982). *Orobanche* L. In: Flora of Turkey and the East Aegean Islands (Ed. Davis P.H.). Edinburgh at the University Press. 7:3-23.

Gressel J, Hanafi A, Head G, Marasas W, Obilana AB, Ochanda J, Souissi T, Tzotzos G (2004). Major heretofore intractable biotic constraints to African food security that may be amanable to novel biotechnological solutions. Crop Protect. 23:661-689.

Hershenhorn J, Eizenberg H, Dor E, Kapulnik Y, Goldwasser Y (2009). *Phelipanche aegyptiaca* management in tomato. Weed Res. 49(Suppl.1):34-47.

Joel DM, Eisenberg H (2002). Three *Orobanche* species newly found on crops in Israel. Phytoparasitica 30(2):187-190.

Kıtıkı A, Acıkgoz N, Cinsoy AS (1993). Baklada (*Vicia faba* L.) orobansın (*Orobanche crenata* Forsk.) kontrolu ve ilaclamanın bazı verim komponentlerine etkisi (Control of broomrape (*Orobanche crenata* Fors.) in broad bean (*Vicia faba* L.) and effect of chemical control on some yield components). Turkiye I. Herboloji Kongresi (I. Turkish Weed Science Congress), 3-5 Şubat (February) 1993-Adana. pp. 297-307.

Koloren O, Uygur FN (2001). Kayısıda zararlı yabancı otlar ve mücadelesi (Harmful weeds and their control in apricot). Kayısı Sempozyumu, 5 Nisan 2001, Malatya, Malatya Kayısı Arastırma Geliştirme ve Tanıtma Vakfı, Sezer Ofset, Malatya.

Linke KH, Sauerborn J, Saxena MC (1989). *Orobanche* field guide. University of Hohenheim, Germany. p. 42.

Odum EP (1971). Fundamentals of ecology. W. B. Saunders Company, Philadelphia, London, Toronto. P. 574.

Parker C, Riches CR (1993). Parasitic weeds of the world: biology and control. CAB International, Wallingford Oxon OX10 8DE, UK., P. 350.

Parker C (2009). Observations on the current status of *Orobanche* and *Striga* problems worldwide. Pest Manage Sci. 65(5):453-459.

Petzoldt K, Nemli Y, Sneyd J (1994). Integrated control of *Orobanche cumana* in sunflower. In: AH Pieterse, JAC Verkleij and SJ ter Borg (eds.), Biology and management of *Orobanche*. proceedings of the 3rd international workshop on *Orobanche* and related *Striga* research. Amsterdam, The Netherlands, Royal Tropical Institute. pp. 442-449.

Qasem JR (2009). Parasitic weeds of the *Orobanchaceae* family and their natural hosts in Jordan. Weed Biol. Manage. 9(2):112-122.

Riches CR, Parker C (1995). Parasitic plants as weeds. Parasitic Plants. Edited by Malcolm C. Press and Jonathan D. Graves. Chapman and Hall, 2-6 Boundary Row, London SE1 8HN, UK. pp. 227-255.

Rubiales D, Fernandez-Apricio M, Wegmann K, Joel DM (2009). Revisiting stategies for reducing the seedbank of *Orobanche* and *Phelipanche* spp. Weed Res. 49 (Suppl.1):23-33.

Sauerborn J (1991). Parasitic flowering plants, ecology and management. Supra-regional project ecology and management of parasitic weeds, gtz-UH. Verlag Josef Margraf, 1991. Scientific Books Mühlstr. 9, P.O. Box 105, D-6992 Weikersheim FR Germany. P. 127.

Uludag A, Demir A (1997). Parasitic weeds of lentil fields in South East Anatolia region. II. Turkish Weed Science Congress. Ege University printery, Bornova, Izmir. pp. 379-384.

Unal MR (2010). Kaysı araştırma raporu (Apricot research report). T.C. Fırat Kalkınma Ajansı, Eylül. P. 63.

Uygur FN (1985). Untersuchungen zu art und bedeutung der verunkrautung der verunkrautung in der Çukurova unter besonderer berücksictigung von *Cynodon dactylon* (L.) Pers. und *Sorghum halepense* (L.) Pers. PLITS, 1985/3 (5), Stuttgard, Germany. P. 169.

7

Effect of seaweed saps on growth and yield improvement of green gram

Biswajit Pramanick[1], Koushik Brahmachari[1] and Arup Ghosh[2]

[1]Department of Agronomy, Bidhan Chandra Krishi Viswavidyalaya, Mohanpur, Nadia, West Bengal, 741252, India.
[2]Central Salt and Marine Chemicals Research Institution (Council of Scientific and Industrial Research), G. B. Marg, Bhabnagar, Gujrat, 364002, India.

A field experiment was conducted during the *pre-kharif* season at Uttar Chandamari village in 2012 to study the effects of seaweed saps on growth, yield and quality improvement of green gram in new alluvial soil of West Bengal. The foliar spray was applied twice at different concentrations (0, 2.5, 5.0, 7.5, 10.0 and 15.0% v/v) of seaweed extracts (namely *Kappaphycus* and *Gracilaria*). Foliar applications of seaweed extract significantly enhanced the growth, yield and quality parameters. The highest grain yield was recorded with applications of 15% *Kappaphykus* sap + recommended dose of fertilizer (RDF), followed by 15% *Gracilaria* - sap + RDF extract resulting in an increase by 38.97 and 33.58% grain yield, respectively compared to the control. The maximum straw yield was also achieved with the application of 15% seaweed extract. Improved crop quality and nutrient uptake [nitrogen (N), phosphorus (P) and potassium (K)] was also observed with seaweed extract applications.

Key words: Seaweed saps, *Kappaphycus*, *Gracilaria*, recommended dose of fertilizer, green gram.

INTRODUCTION

Any improvement in agricultural system that results in higher production should reduce the negative environmental impact and enhance the sustainability of the system. One such approach is the use of bio-stimulants, which can enhance the effectiveness of conventional mineral fertilizers. Long-term indiscriminate use of them invites the crucial problem of soil health disorder vis-à-vis reduced input use efficiency, more precisely, fertilizer use efficiency. Due to these reasons the farmers are being compelled gradually day by day to turn towards various options like organic manures, bio-stimulants, growth regulators etc. One of such options is the use of seaweed extracts as plant nutrient bearing fertilizer. Marine bioactive substances extracted from marine algae are used in agricultural and horticultural crops, and many beneficial effects may be achieved in terms of enhancement of yield and quality. Liquid extracts obtained from seaweeds have recently gained importance as foliar sprays for many crops including various cereals, pulses and different vegetable species. Seaweed extracts contain major and minor nutrients, amino acids, vitamins, cytokinins, auxin and abscisic acid like growth promoting substances and have been reported to stimulate the growth and yield of plants, develop tolerance to environmental stress (Zhang et al., 2003), increase nutrient uptake from soil (Turan and Kose, 2004) and enhance antioxidant properties (Verkleij, 1992). Liquid extracts obtained from seaweeds have recently gained importance as foliar sprays for many crops including various grasses, cereals flowers andvegetable species (Crouch and Van Staden, 1994). In recent years, the use of seaweed extracts have gained in popularity due to their

Table 1. Details of experimental treatment.

Treatment	Details	Time of application
T_1	2.5% *Kappaphycus*-sap + RDF	20 DAS and 40 DAS
T_2	5% *Kappaphycus*-sap + RDF	,,
T_3	10% *Kappaphycus*-sap + RDF	,,
T_4	15% *Kappaphycus*-sap + RDF	,,
T_5	2.5% *Gracilaria*-sap + RDF	,,
T_6	5% *Gracilaria*-sap + RDF	,,
T_7	10% *Gracilaria*-sap + RDF	,,
T_8	15% *Gracilaria*-sap + RDF	,,
T_9	RDF + Water spray	,,
T_{10}	7.5% *Kappaphycus*-sap + 50% RDF	,,

potential use in organic and sustainable agriculture (Russo and Beryln, 1990), especially in rainfed crops, as a means to avoid excessive fertilizer applications and to improve mineral absorption. Unlike chemical fertilizers, extracts derived from seaweeds are biodegradable, non-toxic, non-polluting and non-hazardous to humans, animals and birds (Dhargalkar and Pereira, 2005).

With the advancement of modern agricultural technologies, we somehow achieved food security for our country to feed its ever escalating population. But till date, unfortunately, we could not achieve nutritional security for our people. In West Bengal India, the growing of pulses in the cropping system is nowadays not a common practice, but it is well known that inclusion of any pulse crop in the sequence has all-round advantages. So, if we are to achieve nutritional security we must have to go for crop diversification in cropping pattern introducing crops like pulses for example, green gram, black gram etc. in *Zaid* (Summer) season and lentil, chickpea, garden pea etc. in *Rabi* (Winter) season. On the other hand, to nourish the protein energy malnourished (PEM) people and to maintain soil health properly pulse crops should be included in the cropping system. Green gram or mung bean (*Vigna radiata* L.) is the third most important food legumes grown and consumed in India and is a good source of proteins and minerals and its protein quality is similar to or better than other legumes like chickpea, black gram, peas, pigeon pea etc. (Kataria et al., 1989; Jood et al., 1998a).

Considering each and every corner of the above discussion, an experiment was conducted to study the effect of different seaweed saps on growth and yield of green gram and assess the nutrient removal by seed and stover of this pulse crop.

MATERIALS AND METHODS

Experimental site and soil information

The field experiment was conducted during the *pre-kharif* season of 2012 on inceptisol at Uttar Chandamari village of Nadia district of West Bengal in India. The soil of the site was sandy clay loam with pH 6.45, organic carbon 0.57%, total nitrogen 0.055%, available P_2O_5 26.29 kg ha^{-1} and available K_2O 148.72 kg ha^{-1}. The climate of the region is humid subtropical. The experimental site is located at $22°57'$ N latitude, $88°20'$ E longitude and altitude is 7.8 m.

Experimental designs and treatments

The experiment comprised of ten treatments, the details of the treatments are mentioned Table 1.

Two sprays of *Kappaphycus* and *Gracilaria* extract were applied; one at the seedling stage [20 days after sowing (DAS)] and the other at the pre-flowering stage (40 (DAS). For proper adherence, extracts were mixed with proper surfactant (Active 80 at 0.5 ml L^{-1} of water) at the time of spraying. The total spray volume was 650 L ha^{-1} in each application. The treatments were distributed in a randomized block design with three replications. The plot size was 5 × 6 m. The recommended dose of fertilizer (RDF) for green gram was 20:40:40 kg ha^{-1} N, P_2O_5 and K_2O, respectively and all fertilizers were applied as basal. Date of sowing was 25th of February, 2012 and the crop was harvested on 5th May, 2012.

Preparation and chemical composition of liquid sea weed extract

The seaweed extract used in this study was obtained from *Kappaphycus* sp. and *Gracilaria* sp. The algae were handpicked from the coastal area of Rameswaram, T. N., India during September, 2011. It was washed with seawater to remove unwanted impurities and transported to the field station at Mandapum, Rameswaram. Here, samples were thoroughly washed using tap water. After that, fresh seaweed samples were homogenized by grinder with stainless steel blades at ambient temperature, filtered and stored (Eswaran et al., 2005). The liquid filtrate was taken as 100% concentration of the seaweed extract and further diluted as per the treatments. The nitrogen (N) content of seaweed extract (100% concentrate) was determined by taking 20 ml of filtrate which was oxidized and decomposed by concentrate sulphuric acid (10 ml) with digestion mixture (K_2SO_4 : $CuSO_4$ = 5:1) heated at 400°C temperature for 2½ h as described in the semi-micro Kjeldahl method [AOAC International, 1995, method No. Ba 4b-87(90)], and other nutrient elements were analyzed by inductively coupled plasma-optical emission spectroscopy (ICP-OES), after wet digestion of filtrate (20 ml) with HNO_3-$HClO_4$ (10:4) di-acid mixture (20 ml) and heated at 100°C for 1 h and then raise the temperature to about 150°C (Richards, 1954)

Table 2. Chemical composition of *Kappaphycus* sap.

Nutrient	Amount present	Nutrient	Amount present
Moisture	94.38 g/100 ml	Iron	8.58 mg/100 ml
Protein	0.085 g/100 ml	Manganese	0.22 mg/100 ml
Fat	0.0024 g/100 ml	Nickel	0.35 mg/100 ml
Crude fibre	0.01 g/100 ml	Copper	0.077 mg/100 ml
Carbohydrate	1.800 g/100 ml	Zinc	0.474 mg/100 ml
Energy	7.54 Kcal/100 ml	Chromium	3.50 mg/100 ml
Sodium	18.10 mg/100 ml	Lead	0.51 mg/100ml
Potassium	358.35 mg/100 ml	Thiamine	0.023 mg/100 ml
Magnesium	116.79 mg/100 ml	Riboflavin	0.010 mg/100 ml
Phosphorous	2.96 mg/100 ml	B-Carotene	0.0 mg/100 ml
Calcium	32.49 mg/100 ml	Iodine	160 mg/100 ml
Indole acetic acid	23.36 mg/L	Kinetin + Zeatin	31.91 mg/L
Gibberelin GA_3	27.87 mg/L		

Data courtesy: National Institute of Nutrition, Hyderabad, India (except growth hormone data generated by CSMCRI using quantitative MS-MS and LC-MS techniques).

Table 3. Chemical composition of *Gracilaria* sap.

Nutrient	Amount present	Nutrient	Amount present
Ash	38.91 g/100 g	Calcium	295.50 mg/100 g
Crude protein	9.58 g/100 g	Copper	0.20 mg/100 g
Crude fibre	10.40 g/100 g	Zinc	1.00 mg/100 g
Crude lipid	2.00 g/100 g	Iron	67.35 mg/100 g
Saturated fatty acid	48.92% of total fatty acids	Manganese	4.16 mg/100 g
Total amino acids	889.78 mg/g of protein	Nickel	0.92 mg/100 g
Moisture	88.88%	Cobalt	0.24 mg/100 g
Vitamin C	28.50 mg/100 g	Sulphate	106.20 mg/100 g
Carbohydrate	45.92%	Chlorine	1170.00 mg/100 g
Potassium	8633.00 mg/100 g	Lead	1.11 mg/100 g
Magnesium	549.50 mg/100 g	Cadmium	0.14 mg/100 g
Phosphorus	278.50 mg/100 g	Sodium	158.50 mg/100 g

Source: Benjama and Masniyom (2012), and Sivakumar et al. (2009). *Gracilaria* extract also contains variable amount of phytohormones like Auxin, Cytokinin, Abscisic acid etc. (Yokoya et al., 2010).

(Tables 2 and 3).

Plant sampling

Data were taken through random sampling at 21 DAS, 42 DAS and 63 DAS to measure plant height, dry matter accumulation, crop growth rate (CGR) and leaf area index (LAI). CGR was computed with the help of the formula: $[(W_2 - W_1)/ (t_2 - t_1)]$ where, W_1= dry weight per unit area at t_1, W_2 = dry weight per unit area at t_2, t_1 = first sampling and t_2 = second sampling. LAI was computed by the ratio of leaf area to the area of ground cover. For measuring nodule number per plant, data were taken at 21 and 42 DAS. Data on yield attributes were taken randomly before harvesting. At maturity, green gram seeds and stover samples were collected from each plot, oven dried at 70°C to constant weight and ground to pass through a 0.5 mm sieve for chemical analysis. The N content was determined by the semi-micro Kjeldahl method [AOAC International, 1995, method No. Ba 4b-87(90)]. Phosphorus (P) content was

determined by the Vanado-Molybdate yellow method (Jackson, 1973) and Potassium (K) content by flame photometry (Jackson, 1973).

Statistical analysis

Data were analyzed using analysis of variance (ANOVA) following randomized block design (Gomez and Gomez, 1984). Differences were considered significant at 5% level of probability.

RESULTS

Effect of treatments on growth of green gram

Foliar application of different sea weed saps along with RDF increased growth attributes of greengram

Table 4. Effect of treatments on plant height (cm), dry matter accumulation (g m^{-2}), CGR (g m^{-2} day^{-1}), no. of nodule per plant and LAI of green gram.

Treatment	Plant height (cm)			Dry matter accumulation (g m^{-2})			CGR (g m^{-2} day^{-1})		Nodule no. plant^{-1}		LAI		
	21 DAS	42 DAS	63 DAS	21 DAS	42 DAS	63 DAS	21 - 42 DAS	42 - 63 DAS	21 DAS	42 DAS	21 DAS	42 DAS	63 DAS
T$_1$	15.63	38.33	52.30	19.67	150.31	280.41	6.23	6.19	2.80	18.5	0.65	2.90	2.87
T$_2$	16.71	38.93	52.87	19.79	155.83	288.23	6.48	6.30	3.11	19.3	0.70	2.97	2.99
T$_3$	18.92	42.57	54.67	21.41	170.23	299.57	7.09	6.16	3.82	20.2	0.84	3.43	3.31
T$_4$	19.13	45.52	59.13	27.93	190.78	320.78	7.75	6.19	3.95	23.1	1.08	5.09	6.09
T$_5$	14.25	39.77	50.17	19.17	145.57	271.61	6.04	6.00	2.47	17.8	0.58	2.85	2.74
T$_6$	15.97	40.11	49.33	23.67	149.91	280.19	6.02	6.20	2.67	18.3	0.61	2.93	2.81
T$_7$	16.48	43.77	53.47	24.25	171.33	289.72	7.00	5.64	3.23	19.2	0.88	3.21	3.07
T$_8$	17.27	44.67	55.87	25.69	188.97	308.56	7.78	5.70	3.78	22.9	0.97	4.87	5.57
T$_9$	14.11	39.19	51.23	18.23	142.27	265.57	5.92	5.87	2.39	17.7	0.50	2.81	2.77
T$_{10}$	17.44	37.35	51.37	24.25	161.23	291.46	6.52	6.20	2.71	18.3	0.75	3.48	3.59
SEm(±)	0.93	1.74	1.41	1.90	8.51	6.70	0.56	1.25	0.34	1.26	0.08	0.18	0.06
CD at 5%	2.71	5.08	4.18	5.81	25.73	19.81	1.65	3.69	1.01	3.74	0.24	0.53	0.18

T$_1$: 2.5% *Kappaphycus* - sap + RDF; T$_2$: 5 % *Kappaphycus* - sap + RDF; T$_3$: 10% *Kappaphycus* - sap + RDF; T$_4$: 15% *Kappaphycus* - sap + RDF; T$_5$: 2.5% *Gracilaria* - sap + RDF; T$_6$: 5% *Gracilaria* - sap + RDF; T$_7$: 10% *Gracilaria* - sap + RDF; T$_8$: 15% *Gracilaria* - sap + RDF; T$_9$: RDF + Water spray; T$_{10}$: 7.5% *Kappaphycus*- sap + 50% RDF.

significantly over control (Table 4). In general, a gradual increase in plant height, dry matter accumulation and LAI was observed with increasing seaweed extract application.

Though these parameter are not significantly affected by foliar applications of seaweed extracts up to 5% concentration. Maximum plant height, dry matter accumulation, number of nodules per plant was recorded with 15% *Kappaphycus*- sap + RDF and was statistically at par with 15% *Gracilaria* - sap + RDF treated plot regarding all the observations taken at different DAS. In case of CGR during 21 to 42 DAS, the best result (7.75 g m^{-2} day^{-1}) was recorded with the treatment T$_8$ which was closely followed by T$_4$. Highest value of LAI was recorded with T$_4$ for all the observations.

Effect of treatments on yield attributes and yields of green gram

As per the data depicted in Table 5 the maximum numbers of branches per plant, pods per plant and seeds per pod were observed under the treatment T$_4$ (highest dose, that is, foliar application of 15% *Kappaphycus* sap along with RDF) which was closely followed by 15% *Gracilaria* - sap + RDF (T$_8$). The treatment T$_4$(15% *Kappaphycus*-sap + RDF) showed the maximum increase in yield over control to the extent of 38.97% and this treatment was followed by the treatments T$_8$(15% *Gracilaria* - sap + RDF), T$_3$ (10% *Kappaphycus*- sap + RDF), T$_7$ (10% *Gracilaria* - sap + RDF), T$_2$ (5% *Kappaphycus*- sap + RDF) and T$_6$ (5% *Gracilaria* - sap

+ RDF) recording 33.58, 27.28, 21.17, 19.77 and 13.86% yield increase, respectively over control. Similar kind of results reported for *Phaseolus aureus* (Bai et al., 2008). Increase in yield of several other crops like *Capsicum annuum* (Arthur et al., 2003), black gram (Venkataraman and Mohan, 1997) and canola plants (*Brassica napus*) (Ferreira and Lourens, 2002) are reported with the foliar application of seaweed extract.

Effect of treatments on the uptake of nutrients by greengram

The use of the seaweed extracts significantly increased N, P and K uptake by grains at higher concentrations (10% and above) and reached maximum at 15% seaweed extract compared with control (Table 6). The highest N and K uptake by grain was recorded with the treatment T$_4$(15% *Kappaphycus*- sap + RDF) which was statistically at par with 15% *Gracilaria*- sap + RDF (T$_8$), 10% *Kappaphycus*- sap + RDF (T$_3$) and 7.5% *Kappaphycus*- sap + 50% RDF (T$_{10}$). 15% *Kappaphycus*-sap + RDF showed the maximum uptake of P by grain. In case of nutrient uptake by stover, 15% *Kappaphycus* - sap + RDF was observed to be the best and it was closely followed by 15% *Gracilaria* - sap + RDF and 10% *Kappaphycus* - sap + RDF. Our results confirm those findings previously reported by Crouch et al. (1990) who noted an increased uptake of magnesium (mg), K and calcium (Ca) in lettuce with seaweed concentrate application. Turan and Köse (2004), Nelson and Van Staden (1984), and Mancuso et al. (2006) also observed

Table 5. Effect of treatments on yield components and seed and stover yield of green gram.

Treatment	No. of branches plant^{-1}	No. of pod plant^{-1}	No. of seed pod^{-1}	Test weight (g)	Seed yield (kg ha^{-1})	Stover yield (kg ha^{-1})
T$_1$	14.3	58.0	9.4	30.24	1085.6	4157.9
T$_2$	14.8	59.5	9.5	30.88	1090.3	4199.5
T$_3$	16.4	69.4	9.8	30.03	1158.6	4372.1
T$_4$	19.0	76.0	10.5	31.25	1265.0	5220.3
T$_5$	13.7	57.3	9.3	30.18	995.3	3909.7
T$_6$	14.6	58.9	9.8	29.67	1036.5	4125.3
T$_7$	16.4	70.2	10.1	30.99	1103.0	4657.8
T$_8$	18.5	72.1	10.7	30.61	1216.1	5107.2
T$_9$	12.3	50.0	8.6	28.37	910.3	3712.7
T$_{10}$	15.9	62.8	9.9	30.37	1101.7	4298.3
SEm(±)	0.4	1.7	0.4	1.26	8.12	9.07
CD at 5%	1.2	5.1	NS	NS	24.33	27.12

T$_1$: 2.5% *Kappaphycus* - sap + RDF; T$_2$: 5 % *Kappaphycus*- sap + RDF; T$_3$: 10% *Kappaphycus* - sap + RDF; T$_4$: 15% *Kappaphycus* - sap + RDF; T$_5$: 2.5% *Gracilaria* - sap + RDF; T$_6$: 5% *Gracilaria* - sap + RDF; T$_7$: 10% *Gracilaria* - sap + RDF; T$_8$: 15% *Gracilaria* - sap + RDF; T$_9$: RDF + Water spray; T$_{10}$: 7.5% *Kappaphycus*- sap + 50% RDF.

Table 6. Effect of treatments on nutrient removal by seed and stover of green gram.

Treatment	Nutrient removal by seed (kg ha^{-1})			Nutrient removal by stover (kg ha^{-1})		
	N	P	K	N	P	K
T$_1$	45.33	3.35	17.93	85.47	4.33	58.37
T$_2$	46.31	3.89	18.05	91.93	4.67	60.11
T$_3$	50.97	4.51	19.67	101.67	5.67	70.67
T$_4$	57.67	5.85	22.97	143.33	7.83	80.95
T$_5$	41.65	3.22	16.88	81.23	4.03	55.53
T$_6$	45.23	3.51	17.53	89.77	4.52	59.82
T$_7$	48.82	3.93	17.97	98.05	5.25	66.85
T$_8$	55.09	5.21	21.69	133.33	6.93	76.64
T$_9$	38.11	2.78	14.37	73.11	3.35	48.89
T$_{10}$	47.13	4.03	18.99	97.89	4.99	64.50
SEm(±)	3.81	0.33	1.41	4.89	0.27	1.76
CD at 5%	11.23	0.97	4.13	14.47	0.85	5.29

T$_1$: 2.5% *Kappaphycus*- sap + RDF; T$_2$: 5 % *Kappaphycus*- sap + RDF; T$_3$: 10% *Kappaphycus*- sap + RDF; T$_4$: 15% *Kappaphycus*- sap + RDF; T$_5$: 2.5% *Gracilaria* - sap + RDF; T$_6$: 5% *Gracilaria* - sap + RDF; T$_7$: 10% *Gracilaria* - sap + RDF; T$_8$: 15% *Gracilaria* - sap + RDF; T$_9$: RDF + Water spray; T$_{10}$: 7.5% *Kappaphycus*- sap + 50% RDF.

increased uptake of N, P, K and Mg in grape vines and cucumber with the application of seaweed extract. The presence of marine bioactive substances in seaweed extract improves stomata uptake efficiency in treated plants compared to non-treated ones (Mancuso et al., 2006).

DISCUSSION

Seaweed extract a the rich source of several primary nutrients like K, P; secondary nutrients like Ca, Mg; trace elements like zinc (Zn), copper (Cu), iron (Fe), manganese (Mn) and beneficial elements like nickel (Ni), sodium (Na) etc. Sea weed extracts stimulate various aspects of growth and development resulting in around good health of the plants, while deliberating the effect of sea weed extracts on crops the aspects of root development and mineral absorption, shoot growth and photosynthesis and ultimately crop yield, even vegetative propagation can also be taken into consideration. Due to the presence of good amount of P in it, the liquid seaweed fertilizers (LSF) proliferate root development, enhance root to shoot ratio, thereby, making the plants more able to mine adequate nutrients from the deeper layer of soil and influence crop maturity as a whole. As P is the important constituent of Nitrate reductase (NADP), the niacin component of Vitamin-B complex, helps in photosystem-I to produce NADPH. As LSF is a very good source of K, it helps in regulating the water status of the plants, controls the opening and closing of stomata and thereby the photosynthesis to a large extent. The meristematic growth, translocation of photosynthates and disease resistance are also influenced by it due to the manifestation of good impact of K. Ca being present in seaweed extracts helps in enzyme activation, cell elongation and cell stability. LSF is the opulent source of secondary nutrients like Mg; hence, it helps in photosynthesis, phloem export, root growth and nitrogen metabolism. It also influences the N-fixation in legumes as it contains Mn. Mn is a constituent of several cation activated enzymes like decarboxylase, kinase, oxidase etc., and hence, essential for the formation of chlorophyll, reduction of nitrates and for respiration. The trace elements like Fe, Cu and Zn being present in considerable amount in seaweed extracts inspire redox reaction of respiration and photosynthesis, promote reduction of nitrates and sulphates and stimulate the cation activated enzymes. The organic constituents of seaweed extract include plant hormones which elicit strong physiological responses in low doses. A panorama of phytohormones and plant growth regulators are found in different seaweed concentrates and marine macroalgal extracts viz. Auxins, Gibberellins, Cytokinins etc. which simulate rooting, growth, flower initiation, fruit set, fruit growth, fruit ripening, abscission and senescence when applied exogenously. Seaweeds also contain a diverse range of organic compounds which include several common amino acids inter alia aspartic acid, glutamic acid and alanine in commercially important species. Alginic acid, laminarin and mannitol represent nearly half of the total carbohydrate content of commercial seaweed preparations. Seaweeds also contain a wide range of vitamins which might be utilized by the crops. Vitamins C, B, (thiamine), B_2(riboflavin), B_{12}, D_3, E, K, niacin, pantothenic, folic and folinic acids occur in algae. Although vitamin A is not present, the presences of its precursor carotene and another possible precursor fucoxanthin have been found. Apart from the above organic and inorganic constituents, there is an evidence of existence of different other stimulatory and antibiotic substances. These findings are in agreement with Jeannin et al. (1991), Vernieri et al. (2005), Kowalski et al. (1999), Zhang and Ervin (2008), Mancuso et al. (2006), Norrie and Keathley (2006) and Rayorath et al. (2008).

Thus, being a wealthy source of versatile plant nutrients, phytohormones, amino acids, vitamins, stimulatory and antibiotic substances, the liquid sea weed extract enhances root volume and proliferation, bio-mass accumulation, plant growth, flowering, distribution of photosynthates from vegetative parts to the developing fruits and promotes fruit development, reduces chlorophyll degradation, disease occurrence etc. resulting in improved nutrient uptake, water and nutrient use efficiency causing sound general plant growth and vigor ultimately reflecting higher yield and superior quality of agricultural products.

Conclusion

Thus, it can be concluded that the seaweed extracts are effective in increasing the growth parameters, yield attributes, yield vis-à-vis quality of green gram. The saps also enhance nutrient uptake by this grain legume crop. Presence of micro-elements and plant growth regulators, especially cytokinins (Zodape et al., 2009; Zang et al., 2008) in Kappaphycus and Gracilaria extracts is responsible for the increased yield and improved nutrition of green gram receiving foliar application of the aforesaid two saps.

ACKNOWLEDGEMENT

The authors are thankful to Director, CSMCRI, Bhavnagar (Gujarat) for encouragement and providing necessary facilities for the analytical work.

REFERENCES

AOAC International (1995). Official methods of analysis of AOAC international. In: Cunniff, P. (Ed.), Contaminants, Drugs, 16th ed. Agricultural Chemicals, vol. 1. AOAC International, Wilson Boulevard, Virginia, USA.

Arthur GD, Stirk WA, Van Staden J (2003). Effect of seaweed concentrates on the growth and yield of three varieties of Capsicum annuum. S. Afr. J. Bot. 69:207-211.

Bai NR, Banu NRL, Prakash JW, Goldi SJ (2008). Effect of seaweed extracts (SLF) on the growth and yield of Phaseolus aureus L. Indian Hydrobiol. 11:113-119.

Benjama O, Masniyom P (2012). Biochemical composition and physicochemical properties of two red seaweeds (Gracilaria fisheri and G. tenuistipitata) from the Pattani Bay in Southern Thailand. J. Sci. Technol. 34(2):223-230.

Crouch IJ, Beckett RP, Van Staden J (1990). Effect of seaweed concentrate on the growth and mineral nutrition of nutrient stress lettuce. J. Appl. Phycol. 2:269-272.

Crouch IJ, Van Staden J (1994). Commercial seaweed products as biostimulants in horticulture. J. Home Consumer Horticul. 1:19-76.

Dhargalkar VK, Pereira N (2005). Seaweed: promising plant of the millennium. Sci. Cul. 71:60-66.

Eswaran K, Ghosh PK, Siddhanta AK, Patolia JS, Periyasamy C, Mehta AS, Mody KH, Ramavat BK, Prasad K, Rajyaguru MR, Reddy SKCR, Pandya JB, Tewari A. (2005). Integrated method for production of carrageenan and liquid fertilizer from fresh seaweeds. United States Patent no. 6893479.

Ferreira MI, Lourens AF (2002). The efficacy of liquid seaweed extract on the yield of Canola plants. S. Afr. J. Plant Soil. 19:159-161.

Gomez KA, Gomez AA (1984). Statistical Procedures for Agricultural Research. John Wiley and Sons, New York.

Jackson ML (1973). Soil Chemical Analysis.Prentice Hall of India Pvt. Ltd., New Delhi, India.

Jeannin I, Lescure JC, Morot-Gaudry JF (1991). The effects of aqueous seaweed sprays on the growth of maize. Bot. Marina. 34:469-473.

Jood S, Bishnoi S, Sharma S (1998a). Nutritional and physicochemical properties of chickpea and lentil cultivars. Die Nah/Fd. 42:70-73.

Kataria A, Chauhan BM, Punia D (1989). Antinutrients and protein digestibility (*in vitro*) of mungbean as affected by domestic processing and cooking. Food Chem. 32:9-17.

Kowalski B, Jager AK, van Staden J (1999). The effect of a seaweed concentrate on the in vitro growth and acclimatization of potato plantlets. Potato Res. 42:131-139.

Mancuso S, Azzarello E, Mugnai S, Briand X (2006). Marine bioactive substances (IPA extract) improve foliar ion uptake and water tolerance in potted *Vitisvinifera* plants. Adv. Horticul. Sci. 20:156-161.

Nelson WR, Van Staden J (1984). The effect of seaweed concentrate on the growth of nutrient-stressed, greenhouse cucumbers. Horticul. Sci. 19:81-82.

Norrie J, Keathley JP (2006). Benefits of *Ascophyllum nodosum* marine-plant extract applications to 'Thompson seedless' grape production. (Proceedings of the Xth International Symposium on Plant Bioregulators in Fruit Production, 2005). Acta. Horticul. 727:243-247.

Rayorath P, Narayanan JM, Farid A, Khan W, Palanisamy R, Hankins S, Critchley AT, Prithiviraj B (2008). Rapid bioassays to evaluate the plant growth promoting activity of *Ascophyllum nodosum* (L.) Le Jol. using a model plant, *Arabidopsis thaliana* (L.) Heynh. J. Appl. Phycol. 20:423-429.

Richards LA (1954). Diagnosis and Improvement of Saline Alkali Soils. USDA Handbook No. 60. USDA, Washington, D.C.

Russo RO, Beryln GP (1990). The use of organic biostimulants to help low inputs. J. Sustain. Agric. 1:9-42.

Turan M, Köse C (2004). Seaweed extracts improve copper uptake of grapevine. Acta. Agric. Scand. B-S P. 54:213-220.

Venkataraman K, Mohan VR (1997). The effect of liquid seaweed fertilizer on black gram. Phykos 36:43-47.

Verkleij FN (1992). Seaweed extracts in agriculture and horticulture: a review. Biol. Agricul. Horticul. 8:309-324.

Vernieri P, Borghesi E, Ferrante A, Magnani G (2005). Application of biostimulants in floating system for improving rocket quality. J. Food Agric. Environ. 3:86-88.

Yokoya Nair S, Stirk Wendy A, van Staden Johannes, Novák Ondřej, Turečková Veronika, Pěnčík Aleš, Strnad Miroslav (2010). Endogenous cytokinins, auxins, and abscisic acid in red algae from Brazil. J. Phycol. 6:1198-1205.

Zhang X, Ervin EH, Schmidt ER (2003). Plant growth regulators can enhance the recovery of Kentucky bluegrass sod from heat injury. Crop Sci. 43:952-956.

Zhang X, Ervin EH (2008). Impact of seaweed extract-based cytokinins and zeatin riboside on creeping bent grass heat tolerance. Crop Sci. 48:364-370.

Zodape ST, Mukherjee S, Reddy MP, Chaudhary DR (2009). Effect of *Kappaphycus alvarezii* (Doty) Doty ex silva.extract on grain quality, yield and some yield components of wheat (*Triticum aestivum* L.). Int. J. Plant Prod. 3:97-101.

8

Weed management in winter wheat (*Triticum aestivum* L.) influenced by different soil tillage systems

José M. G. CALADO[1,2], Gottlieb BASCH[1,2], José F. C. BARROS[1,2] and Mário de CARVALHO[1,2]

[1]Department of Crop Science, University of Évora, Apartado 94, 7002-554 Évora, Portugal.
[2]Institute of Mediterranean Agricultural and Environmental Sciences (ICAAM), Apartado 94, 7002-554 Évora, Portugal.

In this study, weed management in winter wheat influenced by different soil tillage systems was investigated. The experiment was carried out under Mediterranean conditions on a Luvisol, during two growing seasons (1996/1997 and 1999/2000). Two factors (different soil tillage systems and post-emergence weed control) were studied, with two levels each: Two soil tillage systems and two levels of post-emergence weed control. The conventional soil tillage system, performed for seedbed preparation after the emergence of a high percentage of weeds, increases the appearance of monocotyledons, especially *Lolium rigidum* Gaud. in wheat crops, when compared to their establishment in no-till system. As a consequence of the higher number of monocotyledons there is an increase in weed-crop competition. Without post-emergence herbicide treatment, the wheat crop yield is lower in the treatments with the conventional soil tillage system and yield reduction is less under no-till system compared to the respective treatments with post-emergence herbicide application.

Key words: No-till system, weeds, wheat, herbicides.

INTRODUCTION

It is known that tillage, carried out by mouldboard plough, was accepted, traditionally, as a principal means for guaranteeing a clean seedbed for cereal crops (Attwood et al., 1977). On the other hand, it can also be fundamental for controlling weeds (Attwood et al., 1977; Gruber and Claupein, 2009).

Substituting mouldboard plough by other tillage equipments can, under some conditions, avoid an increase in weed pressure, as observed by Bàrberi and Cascio (2001), with regard to rotary harrow (reduced tillage).

An increase in the weeds in reduced tillage systems, can be attributed to the direct effect of tillage on the weeds, their seed production, the seed bank, survival during certain periods and spreading, as a consequence of the propagating weed parts that are cut by the equipment (Tørresen and Skuterud, 2002). For example, soil tillage for preparation of autumn-winter cereal seedbed can improve the conditions for germination of weeds (Attwood et al., 1977; Fenster and Wicks, 1982; Bräutigam and Tebrügge, 1997; Mirsky et al., 2010; Morris et al., 2010), which will increase the population density of the weeds in the crop (Fenster and Wicks, 1982; Bräutigam and Tebrügge, 1997).

With regard to no-till systems, which are characterized by depositing seeds on the topsoil (Locke et al., 2002; Morris et al., 2010), it is necessary to follow an appropriate procedure, to avoid high weed densities and prevent unacceptable problems (Brainard et al., 2013). Therefore, in no-till and reduced tillage systems, the application of pre-sowing herbicides is a common practice. This application substitutes tillage, which cause

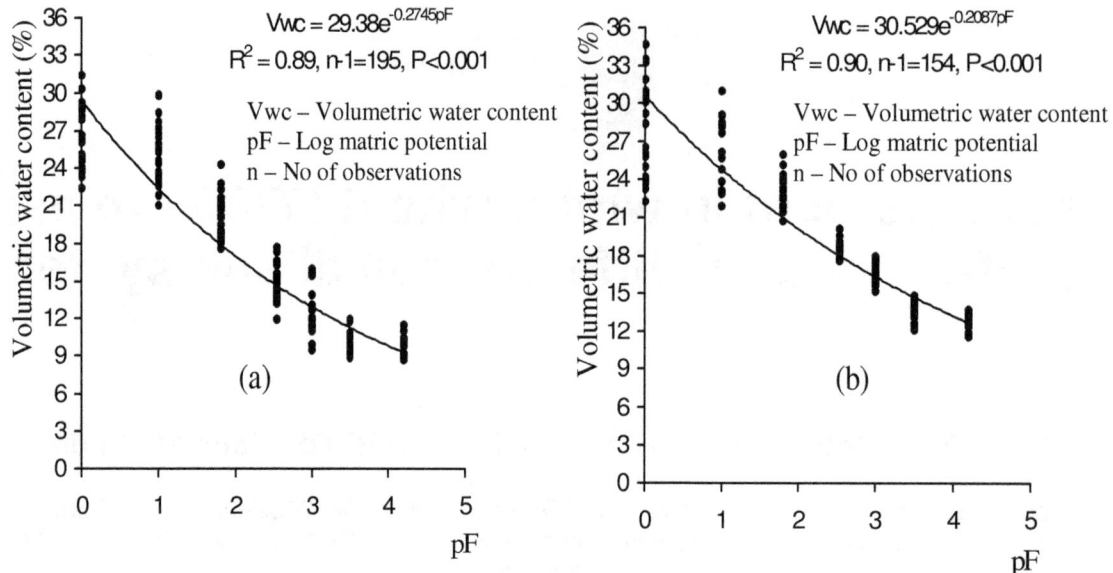

Figure 1. Water retention curve for the topsoil (0 - 10 cm; samples of the Luvisol collected during field work, (a) 1996/1997 and (b) 1999/2000.

great soil disturbances, and can have advantages with regard to water storage and increase wheat grain yield (Fenster and Wicks, 1982; Wicks et al., 1988; Mikha et al., 2013).

During the past two decades Portuguese farmers are searching increasingly for alternatives to the traditional system for the establishment of autumn-sown cereals, based predominantly on plough tillage. In reduced tillage and especially no-till systems they are looking for options able to provide environmental but mainly economical sustainability in a region where natural conditions strongly limit wheat productivity levels. For the adoption of conservation agriculture systems and their wide-spread uptake, weed flora and its dynamics must be understood (Brainard et al., 2013; Han et al., 2013). The knowledge and study of these dynamics is essential to define best management practices for weed control under given environmental conditions. There are differences in the infestation dynamics between traditional tillage based and no-till systems, because soil disturbance can both promote germination of the seeds of the existing seed bank in the soil as well as destroy the weeds that are already established (Forcella, 1986). Thus, this paper studies weed management in common wheat as a function of soil disturbance.

MATERIALS AND METHODS

The experiment was carried out at Revelheira farm in the Évora district of Portugal (38° 27' 54" N; 7° 28' W) on a Luvisol during 1996/1997 and 1999/2000 seasons in a field where winter wheat was direct drilled. The areas were submitted to sheep grazing during the summer period until the establishment of the crop. This type of management was chosen to simulate the conditions of a

wheat crop established in a rotation of cereals and natural pasture, commonly used under rainfed Mediterranean conditions.

Soil and climate conditions

The physical and chemical characteristics of the topsoil (0 to 10 cm) of Luvisol in 1996/1997 were 36.5% sand, 41.6% silt, 21.9% clay, with a bulk density of 1.6 (g cm^{-3}), available P and K of 10.5 ppm P_2O_5 and 69 ppm K_2O, pH 5.8 and 1.1% of organic matter. In 1999/2000 these characteristics were: 68.9% sand, 13.6% silt, 17.5% clay, with a bulk density 1.7 (g cm^{-3}), available P and K of 20.7 ppm P_2O_5 and 64 ppm K_2O, pH 5.6 and 1.2% of organic matter. Water retention characteristics are presented in Figure 1.

The monthly rainfall values and the average minimum, medium and maximum air temperatures are presented in Figure 2. These values were registered at the Reguengos de Monsaraz weather station, located at Revelheira farm, where the field work was carried out. The 30 year monthly average precipitation was obtained through the Instituto Nacional de Meteorologia e Geofísica (1991), based on the values registered at the Reguengos de Monsaraz pluviometric station.

Treatments and experimental design

The experimental model was two factors randomized complete block design for the main factor, which were soil tillage systems with a split with regard to the second factor of post-emergence weed control (subplot factor). There were two soil tillage systems (with and without soil disturbance), two levels of post-emergence weed control (with and without post-emergence weed control) and four replications.

Each main plot was 3 m wide and 25 m long, and divided into 2 subplots of 12.5 m length. Two tillage operations were performed in the treatment with conventional tillage (with soil disturbance), consisting of a tine cultivator followed by an offset disc harrow working at a depth of 15 cm. In the herbicide treatment, a post-emergence mixture of two herbicides (diclofop-methyl + tribenuron-methyl, trade name: Illoxan + Granstar) were applied at a

(a) (b)

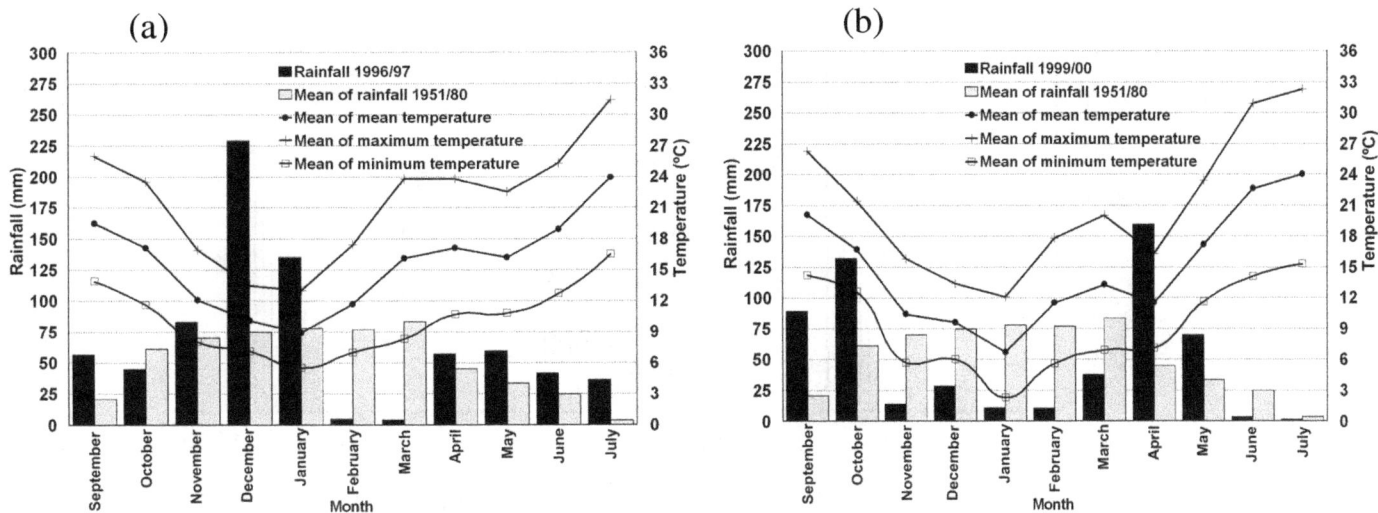

Figure 2. Thermopluviometric conditions in 1996/97 (a) and 1999/00 (b) and the 30-year average rainfall (1951/1980).

Table 1. Cultural practices used in the experiment of weed management in winter wheat influenced by different soil tillage systems.

Cultural practices	Years	
	1996/1997	1999/2000
Soil tillage: tine cultivator followed by an offset disc harrow working at a depth of 15 cm in the treatment with soil disturbance Removing weeds before planting (pre-plant herbicide application, glyphosate 900 g a.i. ha^{-1}) in the treatment without soil disturbance	20/11/1996	04/11/1999
Basic dressing: 36 kg N ha^{-1} and 92 kg P$_2$O$_5$ ha^{-1}	22/11/1996	12/11/1999
Wheat sowing. Variety: Centauro - 170 kg ha^{-1}	22/11/1996	12/11/1999
Application of active ingredient diclofop-methyl (900 g a.i. ha^{-1}) + tribenuron-methyl (13.5 g a.i. ha^{-1}) in the treatment with post-emergence weed control (recommended doses).	22/01/1997	19/01/2000
1st top dressing	27/01/1997 40 kg N ha^{-1}	12/01/2000 42 kg N ha^{-1}
2nd top dressing	10/02/1997 30 kg N ha^{-1}	Not applied
Harvest	June	

recommended doses, normally used by Portuguese farmers in wheat fields (Table 1), for an efficient control of the monocotyledon and dicotyledon weeds. The post-emergence herbicide mixture was applied when about 90% of the monocotyledons weeds, particularly *Lolium rigidum* Gaud., were at the beginning of tillering and dicotyledons were at the two to four true-leaf stages. Sowing density, fertilization and plant protection measures were done according to common local practice (Table 1). Herbicides were applied by a plot sprayer equipped with flat-fan nozzles (110° - 12) and an application volume of 200 L ha^{-1}. Winter wheat was sown with a mechanical drill with single disc openers. The harvest was carried out using a self propelled combine harvester in the small parcels (working width of 1.5 m) and manually in the 0.5 m^2 area.

Observations

In each of the subplots an area of 0.5 m^2 was chosen for the counting and identification of the weeds. This sampling method for

carrying out weed readings was used according to Colbach et al. (2000). The total number of monocotyledons, dicotyledons and some dominant weed species on the trial site (individually) were monitored before and after the post-emergence herbicide application for each date and growth stage of wheat (stages 12 to 20 and 53 to 59) according to Zadoks et al. (1974) scale. Wheat yield and the respective components (number of ears, number of grains ear^{-1} and kernel weight) and the dry matter weight of weeds at wheat harvest time were recorded. Wheat grain and the weed biomass were dried in an oven at 65°C, until a constant weight was obtained (approximately 48 h).

The average values and standard errors of differences between means for the parameters observed were calculated using the MSTAT-C programme (v. 1.42) (Freed et al., 1991). Due to the not "normal distribution" pattern of weed species' populations in the field (Gonzalez-Andujar and Saavedra, 2003), the standard error of the means was used as it helps to estimate the uncertainty of the true value of the population mean, indicating the accuracy of the mean. The relative wheat yield was calculated in relation to the

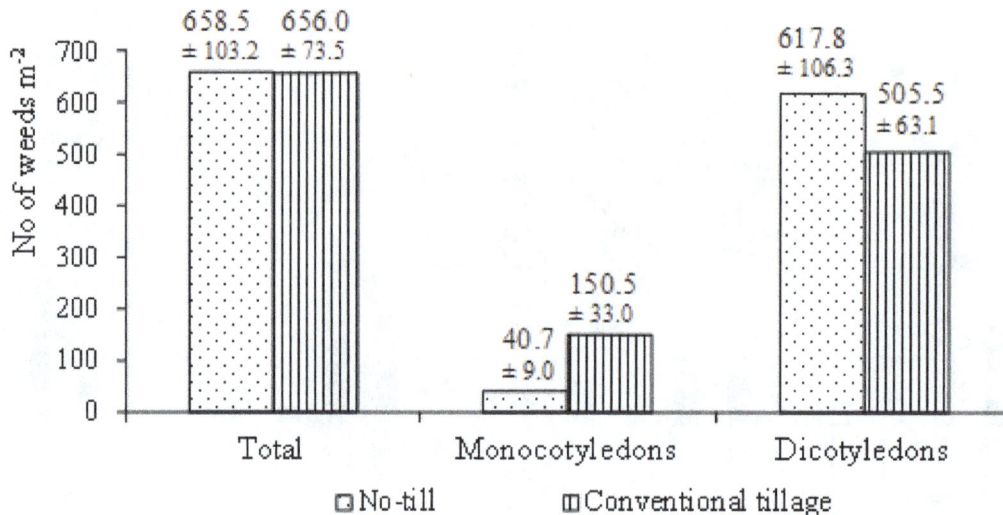

Figure 3. Number of weeds per square metre at tillering stage of the wheat under conventional tillage and no-till (average values of two years (1996/97 and 1999/00) ± standard error, region - Évora, Portugal).

yield obtained from disturbed soil with post-emergence herbicide.

Linear regressions were calculated between monocotyledons at heading stage of the wheat and the grain yield of the wheat as well as the weed dry matter at harvest and the wheat grain yield using Microsoft Excel (v. 2003) and Statistical Package for Social Sciences (SPSS) 15.0.

RESULTS

Effects of different soil tillage systems

The conventional soil tillage system increased the number of grass weeds at the beginning of the tillering stage of the wheat (stage 15). During this period, the no-till system is characterised by the occurrence of less monocotyledons (Figure 3). The counting of the weeds carried out at the heading stage of the wheat (stages 53 to 59), showed that the average number of the monocotyledon weeds continues to be less under no-till, with 13.0 plants m^{-2} in 1996/1997 and 38.8 plants m^{-2} in 1999/2000, against 26.8 plants m^{-2} in 1996/1997 and 49.3 plants m^{-2} in 1999/2000, in the conventional tillage system (Table 2). Therefore, conservation agriculture, especially no-till system, can contribute to decrease grass weeds in the cereal crops.

As we can see from Table 3, the weeds dry matter was less at harvest in no-till system, as compared to that in disturbed soil, especially in 1996/1997 (57.0 g m^{-2} in no-till and 157.0 g m^{-2} in conventional tillage, without post-emergence herbicide application). With regard to wheat yield and the respective components, it is noted that grain yield was greater when there was no-till (no soil disturbance) (Table 3). Also, there was a smaller difference between yield with post-emergence herbicide application and yield without post-emergence herbicide

application in no-till system (413 kg ha^{-1} in 1996/1997 and 570 kg ha^{-1} in 1999/2000) than in conventional tillage system (762 kg ha^{-1} in 1996/1997 and 820 kg ha^{-1} in 1999/2000). Higher grain yield achieved in no-till systems resulted, mainly, in formation of a larger number of grains. The conventional tillage also affected negatively the grain weight when post-emergence herbicide was not applied (Table 3).

Influence of the post-emergence herbicide application on weeds and wheat yield components

As it is demonstrated in Table 4, application of post-emergence herbicides affect the number of weeds observed at the heading stage of the crop (stages 53 to 59). As a consequence of the decrease in the number of weeds with application of post-emergence herbicides, a lower weed dry matter weight was obtained at harvest, which was different from the weight obtained in the treatments without post-emergence herbicide application (Table 3). Due to the influence of herbicide application on weeds, it was noted that without post-emergence herbicide application, there was a decrease in the number of wheat grains, the determining component of wheat yield, under Mediterranean conditions. In turn, there was a decrease in the weight of the grain, without herbicide application in conventional tillage, but in the no-till system there was no difference in the average values of the kernel weight in the two treatments, with and without post-emergence herbicide (Table 3).

Wheat yield influenced by weeds infestation

As mentioned previously, the density of monocotyledon

Table 2. Effect of soil tillage systems on weed plant density (No m⁻²) at heading stage of the wheat (average values (1996/1997 and 1999/2000) ± standard error, region - Évora, Portugal).

No. of plants m⁻²	Year 1996/1997		Year 1999/2000	
	No-till	Conventional tillage	No-till	Conventional tillage
Total	129.3 ± 47.3	71.5 ± 25.1	52.5 ± 20.6	60.0 ± 23.2
Monocotyledons	13.0 ± 6.6	26.8 ± 8.7	38.8 ± 15.9	49.3 ± 19.4
Dicotyledons	116.3 ± 42.3	44.8 ± 17.9	13.8 ± 8.5	10.8 ± 5.1
Lolium spp.	0.5 ± 0.3	6.0 ± 3.5	38.3 ± 15.7	48.5 ± 19.0
Phalaris spp.	11.3 ± 5.9	15.8 ± 8.7	0.3 ± 0.3	0.3 ± 0.3
Coleostephus myconis (L.) Reincheb.	0.0 ± 0.0	11.8 ± 8.7	0.0 ± 0.0	0.0 ± 0.0
Chamaemelum spp.	1.3 ± 0.8	0.0 ± 0.0	0.0 ± 0.0	0.0 ± 0.0
Calendula arvensis L.	0.8 ± 0.8	3.3 ± 2.2	0.0 ± 0.0	0.0 ± 0.0
Anagallis arvensis L.	74.3 ± 32.1	16.8 ± 6.5	0.0 ± 0.0	0.0 ± 0.0
Polygonum aviculare L.	18.3 ± 8.9	7.5 ± 3.0	0.8 ± 0.8	2.3 ± 1.5
Spergularia purpurea (Pers.) G. Don fil.	20.3 ± 11.9	2.0 ± 1.3	8.5 ± 5.7	3.3 ± 3.3

Table 3. Wheat yield, yield components and weed dry matter at harvest, under conventional tillage and no-till and with and without post-emergence herbicide (average values (1996/1997 and 1999/2000) ± standard error, region - Évora, Portugal).

Parameter	Year 1996/1997				Year 1999/2000			
	No-till		Conventional tillage		Post-emergence herbicide			
					No-till		Conventional tillage	
	With	Without	With	Without	With	Without	With	Without
Tillers plant⁻¹	0.365 ± 0.10	0.396 ± 0.11	0.293 ± 0.10	0.069 ± 0.11	0.719 ± 0.14	0.273 ± 0.16	0.301 ± 0.09	0.162 ± 0.06
No of ears m⁻²	356.0 ± 21.0	299.5 ± 22.9	359.0 ± 28.4	286.5 ± 23.5	369.5 ± 15.1	310.0 ± 20.1	346.0 ± 29.0	331.0 ± 29.2
No of grains m⁻²	7836 ± 526	6619 ± 494	6897 ± 619	4773 ± 1091	7981 ± 431	6113 ± 152	7278 ± 255	5385 ± 262
Kernel weight (mg)	31.85 ± 0.35	31.53 ± 0.77	31.32 ± 0.56	29.31 ± 0.15	31.28 ± 0.81	31.64 ± 0.89	31.33 ± 0.39	29.77 ± 0.71
Yield (kg ha⁻¹)	2492 ± 149	2079 ± 129	2162 ± 201	1400 ± 322	2503 ± 185	1933 ± 63	2424 ± 77	1604 ± 91
Harvest index	0.425 ± 0.01	0.400 ± 0.02	0.380 ± 0.0	0.325 ± 0.04	0.352 ± 0.01	0.360 ± 0.03	0.356 ± 0.02	0.352 ± 0.03
Dry matter of weeds (g m⁻²)	2.0 ± 1.4	57.0 ± 16.2	39.00 ± 33.1	157.0 ± 27.8	0.3 ± 0.2	222.0 ± 19.4	1.9 ± 0.8	235.6 ± 17.2

weeds, especially *Lolium rigidum* Gaud., was higher under conventional tillage carried out for seedbed preparation for wheat when compared to no-till. Increasing of weeds population caused an increase in competition for limited resources such as water and consequently a negative influence on yield and the respective components of the wheat crop. This can be verified in Figure 4, which shows the relationship between the number of monocotyledon weeds and wheat yield. In fact, there is a clear tendency towards a drop in the average yield values when the average number of these weeds increases. Figure 4 demonstrates that a weed population of up to 40 monocotyledon plants per square metre is related to wheat grain yields of over 2000 kg ha⁻¹. When the number of monocotyledon weeds moves towards zero, grain production increases to 2500 kg ha⁻¹, while a

Table 4. Effect of post-emergence mixture herbicide application on weed plant density (No m^{-2}) at heading stage of the wheat (average values (1996/97 and 1999/00) ± standard error, region - Évora, Portugal).

No. of plants m^{-2}	Year 1996/1997		Year 1999/2000	
	Post-emergence herbicide			
	With	Without	With	Without
Total	11.3 ± 4.0	189.5 ± 28.6	2.5 ± 0.8	110.0 ± 11.9
Monocotyledons	5.0 ± 3.4	34.8 ± 7.5	0.3 ± 0.3	87.8 ± 9.6
Dicotyledons	6.3 ± 1.7	154.8 ± 29.9	2.3 ± 0.8	22.3 ± 8.3
Lolium spp.	0.0 ± 0.0	6.5 ± 3.4	0.3 ± 0.3	86.5 ± 9.2
Phalaris spp.	3.0 ± 2.1	24.0 ± 8.7	0.0 ± 0.0	0.5 ± 0.3
Coleostephus myconis (L.) Reincheb.	0.0 ± 0.0	11.8 ± 8.7	0.0 ± 0.0	0.0 ± 0.0
Chamaemelum spp.	0.0 ± 0.0	1.3 ± 0.8	0.0 ± 0.0	0.0 ± 0.0
C. arvensis L.	0.0 ± 0.0	4.0 ± 2.2	0.0 ± 0.0	0.0 ± 0.0
Anagallis arvensis L.	6.0 ± 1.8	85.0 ± 29.3	0.0 ± 0.0	0.0 ± 0.0
P. aviculare L.	0.3 ± 0.3	25.5 ± 7.2	0.5 ± 0.5	2.5 ± 1.5
S. purpurea (Pers.) G. Don fil.	0.0 ± 0.0	22.3 ± 11.5	0.0 ± 0.0	11.8 ± 5.9

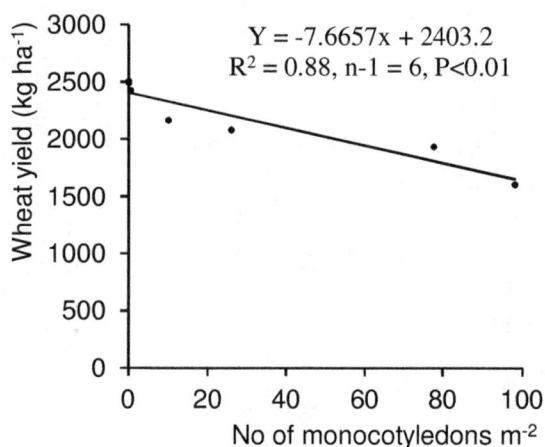

$$Y = -7.6657x + 2403.2$$
$$R^2 = 0.88, n-1 = 6, P<0.01$$

Figure 4. Relationship between the number of monocotyledons at heading stage of the wheat and the wheat yield (average values of two years).

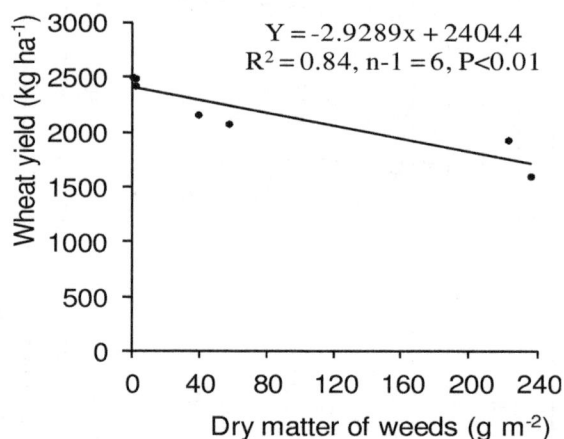

$$Y = -2.9289x + 2404.4$$
$$R^2 = 0.84, n-1 = 6, P<0.01$$

Figure 5. Relationship between the weed dry matter at harvest and the wheat yield (average values of two years).

population of 25 monocotyledon plants per square metre corresponds to a wheat yield of 2250 kg ha^{-1}.

With regard to weed dry matter at wheat harvest period, a negative linear relationship is observed between weed dry matter weight and wheat yield (Figure 5). Weed dry matter weight variations of 0 to 50 gm^{-2} correspond to wheat yields of 2500 to 2000 kg ha^{-1}. Due to the greater density of monocotyledon weeds that are present after soil tillage for seedbed preparation, there is a great decrease in crop yield without herbicide application, compared with post-emergence herbicide application, than that in the no-till system (Figure 6). Therefore, there is a greater relative yield in no-till system, without post-emergence herbicide application, which is another benefit of conservation agriculture systems.

DISCUSSION

Soil tillage enhanced the emergence of monocotyledon weeds, mainly *L. rigidum* Gaud., the dominant species in the region of the study. Therefore, despite the elimination of weeds through tillage performed for seedbed preparation, an increase in weed emergence can be expected after crop establishment (Bräutigam and Tebrügge, 1997; Marginet et al., 2000; Rahman et al., 2000), due to more favourable conditions for weed germination created by the tillage operations (Attwood et al., 1977; Fenster and Wicks, 1982; Bräutigam and Tebrügge, 1997; Mirsky et al., 2010). There are various studies that refer to the negative impact of *L. rigidum* Guad. (the dominant type in the conditions of this study) on wheat grain yield (Lemerle et al., 1995, 1996). According to Lemerle et al. (1996), competition between *L. rigidum* Guad. and wheat plants affected formation of the heading and the crop grain.

In competition between weeds and crops, it is important

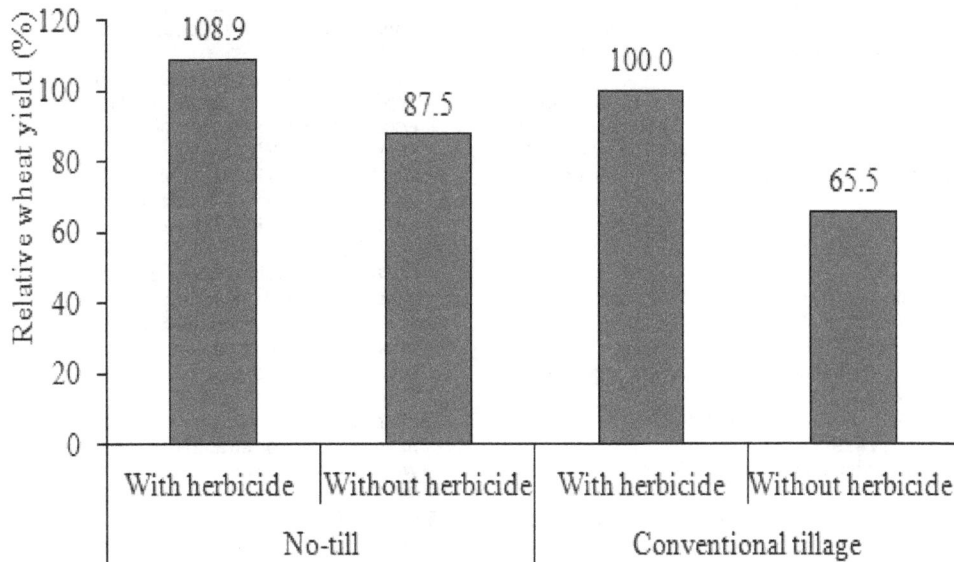

Figure 6. Wheat yield under conventional tillage and no-till and with and without post-emergence herbicide relative to the yield obtained from disturbed soil with post-emergence herbicide (% calculated from average values of two years (1996/1997 and 1999/2000), region - Évora, Portugal).

to analyse the threshold values (Swanton et al., 1999; Munier-Jolain et al., 2002; Gherekhloo et al., 2010), when is it possible to verify and control crop infestation (Swanton et al., 1999; Colbach et al., 2000). As also observed in our field trials, wheat yields under Mediterranean rainfed conditions frequently are around 2500 kg ha^{-1}. From our results we can also conclude that wheat grain losses due to grass weed infestation remain below 10% of this average wheat yield if the number of monocotyledon weeds at wheat heading stage below 25 per square metre.

As different studies have demonstrated that post-emergence herbicide has a significant influence on weed density and as a consequence on wheat yield and its components (Buhler et al., 2000; Han et al., 2013). However, the effect of post-emergence mixture herbicide application on the achievement of acceptable wheat yields depends on the given conditions (Calado et al., 2010). As a result of this study it appears that under Mediterranean conditions the post-emergence herbicide application is more important when crop establishment is based on the conventional tillage system, especially when compared to the no-till system. Considering the relative yield values (Boström, 1999; Colbach et al., 2000; Munier-Jolain et al., 2002), the observed loss in wheat yield without application of post-emergence herbicide, compared to the yield obtained with application of post-emergence herbicide is less in no-till systems than in soil tillage systems. Thus, if the registered autumn precipitation before sowing allows for maximizing emergence of potential weeds (Grundy and Mead, 2000; Froud-Williams, 2001), this is controlled in the no-till system

through application of pre-sowing herbicide (Calado et al., 2010), without significant disturbance of the topsoil layer. In these conditions, the weed population emerging after crop sowing causes less harm to the plants, compared to the soil tillage system, and given the cost benefit results, there is no clear justification for applying post-emergence herbicide. Therefore, it can be concluded that in no-till wheat, in addition to environmental benefits, achieved through the improvement of soil physical, chemical and biological conditions (Mikha et al., 2013), and the benefits resulting from lower costs for crop establishment, it is possible to reduce or avoid the use of post-emergence herbicides.

ACKNOWLEDGMENTS

This study was carried out with the support of the Regional Directorate of Agriculture of Alentejo. The authors expresses their thanks to the Directorate, as well as all those who contributed to the study.

REFERENCES

Attwood PJ, Cussans GW, Holthman ES, Neild JR, Proctor JM, Tottman DR (1977). The principles of weed control in cereals. In: Fryer JD, Makepeace RJ (eds) Weed Control Handbook, 6th Ed., Blackweel Scientific Publications 1:261-281.

Bàrberi P, Lo Cascio B (2001). Long-term tillage and crop rotation effects on weed seedbank size and composition. Weed Res. 41:325-340.

Boström U (1999). Type and time of autumn tillage with and without herbicides at reduced rates in southern Sweden 1. Yields and weed quantity. Soil Till. Res. 50:271-281.

Brainard DC, Haramoto E, Williams II MM, Mirsky S (2013). Towards a no-till no-spray future? Introduction to a symposium on nonchemical weed management for reduced-tillage cropping systems. Weed Technol. 27(1):190-192.

Bräutigam V, Tebrügge F (1997). Influence of long-termed no-tillage on soil borne pathogens and weeds. In: Tebrügge F, Böhrnsen A (eds) Experience with the Applicability of No-tillage Crop Production in the West-European Countries: Proceedings of the EC-Workshoop - III - Évora, pp. 17-29.

Buhler DD, Liebman M, Obrycki JJ (2000). Review - Theoretical and practical challenges to an IPM approach to weed management. Weed Sci. 48:274-280.

Calado JMG, Basch G, Carvalho M (2010). Weed management in no-till winter wheat (Triticum aestivum L.). Crop Prot. 29(1):1-6.

Colbach N, Dessaint F, Forcella F (2000). Evaluating field-scale sampling methods for the estimation of mean plant densities of weeds. Weed Res. 40:411-430.

Fenster CR, Wicks GA (1982). Fallow systems for winter wheat in western Nebraska. Agron. J., 74: 9-13.

Forcella F (1986). Timing of weed control in no-tillage wheat crops. Agron. J. 78:523-526.

Freed R, Eisensmith SP, Goetz D, Reicosky D, Smail VW, Wolberg P (1991). MSTAT-C - A Microcomputer Program for the Design, Management, and Analysis of Agronomic Research Experiments. Michigan State University, East Lansing Michigan, USA.

Froud-Williams RJ (2001). Population dynamics and decision supports for agro-ecosystems: Proceedings The BCPC Conference - Weeds 2001, pp. 455-460.

Gherekhloo J, Noroozi S, Mazaheri D, Ghanbari A, Ghannadha MR, Vidal RA, de Prado R (2010). Multispecies weed competition and their economic threshold on the wheat crop. Planta Daninha 28(2):239-246.

Gonzalez-Andujar JL, Saavedra M (2003). Spatial distribution of annual grass weed populations in winter cereals. Crop Prot., 22(4): 629–633.

Gruber S, Claupein W (2009). Effect of tillage intensity on weed infestation in organic farming. Soil Till. Res. 105(1):104-111.

Grundy AC, Mead A (2000). Modeling weed emergence as a function of meteorological records. Weed Sci. 48:594-603.

Han H, Ning T, Li Z (2013). Effects of tillage and weed management on the vertical distribution of microclimate and grain yield in a winter wheat field. Plant Soil Environ. 59(5):201–207.

Instituto Nacional de Meteorologia e Geofísica (1991). O clima de Portugal - normais climatológicas da região de Alentejo e Algarve correspondentes a 1951/80 - fascículo XLIX. Lisboa, Portugal.

Lemerle D, Verbeek B, Coombes N (1995). Losses in grain yield of winter crops from Lolium rigidum competition depend on crops species, cultivar and season. Weed Res. 35:503-509.

Lemerle D, Verbeek B, Coombes NE (1996). Interaction between wheat (Triticum aestivum) and diclofop to reduce the cost of annual ryegrass (Lolium rigidum) control. Weed Sci. 44:634-639.

Locke MA, Krishna NR, Zablotowicz RM (2002). Weed management in conservation crop production systems. Weed Biol. Manag. 2:123-132.

Marginet A, Van Acker R, Derksen DA, Entz MH, Andrews T (2000). Wild oat (Avena fatua) emergence as affected by tillage and ecodistrit: Proceedings of the 2000 National Meeting - Expert Committee on Weeds, Alberta, Canada, pp. 31-39.

Mikha MM, Vigil MF, Benjamin JG (2013). Long-Term tillage impacts on soil aggregation and carbon dynamics under wheat-fallow in the central great plains. Soil Sci. Soc. Am. J. 77(2):594-605.

Mirsky SB, Gallandt ER, Mortensen DA, Curran WS, Shumway DL (2010). Reducing the germinable weed seedbank with soil disturbance and cover crops. Weed Res. 50(4):341–352.

Morris NL, Miller PCH, Orson JH, Froud-Williams RJ (2010). The adoption of non-inversion tillage systems in the United Kingdom and the agronomic impact on soil, crops and the environment-A review. Soil Till. Res. 10(1-2):1-15.

Munier-Jolain NM, Chauvel B, Gasquez J (2002). Long-term modelling of weed control strategies: analysis of threshold-based options for weed species with contrasted competitive abilities. Weed Res. 42:107-122.

Rahman A, James TK, Mellsop J, Grbavac N (2000). Effect of cultivation methods on weed seed distribution and seedling emergence: Proceedings of the 53rd New Zealand Plant Protection Conference, pp. 28-33.

Swanton CJ, Weaver S, Cowan P, Van Acker R, Deen W, Shrestha A (1999). Weed thresholds: theory and applicability. J. Crop Prod. 2:9-29.

Tørresen KS, Skuterud R (2002). Plant protection in spring cereal production with reduced tillage. IV. Changes in the weed flora and seedbank. Crop Prot. 21:179-193.

Wicks GA, Smika DE, Hergert GW (1988). Long-term effects of no-tillage in a winter wheat (Triticum aestivum)-sorghum (Sorghum bicolor)-fallow rotation. Weed Sci. 42:27-34.

Zadoks JC, Chang TT, Konzak CF (1974). A decimal code for the growth stages of cereals. Weed Res. 14:415-421.

9

Effect of urea and common salt (NaCl) treated glyphosate on parthenium weed (*Parthenium hysterophorus* L.) at Western Hararghe zone, Ethiopia

Zelalem Bekeko

Department of Plant Sciences, Haramaya University Chiro Campus, P. O. Box 335, Chiro, Ethiopia.

Field experiments were conducted from 2008 to 2011, at western Hararghe zone, Eastern Ethiopia to evaluate the effect of urea and common salt treated glyphosate on parthenium weed (*Parthenium hysterophorus*). The experiments were arranged in randomized complete block design with five replications. Glyphosate herbicide at 3 L/ha was applied with different rates (0, 50, 100 and 150 ml) of urea and common salt. From the pooled analysis of variance over years it was observed that there existed a significant difference among treatments (p<0.05) in which the total population of parthenium weed mortality was noted at 25 days after applications across years for 3000 ml of glyphosate treated with 150 ml of urea and 150 ml of common salt. Results from this experiment showed that treating 3000 ml of glyphosate with 150 ml of urea and 150 ml of common salt solutions and spraying at 6 to 8 leaves stages resulted in complete mortality of parthenium weed in short period of time by increasing the phytotoxicity of this herbicide. While spraying this solution at 50 and 75% flowering stages showed poor mortality rates on this weed. This study also elucidated the complete change of parthenium infested plots into soft weed species that suppresses the re-emergency of this noxious weed. This helps in reducing the soil seed bank of parthenium weed, thus its population declines over successive years. This finding helps in reducing the rate and frequency of glyphosate application in conservation tillage and plantation crops like coffee and fruit farms, thus managing cost of parthenium weed can be significantly reduced.

Key words: *Parthenium hysterophorus,* glyphosate, mortality rate, phytotoxicity, soil seed bank.

INTRODUCTION

Parthenium hysterophorus L. (Asteraceae), commonly known as parthenium weed, carrot weed or congress grass, is an aggressive annual herb native to tropical and subtropical America. It has become invasive in North America, South America, the Caribbean, Africa, Asia and Australia (Navie et al., 1996). In many introduced regions (e.g. Australia and India), *P. hysterophorus* has posed serious threats to crop production, natural biodiversity

and human health, because of its prolific growth, rapid spreading and production of toxic allelochemicals (Chippendale and Panetta, 1994; Evans, 1997).

Parthenium is a major new agricultural weed in Ethiopia (Tamado et al., 2002; Taye et al., 2004 and Mohammed, 2010). It is believed to have been introduced into Ethiopia in 1970s during the Ethio-Somali war and has become a serious weed both in arable and grazing lands (Berhanu,

1992; Fasil, 1994; Frew et al., 1996; Tamado et al., 2002). Parthenium can cause severe crop yield losses. In India, a crops yield reduction ranged between 40% up to 90% (Khosla and Sobti, 1981).

Presently the major maize, sorghum, tef, coffee, spice and wheat growing regions in Ethiopia, for example, are being infested by parthenium weed.

In Eastern Ethiopia, parthenium is the second most frequent weed (54%) after *Digitaria abyssinica* (63%) (Tamado and Milberg, 2000) and sorghum grain yield was reduced from 40 to 97% depending on the year and the location (Tamado et al., 2002). Other than direct competition with crops, parthenium poses allelopathic effect on different crops and other plants (Mersie and Singh, 1988; Evans, 1997; Wakjira et al., 2005; Adkins and Naive, 2006). Tadele (2002), Wakjira et al. (2005) and Wakjira (2009) studied the allelopathic effect of parthenium weed in Ethiopia on tef, lettuce and onion, respectively. It was found that this weed has a significant effect on germination capacity and seedling growth of these crops. This in turn has a significant yield reduction effect on these economically important crops. Thus, the spread of parthenium in Ethiopia would be a bigger risk to the expansion and sustainable production of many crops in the country which can potentially interfere with the food self-sufficiency and food security program envisioned (Wakjira, 2009).

Physical control methods of parthenium include manual weeding before flowering, or after flowering leads to increased seed dispersal and germination. Chemical control methods may lead to herbicide resistance by the weed (Adkins et al., 1997; Njoroge, 1991) in addition to their serious ecological problems like groundwater contamination and consequently leading to human health hazards.

Biological control methods like release of insect enemies and rust fungi also have limitations (Taye et al., 2004; Bekeko et al., 2012).

Management of this weed imparts huge economic burden on the countries where it has aggressively invaded (Review of progress towards implementation of Parthenium weed strategic plan 2006-2007). Moreover, allelopathic effect of *P. hysterophorus* on other species makes it difficult for the weed management strategies (Mahadevappa, 1997 and Wakjira et al., 2005). As no single method of control has been successful, an integrated approach is suggested for its effective control (Mahadevappa, 1997; Bekeko et al., 2012).

Although the agricultural methods have been developed for parthenium control has limitation effect. For instance, removing parthenium by slashing or mowing as soon as or before it flowers though it prevents seed production results in regeneration of new shoots leading to a repeated operation. Manual and mechanical uprooting also prove to be of limited value owing to enormous amount of labour and time required (Berhanu, 1992) and vulnerability of workers engaged in the operation to the various kinds of allergies caused by the weed (Kololgi et al., 1997). Chemical control, though effective, is temporary and needs repeated application, besides have problems of residues, selectivity, availability and cost of application (Singh, 1997).

Different chemicals are used in management of parthenium weed such as bromacil, diuron or terbacil, at 1.5 kg/ha (Kanchan and Jaychandra, 1980), diquat at 0.5 kg/ha were reported to effectively control parthenium weed. Spraying 2 kg/ha of 2,4-D sodium salt or 2 l/ha of MCPA were effective to control parthenium at the seedling stage (Bhan et al., 1997). Bhan et al.(1997) also reported 1 to 2% solution of glyphosate with or without surfactant and Metribuzin at 1 to 2 kg/ha gave 90 to 98% visual toxicity on parthenium and advocated the supremacy of chemical control over other control measures on the bases of quick relief, time saving and cost effectiveness.

Chemical pollution of the environment, enormous cost, danger of toxicity to non-target plants, necessity of the chemical application in non-agricultural areas, rapidity of re-invasion of treated areas soon after the effect is diminished are the draw backs of chemical control (Singh, 1997). Similarly, Bhan et al. (1997) reported that chemical control alone is not justifiable as the effect of herbicide will always be of temporary nature and repeated operations are required which will not remain cost effective.

Nevertheless, a number of herbicides were registered to control parthenium (Navie et al., 1996), but for the smallholder farmers of eastern Ethiopia, where the average farm size is 0.65 ha (CSA 2002), use of herbicides to control parthenium is not economically feasible. Instead, parthenium is currently controlled mainly through hand hoeing and hand pulling and sometimes by interrow oxen cultivation. Hoeing by hand involves effort and time (Tamado et al., 2002). Uses of additives such as urea, oils, and common salt in glyphosate help in increasing the efficacy and the phytotoxicity of the herbicide which help overcome the effect of noxious weed species including parthenium weed and increase farmers' productivity (Rao, 1979).

As parthenium is a weed of wasteland, a common man will never invest his money in this venture. Moreover, plants suppressed by chemicals have been observed to regenerate after remaining dormant for a few days (Adkins et al., 2006). Chemical treatment can only kill existing population at the given sites but can not prevent the entry of the seeds from neighbouring places. Use of additives help in increasing phytotoxicity of herbicides through enhancing absorption and translocation of herbicides leading to long term management of weeds by changing the weed spectrum into soft weed species including parthenium weed (Rao, 1979).

Therefore, developing cost effective ways of managing plantation crops such as coffee and fruit farms remains the crux of the matter. So far, very limited information

Effect of urea and common salt (NaCl) treated glyphosate on parthenium weed...

59

Table 1. Mortality percentage of parthenium by glyphosate (3l/ha) tank mixed with urea and common salt at Chiro Campus, at 25, 35, 45 and 55 days observation.

Treatments	25 days	35 days	45 days	55 days
Gly.+ 0 ml of urea/ salt	15	35	20	25
Gly.+ 50 ml of urea/Salt	25	52	38	48
Gly.+ 100 ml of urea / salt	40	84	64	75
Gly.+ 150 ml of urea/ salt	55	72	88	94
CV	8.5	12.2	7.8	8.2
LSD	4.8	8.7	3.4	5.3

Table 2. Effect of glyphosate on parthenium weed leaves discoloration at 25, 35 and 55 days after spraying.

Growth stage (% flowering)	Percentage of leaf discoloration at 25 days (% color change)	35 days	55 days	Total
0	85	15	-	100
25	25	45	5	75
50	15	20	10	45
75	5	15	8	28

exists in Ethiopia regarding parthenium weed parthenium weed for smallholder farmers' field and management through chemicals. Except the investigation made by Tamado and Tamado (2004) using 2.4-D in sorghum fields no experiments were conducted using additives to increase phytotoxicity of glyphosate in controlling parthenium weed in Ethiopia. Therefore, the objective of this experiment was to evaluate effect of glyphosate tank mixed with urea and common salt (NaCl) solutions on parthenium weed at western Hararghe zone, Oromia Regional State, eastern Ethiopia.

MATERIALS AND METHODS

Description of the study area

West Hararghe is located between 7° 55' N to 9° 33' N latitude and 40° 10' E to 41° 39'E longitude. The major crops grown in the study area are sorghum, maize, chat, field beans, potato and tef. The area is characterized by Charcher Highlands having undulating slopes and mountainous in topography. The mean annual rainfall ranges from 850 to 1200 mm/year with minimum and maximum temperatures of 12 and 27 °C, respectively.

Experimental design and procedures

Field experiments were conducted to assess the efficacy of Glyphosate tank mixture with urea and common salt solution against parthenium weed (*P. hysterophorus* L.) at different growth stages in non-cropped area of Chiro Campus, during summer 2008 to 2011. The experiments were arranged in Randomized Complete Block Design (RCBD) with five replications. A plot size of 6 m×6 m (36 m^2) was used in this experiment (Gomez and Gomez, 1984). 3 L/ha of glyphosate was tank mixed with different rates of urea and common salt solution viz, 0 g of urea and 0 g of common salt, 50 ml of urea and 50 ml of salt, 100 ml of urea and 100 ml of salt, and

150 ml of urea and 150 ml of common salt solutions were sprayed to parthenium weed respectively at 6 to 8 leaves, 25% flowering, 50% flowering, and 75% flowering stages using knapsack sprayer (CP 15) on 10 April 2008, 5 May 2009, June 15, 2010 and July 10, 2011 for four years.

Data on mortality percentage of parthenium weed was taken every ten days for two months and percentage of weed spectrum shifted to another weed species were noted on the experimental plots and subjected to analysis of variance using SAS package (Table 1).

RESULTS

Parthenium weed control at different growth stages

The statistical analysis of the data showed that Glyphosate tank mixed with different concentration of urea and common salt solutions had significant effect on parthenium weed mortality under field conditions (p=0.05). The treatments provided 32 to 89% mortality at two weeks after treatment (WAT) and 43 to 96% mortality at 4 WAT (Table 2).

This result exhibited that maximum parthenium weed mortality (96%) at 4 WAT was recorded in glyphosate treated with 150 ml of urea and 150 ml of common salt solutions which was followed by 100 ml of urea and 100 ml of common salt solutions on treated plots scoring 80% mortality (Table 3).

Effects of the solutions on parthenium weed leaf discoloration

Spraying glyphosate treated with 100 ml of urea and 100 ml of common salt solution and 150 ml of urea and 150

Table 3. Field observation on mortality rate of parthenium weed due to effects of the treatments at different growth stages of parthenium weed.

Growth stage	Effect observed 35 days after spraying	Remark
6-8 leaf stage	96	Very rapid
25% flowering	80	Rapid
50% flowering	25	Fair
75% flowering	15	Poor

Table 4. Ground cover shift into different weed species at Chiro. 2008- 2011 after mortality of parthenium weed.

No	Dominant weed species /family	Mean number of seedlings/4 m^2			
		April	May	June	July
1	*Amaranthus hybridatus*	56	68	48	22
2	*Parthenium hysterophorus*	32	14	8	5
3	*Argemonium mexicana*	8	12	24	38
4	*Cynodon dactylon*	42	52	52	52
5	*Bidens pilosa*	46	68	34	12
6	*Digitaria abyssinica*	28	42	42	42
7	*Rumex abyssinica*	14	24	18	18
8	*Commelina bengalensis*	6	12	12	12
9	*Datuta stramonium*	4	10	10	10
10	*Galensoga palviflora*	54	54	28	10
11	*Guziotia scabra*	3	11	11	11
12	*Killinga bulbosa*	18	22	22	22

ml common salt solutions, caused 55 and 85% parthenium mortality, respectively at 0% flowering stage or 6 to 8 leaves stage 25 days after treatments (Table 2). While spraying these solutions at 25, 50 and 75% flowering stages had no significant effect on parthenium mortality at 25, 35 and 55 days after treatments. Spraying the solutions at 6 to 8 leaves stage caused completely (100% mortality) control of parthenium weed at 35 days after treatments (Table 2).

It appeared that spraying this mixture at the 6 to 8 leaves stage of parthenium weed and at 25% flowering stages, resulting in 96 and 80% weed mortality, respectively (Table 3). Observation on the rate of parthenium weed mortality, showed the existence of significant differences among treatments at different growth stages of this aggressive weed over years in which maximum mortality was noted in 2011 (Table 3).

Change on the weed spectrum (ground cover shift) in the treated plots

The experimental plots treated with the glyphosate solution had caused rapid death of parthenium weed. After the mortality of parthenium weed in the experimental plots it appeared that new weed species

had emerged in which the population of *Amaranthus hybridatus* and *Bidens pilosa*, respectively reached 68% in the month of May (Table 4). New weed species such as *Amaranthus hybridatus*, *Bidens pilosa*, *Cynodon dactylon*, *Digitaria abyssinica* and *Galensoga palviflora* emerged dominantly in the treated plots after the death of parthenium weed. In addition another new weed species such as Rumex, Commelina, and Argemonium had emerged in these plots (Table 4). Generally it was observed that in the treated plots new weed species appeared 30 days after glyphosate tank mixture treatment which had resulted into complete change of the weed spectrum into soft weeds in the periods of April to July (Table 4).

DISCUSSION

The results indicated that parthenium plants can effectively be controlled with glyphosate tank mixed with low concentrations of urea and common salt (Table 1). So far studies conducted on management of this weed using other herbicides did not provide satisfactory control when applied at bolted stage, even high rates of herbicides failed to control parthenium weed. Singh et al., (2004) reported that 2,4-D, atrazine, metribuzin,

Effect of urea and common salt (NaCl) treated glyphosate on parthenium weed...

61

metsulfuron, chlorimuron, and glufosinate failed to control parthenium weed, while glyphosate at 2.7 and 5.4 kg ha^{-1} provided greater than 95% control of bolted plants at 18 WAT. Similarly Walia et al. (2002) reported that other herbicides with the exception of glyphosate applied to well establish parthenium weed plants did not provide satisfactory control.

Therefore, this finding is supported by the investigation made by Singh et al. (2004) and Walia et al. (2002). The rapid mortality of parthenium weed was due to enhanced phytotoxicity effect of the herbicide against this weed. Rao (1979) reported that the phytotoxicity of glyphosate can be increased by adding urea, oils and salt solutions to the glyphosate solution. In this study also lower concentrations of urea and common salt enhanced the phytotoxicity level of glyphosate on parthenium weed.

Foliar application of urea and micro nutrients like Na and Cl helps in rapid absorption and translocation of glyphosate so that faster effect is noticed in rapid leaf discoloration or chlorosis leading to decrement of the rate of photosynthesis in parthenium weed and rapid oxidation of the photo assimilate reserved in the leaves and stem of this weed. Some variations in parthenium weed control with treatments of treated glyphosate were recorded in 2008, 2009 and 2010 compared to 2011. This might have been partly due to differences in weather conditions among the years and growth stages of this weed.

In wasteland, non cropped areas, plantation crops and roadsides, the use of glyphosate has shown promising results. The stage/time of parthenium weed for herbicidal control is important. Therefore parthenium weed should be treated at 6 to 8 stage and/or before flowering to get maximum effect of this herbicide solution (Table 3). Generally all weeds were sensitive to the herbicide at vegetative stage particularly at 4 to 6 leaf stage.

Parthenium weed control at 6 to 8 leaves and before flowering stages was highest with glyphosate treated with 150 ml urea and 150 ml of common salt (96%) followed by 100 ml urea and 100 common salt solutions (80%) at 4 WAT and control was lowest with sole glyphosate (without urea and common salt) which resulted in (55 %) mortality at 4 WAT.

This result indicated that parthenium weed can effectively be controlled with glyphosate tank mixed with urea and common salt solution, while glyphosate alone used in the study did not provide satisfactory control when applied at all stages of this weed (5-55%). Parthenium weed is highly sensitive to amino acid synthesis and photosynthesis inhibitors compared to herbicides with other modes of action (Singh et al., 2004).This result has also shown the complete shift of the weed spectrum into weak weed species which can be easily controlled and helped in suppressing the re-establishment of parthenium weed in the treated plots (Table 4) thus, its soil seed bank can decline over successive years.

In this study, glyphosate treated with lower concentrations of urea and common salt is recommended for the control of parthenium weed in non-cropped areas and plantation crops such as coffee and fruit farms in all parts of Ethiopia. It is recommended that spread of parthenium weed should be prevented to avoid its harmful effect on the crop production, biodiversity, the environment and human health.

Therefore, by treating glyphosate (3 L/ha) with with 150 ml of urea and 150 ml common salt solutions can control parthenium weed. Using this combinations and its frequency of application can be help in the long term control of this noxious weed. However, further studies have to be conducted to evaluate the effect of these solutions on the physico chemical properties of the soil and soil micro organisms.

ACKNOWLEDGEMENTS

The author is grateful to the Ministry of Agriculture and Rural Development of the Federal Democratic Republic of Ethiopia Chiro ATVET College for funding this research project, and Professor Temam Hussein for his technical advice during the field experiment.

REFERENCES

Adkins SW, Navie SC (2006). Parthenium weed: A potential major weed for agro-ecosystems in Pakistan. Pak. J. Weed Sci. Res. 12(1-2):19-36.

Bhan VM Kumar S, Raghuwanshi MS (1997). Future strategies for effective Parthenium management. In: Mahadevappa, M. and Patil, V.C. (eds.). Proceedings of the 1st International Conference on parthenium management, 6-8 October 1997, Univ. Agric. Sci. Daharwad, India, pp. 90-95.

Bekeko Z, Temam H, Taye T (2012). Distribution, Incidence, Severity and Effect of the Rust (*Puccinia abrupta* var. *partheniicola*) on *Parthenium hysterophorus* L. in Western Hararghe Zone, Ethiop. Afri. J. Plant Sci. 6(13):337-345.

Berhanu GM (1992). *Parthenium hysterophorus*, a new weed problem in Ethiopia. FAO Plant protection Bulletin 40:49.

Chippendale JF, Panetta FD, (1994). The costs of parthenium weed to the Queensland cattle industry. Plant Protection Q. 9(2):73-76.

Evans HC (1997). *Parthenium hysterophorus*: a review of its weed status and the possibilities for biological control. Biocontrol News and Information 18:89-98.

Fasil R (1994). The biology and control of parthenium In: Rezene Fessahaie (ed.), Proceedings of the 9th annual conference of the Ethiopian weed Science committee, 9-10 April 1991, Addis Ababa, Ethiopia. EWSS, Addis Ababa, pp. 1-6.

Frew M, Solomon K, Mashilla D (1996). Prevalence and distribution of *Parthenium hysterophorus* L. in eastern Ethiopia." Arem 1:19-26.

Gomez KA, Gomez AA (1984). Statistical Procedures for Agricultural Research. 2nd Edition. John and Wiley and Sons, New York. P. 680.

Kanchan S, Jayachandra (1980). Pollen allelopathy - a new phenomenon. New Phytologist 84:739-746.

Khosla SN, Sobti SW (1981). Effective control of *Parthenium hysterophorus* L. Pesticide 13:121-127.

Kololgi PD Kololgi SD, Kololgi NP (1997). Dermatologic hazards of Parthenium in human beings. In: Mahadevappa M,Patil VC,eds. Proceedings of the First International Conference on Parthenium Management, Dharwad, India, 6-9 October 1997. Dharwad, India: University of Agricultural Sciences. P. 19.

Mahadevappa M (1997). Ecology, distribution, menace and management of parthenium. In: Mahadevappa M, Patil VC, eds.

Proceedings of the First International Conference on Parthenium Management, Dharwad, India, 6-9 October 1997. Dharwad, India: University of Agricultural Sciences, pp.1-12.

Mersie W, Singh M (1988). Effect of phenolic acids and rage weed *Parthenium hysterophorus* L. extracts on tomato (*Lycopersicum esculantum*) growth, nutrient and chlorophyll content. Weed Sci. 36:278-281.

Mohammed W (2010). Prevalence and distribution survey of an invasive alien weed (*Parthenium hysterophorus* L.) in Sheka zone, Southwestern Ethiopia. Afr. J. Agric. Res. 5(9):922-927.

Nath R (1988). *Parthenium hysterophrous*-a general account. Agric. Rev. 9:171-179.

Navie SC Panetta FD McFadyen RE, Adkins SW (1996). The biology of Australian Weeds 27: *Parthenium hysterophorus* L. Plant Protect. Q. 11:76-87.

Rao RS (1979). Parthenium - a new record for Assam. J. Bombay. Natural. History Soc. 68:44.

Singh SP (1997). Perspectives in biological control of parthenium in India. In: Mahadevappa M, Patil VC, eds. Proceedings of the First International Conference on Parthenium Management, Dharwad, India, 6-9 October 1997. Dharwad, India: University OF Agricultural Sciences. pp. 22-32.

Singh AY, Balyan RS, Malik RK, Singh M (2004). Control of Ragweed Parthenium (*Parthenium hysterophorus*) and associated weeds. Weed Tech. 18:658-664.

Tadele T (2002). Allelopathic effect of Parthenium hysterophorus extracts on seed germination and seedling growth of Eragrostis tef. J. Agron. Crop Sci. 188:306-310.

Tamado T, Milberg P (2000). Weed flora in arable fields of Eastern Ethiopia with emphasis on the occurrence of *Parthenium hysterophorus*. Weed Res. 40:507-521.

Tamado T, Schutz W, Milberg P (2002). Germination ecology of the weed *Parthenium hysterophorus* in eastern Ethiopia. Ann. Appl. Biol. 140:263-270.

Tamado T, Milberg P (2004). Control of *Parthenium hysterophorus* in grain sorghum (*Sorghum bicolor*) in the smallholder farming system in eastern Ethiopia. Weed Tech. 18:100-105.

Taye T, Gossmann M, Einhorn G, Buttner C, Metz R, Abate D (2004). The potential of rust as biological control of parthenium weed (*Parthenium hysterophorus* L.) in Ethiopia. Pest Manage J.. Eth. 8:83-95.

Wakjira M Berecha G, Bullti B (2005). Allelopathic effect of *Parthenium hysterophorus* extract on germination and seedling growth of lettuce. Trop. Sci. 45:159-162.

Wakjira M (2009). Allelopathic effects of *Parthenium hysterophorus* on germination and growth of onion. Allelopathy J. 24:351-362.

Walia US, Brar SJ, Kler DS (2002). Control of *Sorghum halepense* and *Parthenium hysterophorus* with different brands of glyphosate. Environ. Ecol. 20:540-543.

10

Studies on survivability of field pea rust caused by *Uromyces viciae-fabae* (Pers.) de Bary in Tarai region of Uttarakhand (India)

Singh D.[1], Kumar A.[2], Singh A. K.[3], Prajapati C. R.[4] and Tripathi H. S.[1]

[1]Centre of Advanced Studies in Plant Pathology, G.B. Pant University of Agriculture and Technology, Pantnagar-263145, Udham Singh Nagar (Uttarakhand), India.
[2]Department of Plant Pathology, IARI, Regional Station, Pusa-848125 Samastipur (Bihar), India.
[3]Department of Agronomy, ICAR Research Complex for Eastern Region Patna-800 014, Bihar, India.
[4]Department of Plant Protection, Krishi Vigyan Kendra, Baghpat, S.V.P.University of Agriculture and Technology, Meerut- 250 609, UP, India.

Pea rust, which is caused by *Uromyces viciae-fabae* (Pers.) de Bary has become the important pathogen since last decade and results in significant yield losses in India. The experiment was conducted during the 2006 to 2010 crop seasons on the survivability on rust of field pea under *in vitro* condition at Pantnagar, India. The results indicated that maximum urediospores germination was recorded from 2006 to 2010 in the month of April followed by May and June. However the number of uredinia on assayed plant indicated apparent survival of urediospores increased dramatically in the viable germination test of urediospore in June, July and August in the year 2008 to 2009. The urediospore survived at 10°C for four months in infected crop debris followed by 5 and 20°C, while failed to survive at temperature (30 to 40°C). Crop debris at different depths revealed that survivability of urediospore declined sharply over period of time as well as increased in the depth of placement. Maximum survivability of fungus was recorded in samples buried at 10 cm followed by 5, 15 and 20 cm deep, respectively. However, the urediospore failed to survive at any depth beyond 4 weeks in 2007 and 2008.

Key words: *Uromyces viciae-fabae*, survivability, urediospore, aeciospores.

INTRODUCTION

The rust fungus *Uromyces viciae-fabae* (Pers.) de Bary is one of the important pathogens of field pea (*Pisum sativum* L.). It is worldwide distributed pathogen of pea and also reported from faba bean (*Vicia faba* L.), lentil (*Lens culinaris* Medic.) and sweet pea (*Lathyrus sativus* L.) (Chung et al., 2004; Emeran et al., 2005, 2008; Kushwaha et al., 2006; Shroff and Chand, 2010). In India, Field pea, is grown over an area of 0.77 million hectare with a production 0.71 million tonnes and productivity 1032 kg/ha (Anonymous, 2009). It is being cultivated in Uttar Pradesh, Madhya Pradesh, Orissa, Bihar, Rajasthan, Punjab, Himachal Pradesh, Haryana and Uttarakhand states of India. Pea rust is an important disease in Uttar Pradesh, Uttarakhand and its surrounding areas, resulting adverse effect on grain yield (Singh, 2005). Initial infection is commonly found in areas of fields bordering to faba bean and lentil crops that were infected in previous year or nearly sheltered areas of fields where moisture favour prolonged plant wetness. In general, rust appears late in season and causes an estimated 20% yield loss in faba bean crop (Mohamed, 1982). However, these losses could go up to 45 to 50%, if severe infection occurs early in the season (Rashid and Bernier, 1991). In the last few years, disease has been observed in almost epiphytotic form in India when rust infection recorded early in season in first fortnight of November (Sharma, 1998).

The fungus *U. viciae fabae* (Pers.) de Bary is an autoecious with aeciospores, urediospores and teliospores on the surface of host plant and completes its life cycle on the same host. However, initial inoculum has not been clearly established (Gaumann, 1998). In India under field condition, urediospores are converted to teliospores in the month of March due to the higher temperature. It is assumed to teliospores overwinter in the soil or in association with their alternate host debris (Singh, 2005). Urediospores are short-lived, but the teliospores can survive in plant debris from one season to another (Hebblethwaite, 1983). However urediospores can be preserve for two years under freezing condition has been documented by Cunningham (1973) and Davison and Vaugham (1963). Germination of teliospores takes place between 17 to 22°C temperature and at the start of next season producing basidiospores which initiated new infection cycle (Joseph and Hering, 1997; Joshi and Tripathi, 2012). The basidial stage and initiation of primary infection are not fully understood. The disease is favoured by high humidity, cloudy and rainy weather condition. Disease development in field is favoured between 20 and 22°C (Webb and Hawtin, 1981; Kushwaha et al., 2006). Limited information is available on survival of pea rust fungus *U. viciae-fabae* (Pers.) de Bary in nature. Keeping this fact in view, the present investigation was undertaken to study the effect of temperature and soil depth on the survivability of urediospore of *U. viciae-fabae* in tarai region of Uttarakhand.

MATERIALS AND METHODS

Present investigation was commenced to study the effect of temperature and soil depth on the survivability of urediospore of *U. viciae-fabae* in Tarai region of Uttarakhand (India). The following methods were adopted to study the specific objective of experiment.

Production of aeciospores /urediospores in glasshouse condition

The experiment was conducted during 2006 to 2007 and 2009 to 2010 in glasshouse, Department of Plant Pathology, College of Agriculture, G.B. Pant University of Agriculture and Technology, Pantnagar, India. The susceptible field pea variety "Aparna" was inoculated with rust urediospores and aeciospores collected from the naturally infected leaves of field pea plants. Spores from field infected plants were collected and dried over calcium sulphate (8 mesh) at 4°C for 20 h, then maintained in 1.0 ml cryogenic vials at 15°C. Single plant was grown in 6.5 cm square plastic pots filled with potting medium (sterilized sand, soil and farm yard manure mixture, 1:2:1). One month after planting in late October plants were inoculated by spraying a suspension of 10 ml Soltrol 170 oil carrying 0.03 g rust spores through an atomizer (Browder, 1971). Thereafter evaporation of oil, plants were wetted with a solution of 0.1% Tween 20 and transferred to a humidity chamber (+ 90% relative humidity at 18°C) for 16 h. Field pea plants were incubated in glasshouse at 22°C for 14 days for the development of rust pustules and production of aeciospores/ urediospores.

Exposure of infected leaves

Uniformly infected leaves, each with approximately more than 100 pustules, were removed from arbitrarily randomly selected plants and placed six leaves per bag into each of 16 nylon mesh bags, which were stapled. Bags were randomly divided in two groups and each group contained a wire mesh basket to provide protection from animals. The wire mesh baskets were attached to a small tree in shelterbelt at Crop Research Centre of the G.B. Pant University of Agriculture and Technology Pantnagar, in the third week of March. At each site, one basket was fixed 0.3 m above the ground level and one was placed on the ground. Within the first 10 days of each month a single nylon bags was removed from each wire enclosed and plant tissues were assayed for viable urediospores. Standard germination test method was opted to test the viability of aeciospores/urediospores (Shroff and Chand, 2010). Identical tests were made each year over a four year period.

The spore suspension (aeciospores/urediospores) of *U. viciae-fabae* was prepared. A drop of 20 µl of spore suspension prepared in distilled water was placed on glass cavity slides on moist Petriplates with the help of a Gilson micropipette. The aeciospores/urediospores of *U. viciae-fabae* were incubated at 15 ± 2°C. Three replications were maintained for each treatment. After 12 h incubation period, the slides were taken out and 10 µl lactophonal was added to each slide immediately to check further germination of spores. The observations for spore germination were recorded and the percent spore germination was calculated by using following formula:

$$\text{Spore germination (\%)} = \frac{\text{Number of spores germinated}}{\text{Total number of spores observed}} \times 100$$

Assay of plant tissue

The experimental leaf debris was weighted, crushed, suspended in 0.1% Tween 20 solution within 4 h of recovery from the exposed mesh bags and sprayed onto 20 leaves of susceptible field pea cultivar "Aparna" in the glasshouse. Inoculated plants were gently sprayed with Tween 20 solution and incubated in moist condition in the greenhouse. Rust pustules on each leaf were counted 14 days after incubation in the glasshouse. Infected recovered material were leaves and plant part segregated on separate aluminum foil sheets and conditioned at 4°C temperature and 90% relative humidity for 7 days in a refrigerator. Conditioned material from each exposure was weighted and grinded with mortar and pestle. The crushed material was loaded into a 3 cc syringe and suspended in 1.5 ml Soltrol 170 oil and thoroughly mixed. The syringe was attached to a Swinney 13 mm filter holder fitted with a 300 mesh screen. The suspension was passed through the screen and filtrate was collected in empty gelatin capsules. Each capsule was fitted into an atomizer and the contents were dispersed onto a group of 20 to 25 days old susceptible cultivar of field pea (Aparna) grown in a rust free glasshouse. Each group of plants were placed in a separate humidity chamber and later kept in glass house with growing

Table 1. Mean number of viable urediospore obtained in germination test.

Year	No. of viable urediospore (mean spore germination of 03 sample)				
	April	May	June	July	August
2006-2007	44	8	5	0	NT
2007-2008	56	10	5	1	NT
2008- 2009	67	12	35	17	5
2009-2010	60	13	8	0	NT

NT = Not tested because in month of August due to no viable spore was observed in July.

conditions as previously described.

Survival of urediospores at different temperature

Infected crop debris having urediospores /teliospores were collected from the field during the month of April 2007 and 2008, respectively and they were cut into small pieces and wrapped in butter paper. These pieces were stored in incubator at different temperature viz., 10, 20, 30, 40 and at 5°C in refrigerator. Germination of urediospores /teliospores from crop debris was tested at monthly interval by serial dilution technique to ascertain the survivability of spores.

Survival of aeciospores/urediospores at different soil depth

The experiment was conducted for two consecutive years (2007 and 2008) from April to May under glasshouse condition to observe the survivability of spores. The small pieces of infected plant debris were placed in nylon mesh bags (50 g/bag) and buried at different depths viz., 5, 10, 15 and 20 cm, respectively in plastic pots filled with natural soil. These inoculated pots were replicated 4 times and randomly kept in glasshouse. At weekly interval, one gram of debris was taken out from each sample and germination of spore was tested by serial dilution technique as above to ascertain the survival of urediospore/ teliospores at different soil depths, if any.

RESULTS and DISCUSSION

Production of aeciospores/urediospores in glasshouse

Field pea plants were inoculated with rust urediospores and aeciospores collected from the naturally infected leaves and incubated in glasshouse at 22°C for 14 days for the development of rust pustules and production of aeciospores/urediospores. The survival of viable aeciospores/urediospore from detached debris and standing plant residue was apparent. Therefore data were combined and mean spore germination was calculated for each month. Within the first 10 days of each month plant tissues were sampled three times and assayed for viable urediospores. The table indicated that maximum urediospore germination was recorded during 2006 to 2010 in the month of April followed by May and June (Table 1). For the first 2 years (2006 to 2007 and 2007 to 2008) of the study, the limited number of uredinia

developed on the assayed plant indicated that the urediospore survival was poor or lasted up to the month of July.

In June 2008 to 2009, urediospores were left in the refrigerator for 1 week because susceptible test plant had poor/no germination and had to be replanted. The number of uredinia on assayed plant indicated apparent survival increased dramatically in June, July and August in the year 2008 to 2009 (Table 1). The difference was interpreted as the result of spore conditioning. The work has been duly supported by earlier investigation on bean with rust pathogens as conditioning increases urediospores viability (Beard and Hamlin, 1995). In the year 2008 to 2009, conditioned spores were assayed and urediospores found survived up to August. In general rust survival declined over time from April onwards. Temperature was warmer than normal in 2006 to 2007 and cooler than normal for 2008 to 2009. Although most uredinia are converted to telia on mature and senescent plant tissues, uredinia are also common on field pea at or after harvest. Overwintered urediniospores could contribute to early initiation of disease and also to build-up of new races known to attack on field pea in Pantnagar.

Effect of different temperatures on survival of urediospore

The results revealed in year 2007 that initial survival of urediospore was 65.0%, declined up to 8.67% after lapse of 4 month period when stored at 10°C, while at 5°C temperature only 5% survivability of spore was observed in the month of July. Urediospore incubated at 40°C and 30°C in the month of April and onward, only 4.0% survival was found in the month of April at 30°C whereas no germination was recorded at 40°C temperature. At 20°C by the end of June 14.0% survivability was recorded and thereafter no germination of spores at all (Table 2). Similar trend was observed in respect to effect of different temperature on survival of urediospore during the year 2008. Maximum survivability (67.0%) was recorded at low temperature when urediospore were incubated at 10°C followed by 5°C (65.0%), 20°C (60.33%) and 30°C (8.33%) in the month of April. In the month of July

Table 2. Effect of different temperatures on survival of urediospores of *U. viciae-fabae.*

Month/week	Percent survival									
	2007					2008				
	5°C	10°C	20°C	30°C	40°C	5°C	10°C	20°C	30°C	40°C
April	62.00 (51.99)	65.00 (53.76)	45.00 (42.12)	4.00 (11.35)	0.00 (0.00)	65.00 (53.76)	67.00 (55.01)	60.33 (50.98)	8.33 (16.73)	0.00 (0.00)
May	45.00 (42.12)	55.00 (47.88)	32.00 (34.44)	0.00 (0.00)	0.00 (0.00)	44.00 (41.54)	50.00 (44.99)	37.33 (37.64)	0.00 (0.00)	0.00 (0.00)
June	30.00 (33.16)	36.67 (37.25)	14.00 (20.69)	0.00 (0.00)	0.00 (0.00)	26.67 (30.39)	30.00 (33.16)	13.33 (17.21)	0.00 (0.00)	0.00 (0.00)
July	5.00 (12.92)	8.67 (17.35)	0.00 (0.00)	0.00 (0.00)	0.00 (0.00)	7.67 (16.32)	5.33 (13.32)	0.00 (0.00)	0.00 (0.00)	0.00 (0.00)
RCD (P = 0.01)	3.01	8.70	5.37	1.72	0.00	7.84	7.39	9.68	1.52	0.00
CD (P = 0.05)	1.99	13.14	8.11	2.60	0.00	11.84	11.17	14.63	2.30	0.00
GM±SEM	40.50±0.57	46.33±2.51	27.75±1.55	1.0±0.50	0.00	38.08±2.26	43.08±2.14	33.16±2.80	2.88±0.44	0.00
CV (%)	2.46	9.41	9.70	86.60	0.00	10.31	8.60	14.64	36.66	0.00

Figures given in parentheses are angular transformed values. Average of four replications.

urediospore was incubated at 5 and 10°C temperatures. Only 7.67 and 5.33% survivability was recorded, while at higher temperature, it was also observed that at 30°C temperature, spore survived only for one month in both the years and no germination was observed at 40°C. Similarly, optimum germination of urediospore and other spores of *U. viciae-fabae* were recorded at 20°C temperature by Joseph and Hering (1997) and Joshi and Tripathi (2012). Experimental result indicated that constant exposure of urediospore at 30 and 40°C temperature was lethal to the survival of urediospore as they cannot survive in hot months of May, June and July. The results of present study revealed that the fungus survived in

infected crop debris in the form of urediospores at different temperature (Table 2). Similarly Singh (2005) reported that optimum temperature for germination of urediospores is 16 and 25°C. However, there was no germination observed at 28 and 29°C. Batra and Stavely (1994) reported that *U. appendiculatus* urediospore germinates best at 15 to 24°C, while the teliospore in crop debris germinates best at 10 and 15°C.

Magsi (2000) observed that germination of *U. striatus* urediospore causes the rust of Lucerne at 5, 10, 15, 20, 25, 30 and 35°C under 24 h in dark condition. It was reported that 20°C favoured maximum spore germination, whereas minimum germination occurred at 10 and 25°C. Singh

(2005) reported in rust of chickpea caused by *U. ciceris-arietini* that the urediospore can germinate at any temperature between 5 and 30°C but the optimum is 11 and 12°C. No germination occurs at 35°C and above temperatures.

Effect of different soil depth on urediospore survival

The experiment was conducted for two consecutive years (2007 and 2008) from April to May under glasshouse condition to observe the survivability of urediospore, buried at 5, 10, 15 and 20 cm depth in natural soil. The urediospore

Table 3. Survival of *U. viciae fabae* in infected crop debris buried in soil at different depth during the month of April to May in 2007and 2008 crop seasons.

Months/week	Percent survival							
	2007				2008			
	5 cm	10 cm	15 cm	20 cm	5 cm	10 cm	15 cm	20 cm
April								
1st week	44.00 (41.55)	54.00 (47.29)	16.33 (23.77)	11.33 (19.66)	51.00 (45.57)	61.67 (51.75)	21.67 (27.71)	12.67 (20.80)
2nd week	36.67 (37.20)	45.00 (42.12)	11.30 (19.21)	7.67 (16.07)	39.67 (39.03)	47.67 (43.66)	16.67 (24.08)	9.67 (18.07)
3rd week	2.400 (29.66)	35.00 (38.92)	6.67 (14.72)	5.33 (13.34)	32.67 (34.82)	36.67 (37.25)	9.67 (18.07)	6.67 (14.89)
4th week	18.00 (21.32)	20.33 (26.45)	5.00 (12.75)	0.00 (0.00)	21.67 (27.71)	23.33 (28.78)	6.33 (14.50)	0.00 (0.00)
May								
1st week	11.33 (19.56)	31.66 (34.23)	0.00 (0.00)	0.00	11.67 (19.74)	21.00 (27.70)	0.00 (0.00)	0.00
2nd week	0.00	0.00	0.00	0.00	0.00	0.00	0.00	0.00
3rd week	0.00	0.00	0.00	0.00	0.00	0.00	0.00	0.00
4th week	0.00	0.00	0.00	0.00	0.00	0.00	0.00	0.00
CD (P = 0.01)	4.36	9.89	3.94	1.14	4.56	7.20	3.33	2.42
GM ± SEM	10.45 ± 1.98	20.16 ± 3.77	5.20 ± 0.93	3.33 ± 0.27	20.50 ± 1.93	28.95 ± 2.16	7.16 ± 0.79	3.62 ± 0.57
CV (%)	10.69	20.05	31.17	14.07	13.13	10.37	19.13	27.58

Figures given in parentheses are angular transformed values. Average of four replications.

of *U. viciae-fabae* survived up to 5 weeks in natural soil at 10 cm deep. However, the survivability was significantly declined with increase in soil depth and exposure period. The survivability of urediospore was maximum (54.0%) when it was buried at 10 cm deep in the first week of April followed by 5 cm (44.0%), 15 cm (16.33%) and 20 cm (11.33%) (Table 3). The survivability of urediospore was significantly higher (31.66%) when it was buried at 10 cm depth followed by 5 cm (11.33%) depth in the first week of May. The survivability declined from 11.33 to 0.00% by 6th week at 5 cm depth, while no survivability was recorded at 15 and 20 cm depth from 6th to 8th week burial in soil during both the seasons. The similar trend was observed during the next year also where highest survivability was recorded when urediospore were buried at 10 and 5 cm deep. There was gradual decrease in survivability with increase in soil depth as well as lapse of time. Likewise, percent survivability of *U. viciae-fabae* declined sharply with increase in depth reported by Joseph and Hering (1997) and Joshi and Tripathi (2012).

Loss of survivability of urediospores of *U. viciae-fabae* through infected crop debris at 15 and 20 cm soil depth might be due to decomposition of crop debris by saprophytic fungi like Aspergillus, Rhizopus, Mucor, Penicillium present in soil. It is presumed that maximum survivability at 10 cm due to less competition with other

microorganism and comparatively lower temperature than other soil depths tried. The results and assumption was duly supported by various workers (Pandey and Sinha, 2008; Thorman et al., 2003). The role of environmental factors is more complex with air borne disease than soil borne. Apart from the direct effect of moisture and temperature on the germination, dissemination and pathogenicity of organism the complexity of soil itself is highly significant as soil environment present a physical, chemical and biological barrier for survival. Obviously all these factors interact, as has been hypothesized as an explanation for the survivability of the fungus. Therefore, this information will be apparently useful for epidemiological studies and development of integrated disease management module for pea rust disease.

REFERENCES

Anonymous (2009). Project Coordinator's Report. All India Coordinated Research Project on MULLaRP. Published by IIPR (ICAR) Kanpur. P. 2.

Batra LR, Stavely R (1994). Attraction of two spotted spider mites to bean rust uredinia. Plant Dis. 78:282-284.

Beard LW, Hamlin WG (1995). North Dakota Agricultural Statistics. No. 64.

Browder LE (1971). Pathogenic specialization in cereal rust fungi especially *Puccinia recondita* f.sp. *tritici*: concepts methods of study and application. U.S. Deptt. Agric. Tech. Bull. P. 1432.

Chung WH, Tsukiboshi TOY, Kakishma M (2004). Phylogenic analyses of *Uromyces viciae-fabae* and its varieties on Vicia, Lathyrus and Pisum in Japan. Mycoscience pp. 45:1-8.

Cunningham JL (1973). Longevity of rust spores in liquid nitrogen. Plant Dis. Rep. 57:736-737.

Davison A, Vaugham EK (1963). Longevity of urediospores of race 33 of *Uromyces phaseoli* var. *phaseoli* in storage. Phytopathology 53:736-737.

Emeran AA, Roma NB, Sillero JC, Satovic Z, Rubiales D (2008). Genetic variation among and within *Uromyces* species infecting legumes. J. Phytopathol. 156:419-424.

Emeran AA, Sillero JC, Niks RE, Rubiales D (2005). Infection structure of host-specialized isolates of *Uromyces viciae-fabae* and of other species of *Uromyces* infecting leguminous crops. Plant Dis. 89:17-22.

Gaumann EA (1998). Comparative morphology of fungi, Translated by Caroll William Dodge, Biotech Book, Delhi. P. 563.

Hebblethwaite PD (1983). The faba bean. Butterworths London. U.K. P. 573.

Joseph ME, Hering TF (1997). Effect of environment on spore germination and infection by broad bean rust (Uromyces *viciae fabae*). J. Agric. Sci. 128:73-78.

Joshi A, Tripathi HS (2012). Studies on epidemiology of lentil rust (*Uromyces viciae fabae*). Indian Phytopathology. 65(1):67-70.

Kushwaha C, Chand R, Srivastava CP (2006). Role of aeciospores in outbreak of pea (*Pisvum sativum*) rust (*Uromyces fabae*). Eur. J. Plant Path. 115:323-330.

Magsi MR (2000). Studies on physiology of *Uromyces striatus* Schroet. Pakistan J. Phytopathol. 12(2):130-133.

Mohamed HAR (1982). Major disease problems of faba beans in Egypt. In Faba Bean Imorovement (Hawtin, GC, Webb C, eds.). Martinus Nijhoff Publishers, The Hague, Netherlands. pp. 213-225.

Pandey V, Sinha A (2008). Mycoflora associated with decomposition of rice stubble mixed with soil. J. Plant Prot. Res. 48(2):247-253.

Rashid KY, Bernier CC (1991). The effect of rust on yield of faba bean cultivars and slow rusting populations. Can. J. Plant Sci. 71:967-972.

Sharma AK (1998). Epidemiology and management of ruts disease of french bean. Vegetable Sci. 25(1):85-88.

Shroff S, Chand R (2010). Pre-infection Biology of aeciospores of *Uromyces fabae*. Int. J. Curr. Trends Sci. Tech. 1(2):1-10.

Singh RS (2005). Plant Diseases. Oxford and IBH, New Delhi. P. 396.

Thorman MN, Currah RS, Bayley SE (2003). Succession of microfungal assemblages in the decomposing peat land plants. Plant Soil. 250(3):323-333.

Webb C, Hawtin G (1981). Lentils. Commonwealth Agricultural Bureaux London, U.K. P. 216.

11

Effect of tillage and nutrient management practices on weed dynamics and yield of fingermillet (*Eleusine coracana* L.) under rainfed pigeonpea (*Cajanus cajan* L.) fingermillet system in *Alfisols* of Southern India

Vijaymahantesh, Nanjappa H. V. and Ramachandrappa B. K.

Department of Agronomy, University of Agricultural Sciences, Bangalore, Karnataka, India.

Field experiments were conducted during the *Kharif* seasons of 2010 and 2011 at GKVK, UAS, Bangalore, Karnataka, India to study the influence of three tillage practices *viz.*, conventional tillage (3 ploughings + 3 inter cultivations), reduced tillage (2 ploughings + 2 inter cultivations) and minimum tillage (1 ploughing + 1 inter cultivation) and three nutrient management practices *viz.*, 100% N through Urea, 100% N through integrated supply (50% N through urea+ 25% N through FYM+ 25% N through Glyricidia) and 100% N through organic source (50% N through FYM+ 50% N through Glyricidia) on weed dynamics and yield of fingermillet (*Eleusine coracana* L.) under rainfed pigeonpea-fingermillet system in *Alfisols*. The results showed that conventional tillage reduced the infestation of *Borreria articularis*, *Cynodon dactylon* and *Cyperus rotundus* compared to other tillage practices. However, nutrient management practices did not influence weeds significantly. Among tillage practices conventional tillage recorded significantly higher fingermillet yield (3030 kg ha^{-1}) compared to other tillage practices and among nutrient management practices integrated supply of N recorded higher yield of 2666 kg ha^{-1} compared to other nutrient management practices. More weed seeds were distributed in upper 10 cm soil depth in minimum tillage where as in conventional tillage weed seed distribution was more or less uniform in the soil profile studied.

Key words: Tillage, nutrient management, glyricidia, weed dynamics, grain yield, weed seed bank.

INTRODUCTION

Pigeonpea (*Cajanus cajan* L.) is one of the important pulse crop rich in protein content. Pigeonpea being a legume fixes atmospheric nitrogen and also it sheds lot of leaves at maturity which improves organic matter content of soil. Fingermillet (*Eleusine coracana* L.) is a staple food for working class and also an ideal food for patients suffering from diabetes. The grains are rich in calcium and iron besides being rich in carbohydrate and protein.

Under rainfed conditions in southern India particularly in Karnataka state, pigeonpea-fingermillet cropping system is predominantly followed. Both crops are largely grown under rainfed conditions, experiencing moisture deficiency at different growth stages. Deficiency of moisture for both the crops affects normal growth and development resulting in lesser yield under rainfed conditions. Further nutrient deficiency particularly N and unchecked weed growth inflict considerable reduction in fingermillet yield. The role of tillage in conserving soil moisture and its subsequent beneficial effect on crop productivity has long been recognized. Adequate tillage operations controlled weeds and resulted in higher crop productivity, but caused more soil loss and were more capital intensive (Dogra et al., 2002). Tillage influences the vertical distribution of weed seeds in soil layer and weed diversity. No till cropping systems leave most seeds in top 1.0 cm layer of soil profile (Yenish et al., 1992). Differential distribution of seeds in the soil profile subsequently leads to change in weed population dynamics (Buhler, 1991). Use of organic manure is inevitable for sustained agricultural production by reducing dependence on inorganic fertilizers and to build the soil fertility and improve the soil biological activity. Keeping this in view, a study was under taken to find out the combined influence of tillage and nutrient management practices on weed dynamics and yield of fingermillet under pigeonpea-fingermillet system in *Alfisols* of Southern India.

MATERIALS AND METHODS

Field experiments were conducted during *Kharif* seasons of 2010 and 2011 at the University of Agricultural Sciences, G.K.V.K, Bangalore. The soil of the experimental field was red sandy clay loam having a pH 5.5, with 0.36% organic carbon, available nitrogen 175 kg ha^{-1}, phosphorus 68.40 kg ha^{-1} and potassium 160 kg ha^{-1}. The treatments consisting of three tillage practices *viz.*, conventional tillage (3 ploughings + 3 inter cultivations), reduced tillage (2 ploughings + 2 inter cultivations) and minimum tillage (1 ploughing + 1 inter cultivation) in main plots combined with three nutrient management practices *viz.*, 100% N through Urea, 100% N through integrated supply (50% N through urea+ 25% N through farm yard manure (FYM) + 25% N through Glyricidia) and 100% N through organic sources (50% N through FYM+ 50% N through Glyricidia) in sub plots were replicated thrice in split- plot design. Tillage practices were done as per treatment details *viz.*, conventional tillage: three ploughings (15 to 20 cm deep) and three inter cultivations during the crop period was done first after 30 days after sowing and remaining two at an interval of fifteen days. reduced tillage: two ploughings (15 to 20 cm deep) and two inter cultivations during the crop period was done first inter cultivation after 30 days after sowing and remaining one after an interval of fifteen days. Minimum tillage: one ploughing (15 to 20 cm deep) and one inter cultivations during the crop period was done after 30 days after sowing. Nutrient management practices were followed as per the treatment details *viz.*, 100% N through Urea: entire dose of nitrogen was applied through urea as basal in pigeonpea where as in fingermillet 50% N as basal and remaining 50% after 30 days after sowing, 100% N through organic sources: 50% of N through FYM and 50% of N through glyricidia was supplied to the crop by incorporating to the field 20 days before sowing of crop, 100% N through integrated supply: 50% N through urea, remaining nitrogen was supplied through farm yard manure and glyricidia in equal proportion to meet the remaining nitrogen based on their nitrogen equivalent before 20 days of sowing. Whereas recommended phosphorus and potassium was supplied through Single Super Phosphate and muraite of potash respectively to all the treatments as basal. Sowing of fingermillet was done at a spacing of 30 × 10 cm with a plant density of 3,33,333 per ha. Ear heads in the plot were harvested on maturity and were dried separately then hand threshed by beating with wooden sticks, winnowed and cleaned treatment wise and expressed as kg ha^{-1}.

In each treatment a quadrant of 0.5 × 0.5 m was selected at random for recording weed count. Accordingly the number of monocots, dicots and sedges present within quadrant were counted and expressed as No. m^{-2}. Later the original values were subjected to suitable transformations (square root or logarithmic) depending on the variation in the data and subjected to statistical analysis. For weed seed bank analysis, soil samples were collected two times, before sowing of pigeonpea in May 2010 and after harvest of fingermillet in November 2011 to determine the weed seed bank composition. Samples were taken from three soil depths (0-10, 10-20 and 20-30 cm) in the field. From each plot five samples of soils with a core auger were taken as at random. Soil samples from each plot were pooled within the same depth. Soil samples from each plot were thoroughly mixed air dried under shade and ground gently with hands in to the small pieces. Thereafter 1 kg each of soil, devoid of large rocks and root fragments, was transferred into 20 × 35 cm plastic trays and with 2 cm soil thickness, placed in light screen house. These were watered as and when needed to maintain adequate moisture. After weed seedlings emerged were identified, counted and removed. Seedlings of unidentified weeds were transplanted in to other pots and grown until their identities could be verified. After this, the soil was thoroughly mixed watering was continued for next flush of weed seed germination. The cycle of operation was repeated after every flush of germination, identification and removal of seedlings. Watering was continued for three weeks after weed seed germination ceased. The data were subjected to statistical analysis to determine differences among tillage, nutrient management practices and different soil depths.

RESULTS AND DISCUSSION

The dominant weed species observed both in experimental field and weed seed bank studies were *Borreria articularis*, *Cynodon dactylon* (L.) and *Cyperus rotundus* (L.).

Effect of tillage on weed dynamics and yield of fingermillet

Different tillage practices significantly influenced weed population. Irrespective of the weed species, conventional tillage significantly reduced the population of weeds compared to reduced tillage and minimum tillage. The inversion of soil by following conventional tillage resulted in deeper placement of weed seeds which could not emerge out, causing a significant reduction in the population of weeds. Similar result was observed by Chahal et al. (2003). In both years of experimentation the trend remained the same only dominance of perennial grass and sedge started to increase in the minimum tillage. In minimum tillage is due to less disturbance and falling of weed seeds on to the surface of soil both weed population and weed dry weight was significantly higher compared to reduced and conventional tillage treatments (Table 2). Satisfactory weed control in conventional tillage treatment may be attributed to the stimulatory effect of tillage in inducing weed seed germination and it might be due to the greater deposition of weed seed at

Table 1. Effect of Tillage and Nutrient management practices on growth and yield of fingermillet.

Treatments	Plant height (cm) at harvest	Number of tillers plant^{-1}	Grain weight plant^{-1}(g)	Grain yield (kg ha^{-1})	B:C ratio
Tillage practices (M)					
M$_1$:3Ploughings + 3Inter cultivations	79.16	6.16	12.00	3030	3.57
M$_2$:2Ploughings + 2 Inter cultivations	70.26	5.37	10.73	2256	2.83
M$_3$:1Ploughing + 1 Inter cultivation	61.21	4.47	8.27	1112	1.56
S.Em±	0.72	0.15	0.30	124.37	
CD at 5%	2.81	0.61	1.18	488.35	
Nutrient management practices(F)					
F$_1$: 100% N through urea	70.00	5.33	10.40	2067	3.02
F$_2$:25% N through FYM + 25% N through Glyricidia + 50% N through Urea	74.41	5.77	10.74	2666	3.23
F$_3$: 50% N through FYM + 50% N through Glyricidia	66.23	4.91	9.86	1665	1.71
S.Em±	1.49	0.14	0.17	81.35	
CD at 5%	3.60	0.40	0.31	250.65	
Interaction (M XF)					
F at same level of M					
S.Em±	2.58	0.24	0.30	140.90	
CD at 5%	NS	NS	NS	NS	
M at same or different levels of F					
S.Em±	2.23	0.25	0.39	169.42	
CD at 5%	NS	NS	NS	NS	

soil surface and ploughing each time might kill the germinated weeds. This research had a general agreement with several previous studies of Ball and Miller (1990), Amanuel and Tanner (1991) and Mohler (1993).

In conventional tillage, weed seeds were distributed uniformly among different soil depths compared to minimum tillage and reduced tillage (Tables 3 and 4). In minimum tillage all three types of weed species viz., broad, grass and sedges were significantly higher than other two tillage practices and most of the weed seeds were concentrated in top layers of soil. Plant height was significantly higher under conventional tillage (79.16 cm) than reduced tillage (70.26 cm) and minimum tillage (61.21 cm). Similarly conventional tillage produced significantly higher tiller number (6.16 per plant), grain weight (12 g per plant) and grain yield (3030 kg ha^{-1}) compared to other tillage practices (Table 1). This may be due to creation of favourable physical condition for seed germination, seedling emergence, stand establishment and subsequent growth which contributed for better growth, yield attributes and yield of fingermillet.

Soil depth and tillage interactions

The interaction between soil depth and tillage was significant. All the dominant weed species observed *viz.,*

Borreria articularis, Cyperus rotundus and *Cynodon dactylon* were significantly reduced by conventional tillage (Table 3). The weed seed bank distribution differed between tillage practices. In the minimum tillage practices large number of weed seed was found in the depth of 0 to 10 cm followed by reduced tillage and conventional tillage. This may be attributed to greater deposition of weed seed at the soil surface due to fewer disturbances to weeds in minimum tillage which resulted in more addition of weed seeds at the end of their life cycle. In conventional tillage weed seeds were distributed more or less uniform compared to reduced and minimum tillage. The total weed seed bank was higher in minimum tillage and lowest weed seed bank was observed in conventional tillage (Table 4) this may be attributed satisfactory control of weeds by intensive tillage practices. Similar findings were reported by Ball and Miller (1990).

Effect of nutrient management practices on weed dynamics and yield of fingermillet

Nutrient management practices did not influence weed seed bank and weed dynamics significantly. However, nutrient management practices significantly influenced on growth and yield of fingermillet in the present study. The

Table 2. Effect of Tillage and Nutrient management practices on weed density (No. m^2) and weed dry weight (g 0.25 m^2) in finger millet at harvest.

Treatments	Monocots (+)	Dicots (+)	Sedges (+)	Total weed density (≠)	Total weed dry weight (≠)
Tillage practices (M)					
M$_1$: 3 Ploughings + 3 Inter cultivations	3.52 (11.53)	2.76 (6.66)	2.81 (6.98)	1.43 (25.17)	1.26 (16.6)
M$_2$: 2 Ploughings + 2 Inter cultivations	4.94 (23.50)	3.83 (13.75)	4.22 (16.91)	1.75 (54.16)	1.69 (46.87)
M$_3$: 1 Ploughing + 1 Inter cultivation	5.78 (32.59)	4.70 (21.15)	4.96 (23.78)	1.89 (77.52)	1.80 (61.85)
S.Em±	0.15	0.10	0.12	0.01	0.03
CD at 5%	0.59	0.41	0.37	0.08	0.10
Nutrient management practices (F)					
F$_1$: 100% N through Urea	4.64 (21.64)	3.71 (13.34)	3.90 (15.12)	1.67 (50.11)	1.56 (39.46)
F$_2$:25% N through FYM + 25% N through Glyricidia + 50% N through Urea	4.76 (22.64)	3.77 (13.93)	4.0 (15.89)	1.69 (52.46)	1.59 (41.84)
F$_3$: 50% N through FYM + 50% N through Glyricidia	4.84 (23.34)	3.81 (14.28)	4.10 (16.65)	1.71 (54.28)	1.61 (44.02)
S.Em±	0.10	0.04	0.09	0.01	0.02
CD at 5%	NS	NS	NS	NS	NS
Interaction (M XF)					
F at same level of M					
S.Em±	0.18	0.07	0.20	0.03	0.03
CD at 5%	NS	NS	NS	NS	NS
M at same or different levels of F					
S.Em±	0.21	0.12	0.19	0.02	0.03
CD at 5%	NS	NS	NS	NS	NS

*Figures in parenthesis indicate original values, NS- Non significant, Data subjected to, + = square root of X+1, ≠ = log (X + 2) transformations.

Table 3. Total weed no. per kg of soil as influenced by tillage, Soil depth and nutrient management practices.

Treatments	At 15 days		At 30 days		At 45 days		At 60 days	
	Tillage practices							
M$_1$	1.16(13.05)		1.27(18.31)		0.99(8.93)		0.69(3.94)	
M$_2$	1.56(36.22)		1.66(49.89)		1.48(34.68)		1.06(11.70)	
M$_3$	1.63(44.78)		1.81(66.80)		1.64(46.19)		1.26(19.70)	
Nutrient management practices								
F$_1$	1.42(31.44)		1.55(41.49)		1.35(26.63)		0.97(9.89)	
F$_2$	1.45(33.04)		1.58(43.31)		1.37(28.33)		1.01(11.32)	
F$_3$	1.48(34.68)		1.60(45.07)		1.41(30.22)		1.04(10.52)	
Depth (cm)								
D$_1$-10	1.65(53.16)		1.74(60.52)		1.56(40.81)		1.26(20.33)	
D$_2$- 20	1.44(28.07)		1.59(41.73)		1.38(27.89)		1.01(10.04)	
D$_3$- 30	1.26(17.93)		1.40(27.3)		1.19(16.48)		0.75(4.98)	
	SEm±	CD (0.05%)	SEm±	CD (0.05%)	SEm±	CD (0.05%)	SEm±	CD (0.05%)
M	0.02	0.07	0.02	0.06	0.02	0.07	0.03	0.08
F	0.02	NS	0.02	NS	0.02	NS	0.03	NS
D	0.02	0.07	0.02	0.06	0.02	0.07	0.03	0.08
MF	0.03	NS	0.03	NS	0.03	NS	0.05	NS

Table 3. Contd.

MD	0.03	0.09	0.03	0.07	0.03	0.09	0.05	0.14
FD	0.03	NS	0.03	NS	0.03	NS	0.05	NS
MFD	0.05	NS	0.05	NS	0.06	NS	0.09	NS

M_1- 3 ploughings + 3 Inter cultivations, M_2- 2 ploughings + 2 Inter cultivations, M_3- 1 ploughing + 1 Inter cultivation, F_1-100% N through Urea, F_2-25% N through FYM, 25% N through Glyricidia and 50% N through Urea, F_3-50% N through FYM and 50% N through Glyricidia, NS - Non significant, *Figures in parenthesis indicate original values, Data subjected to log (X + 2) transformation.

Table 4. Total weed number per kg of soil as influenced by tillage at different soil depths.

Treatments	At 15 days	At 30 days	At 45 days	At 60 days
M_1D_1	1.26(16.59)	1.46(27.56)	1.21(14.89)	0.95(7.31)
M_1D_2	1.19(13.56)	1.29(18.15)	0.97(7.33)	0.74(4.00)
M_1D_3	1.03(9.00)	1.04(9.22)	0.81(4.56)	0.38(0.50)
M_2D_1	1.71(49.56)	1.77(57.56)	1.63(41.22)	1.25(17.67)
M_2D_2	1.55(33.67)	1.67(44.89)	1.46(27.33)	1.11(12.56)
M_2D_3	1.43(25.44)	1.53(31.89)	1.36(21.67)	0.82(4.89)
M_3D_1	1.98(93.33)	1.99(96.44)	1.83(66.33)	1.57(36.00)
M_3D_2	1.57(37.00)	1.80(62.17)	1.70(49.00)	1.17(13.56)
M_3D_3	1.33(19.33)	1.64(41.78)	1.39(23.22)	1.05(9.56)
SEm±	0.03	0.03	0.03	0.05
CD (0.05%)	0.09	0.07	0.09	0.14

M_1-3ploughings + 3 Inter cultivations, M_2-2 ploughings + 2 Inter cultivations, M_3- 1 ploughing +1 Inter cultivation, D_1- 10 cm depth, D_2- 20 cm depth, D_3- 30 cm depth,*Figures in parenthesis indicate original values, Data subjected to log (X + 2) transformation.

grain yield of fingermillet in 100% N supplied through urea was 2067 kg per ha, which increased to 2666 kg per ha due to 50% substitution of N with farm yard manure and Glyricidia leaf manure (Table 1). This has accounted for 28.97% increase in grain yield over 100% N supply through urea. Further, increasing the level of substitution of N by 100% with organics (FYM and Glyricidia) did not influence the grain yield rather result in reduction in yield and it was significant. Combined application of both the source of nitrogen has resulted in better availability of nitrogen throughout the crop growth period. Fertilizer source of N has met the nutrient requirement of the plant in the early growth stages and the mineralized nitrogen from FYM and Glyricidia could supply the nutrient in the later growth stages of the crop. Hence, there was continuous supply of nutrients throughout the crop growth period. Whereas, in 100% N substitution by farm yard manure and Glyricidia, mineralization occurs slowly and the supply of nitrogen in the early stages of crop growth was delayed and thus the crop has starved of nitrogen which has affected crop growth and yield. Similar results were obtained by Dass and Patnaik (2007), Aruna and Mohammad (2006) and Kumar et al. (2007).

Conclusion

Tillage and soil depth had significant effects on weed

dynamics and weed seed bank. Weed seed bank size was greater in minimum tillage than conventional tillage or reduced tillage. This resulted in better performance of finger millet crop in southern India. Among nutrient management practices integrated supply of N found promising in getting better yield of fingermillet however, there was no effect of nutrient management practices on weed dynamics and weed seed bank. From the research results it could conclude that tillage is the limiting factor for weed seed bank size in the soil. This suggested that by intensive tillage practices could make considerable weed seed bank reduction in the soil. However, further research will be required to confirm these initial findings and to determine whether dynamics of individual species follows the same pattern as total weed density and which can be use for more accurate future predictions related to the population dynamics of the weed seed in the soil.

ACKNOWLEDGMENT

The authors wish to acknowledge the support provided by the AICRPDA, UAS, GKVK, Bangalore for this study.

REFERENCES

Amanuel G, Tanner DG (1991). The effect of crop rotation in two wheat production zones of southern Ethiopia. Weed Sci. 39:205-209.

Aruna E, Mohammad S (2006). Nutrient management options in rice-sunflower cropping system to sustain productivity and conserve the soil fertility. Int. J. Agric. Sci. 2(1):122-129.

Ball DA, Miller SD (1990).Weed seed population response to tillage and herbicide use in three irrigated cropping sequences. Weed Sci. 38:511-517.

Buhler DD (1991). Influence of tillage systems on weed population dynamics and control in the northern corn belt of the United States. Advan. Agron. 1:51-60.

Chahal PS, Brar HS, Walia US (2003). Management of *phalaris minor* in wheat through integrated approach. Indian J. Agron. 35(1,2):1-5.

Dass A, Patnaik US (2007). On-farm evaluation of integrated nutrient management in Ragi (*Eleusine coracana*). Indian J. Soil Cons. 35(1):79-81.

Dogra P, Joshi BP, Sharma NK (2002). Economic analysis of tillage practices for maize cultivation in the Himalayan humid subtropics. *Indian* J. Soil conserv. 30(2):172-178.

Kumar O, Basavaraj naik T, Palaiah P (2007). Effect of weed management practices and fertility levels on growth and yield parameters in fingermillet. Karnataka J. Agric. Sci. 20(2):230-233.

Mohler CL (1993). A model of the effect of tillage on emergence of weed seedlings. Ecological Applications 3:53-73.

Yenish JP, Doll JD, Buhler DD (1992). Effect of tillage on vertical distribution and viability of weed seeds in soil. Weed Sci. 40:429-433.

12

Pharmacodynamic and ethnomedicinal uses of weed speices in nilgiris, Tamilnadu State, India: A review

M. V. N. L. Chaitanya[1], Dhanabal S. P.[1], Rajendran[1] and Rajan S.[2]

[1]Department of Pharmacognosy and Phytopharmacy, JSS College of Pharmacy (A constituent College of JSS University, Mysore), Rock lands, Ootacamund-643001, Tamilnadu, India.
[2]Survey of Medicinal plants and collection Unit, Indira Nagar, Emerald – 643209, India.

Generally, weeds are considered as nuisances in the garden and enemies to the farmer, as there is a misconception that they are useless. Many of the herbs used in Indian traditional medicine and tribal medicine are considered weeds by agriculturists and field botanists (for example, *Phyllanthus amarus L.*, *Eclipta alba L.*, *Centella asiatica* (L.) etc.). Even though many of these weeds have high ethnopharmacological importance, they are being destroyed and there is a lack of scientific knowledge and guidance. In the Nilgiris many medicinally valuable weeds like *Achyranthes bidentata* Blume., *Artemisia nilagirica* Clarke., *Centella asiatica* L., are very prominent having good therapeutic values like diuretic, antimalarial and brain tonic. The main aim of this review is to expose the important pharmacodynamic and ethnomedicinal values of 50 prominent weeds belongs to 26 different families that grow wild in the Nilgiris. It is possible that some of these weeds could provide an additional income to farmers. There is increasing evidence to support that weeds are relatively high in bioactive molecules thus very important for new drug discovery. Innovative research should be encouraged and scientific workshops conducted by government bodies to communicate the medicinal value of weeds, make weeds economically important and to fill the gap between weeds, farmers and the economy.

Key words: Weeds, pharmacological importance, pharmacodynamic uses, *Silybum marianum L., Artemisia parviflora Roxb.*

INTRODUCTION

Weed

A weed may be defined as any plant or vegetation that interferes with the objectives of farming or forestry, such as growing crops, grazing animals or cultivating forest plantations. A weed may also be defined as any plant growing where it is not wanted. For example, a plant may be valuable or useful in a garden, or on a farm or plantation – but if the same plant is growing where it reduces the value of agricultural produce or spoils aesthetic or environmental values, then it is considered a weed. However, some plants are weeds regardless of where they grow.

There are numerous definitions of a weed. Some common definitions include:

1. A plant that is out of place and not intentionally sown
2. A plant that grows where it is not wanted or welcomed
3. A plant whose virtues have not yet been discovered
4. A plant that is competitive, persistent, pernicious, and

interferes negatively with human activity. No matter which definition is used, weeds are plants whose undesirable qualities outweigh their good points, at least according to humans. Human activities create weed problems since no plant is a weed in nature. Though we may try to manipulate nature for our own good, nature is persistent. Through manipulation, we control certain weeds, while other more serious weeds may thrive due to favorable growing conditions. Weeds are naturally strong competitors, and those weeds that can best compete always tend to dominate.

Both humans and nature are involved in plant-breeding programs. The main difference between the two programs is that humans breed plants for yield, while nature breeds plants for survival.

Characteristics of weeds

There are approximately 250,000 species of plants worldwide; of those, about 3%, or 8,000 species, behave as weeds. Of those 8,000, only 200 to 250 are major problems in worldwide cropping systems. A plant is considered as weed if it has certain characteristics that set it apart from other plant species.

Weeds possess one or more of the following characteristics that allow them to survive and increase in nature:

1. Abundant seed production
2. Rapid population establishment
3. Seed dormancy
4. Long-term survival of buried seed
5. Adaptation for spread
6. Presence of vegetative reproductive structures
7. Ability to occupy sites disturbed by humans

Problems with weeds

Weeds are troublesome in many ways. Primarily, they reduce crop yield by competing for: water, light, soil nutrients, space, CO_2, reducing crop quality by contaminating the commodity, interfering with harvest, serving as hosts for crop diseases or providing shelter for insects to overwinter, limiting the choice of crop rotation, sequences and cultural practices, producing chemical substances that can be allergins or toxins to humans, animals, or crop plants (allelopathy), producing thorns and woody stems that cause irritations and abrasions to skin, mouths, or hooves of livestock being unsightly, dominant, aggressive, or unattractive obstructing visibility along roadways, interfering with delivery of public utilities (power lines, telephone wires), obstructing the flow of water in water ways, and creating fire hazards, accelerating deterioration of recreational areas, parking lots, buildings and equipment, invading exotic weed species that can displace native species in stabilized natural areas.

Benefits of weeds

Despite the negative impacts of weeds, some plants usually thought of as weeds may actually provide some benefits, such as: Stabilizing and adding organic matter to soils, providing habitat and feed for wildlife, providing nectar for bees, offering aesthetic qualities, serving as a genetic reservoir for improved crops, providing products for human consumption and medicinal use, creating employment opportunities.

Controversial nature of weeds

Weeds have a controversial nature. But to the agriculturist, they are plants that need to be managed in an economical and practical way in order to produce medicine, food, feed, and fiber for humans and animals. In this context, the negative impacts of weeds indirectly affect all living beings. (http://www.weeds.psu.edu).

The term weed in variety of senses

The term weed is used in a variety of senses, generally centering around a plant that is not desired within a certain context. The term weed is a subjective one, without any classification value, since a plant that is a weed in one context is not a weed when growing where it belongs or is wanted. Indeed, a number of plants that many consider "weeds" are often intentionally grown by people in gardens or other cultivated-plant settings. Therefore, a weed is a plant that is considered by the user of the term to be a nuisance. The word commonly is applied to unwanted plants in human-controlled settings, especially farm fields and gardens, but also lawns, parks, woods, and other areas. More vaguely, "weed" is applied to any plants that grow and reproduce aggressively and invasively (Vjanick and Jules, 1979).

Beneficial weeds

Even though weeds may be considered as unwanted for a number of reasons, the most important one is that they interfere with food and fiber production in agriculture, but there are many weeds having ethnomedicinal and pharmacological value, like the phrases in the poem wrote by Gerard Manley Hopkins' "What would the world be, once bereft, of wet and wildness? Let them be left .O let them be left; wildness and wet; Long live the weeds and the wilderness yet." A number of weeds, such as the dandelion (*Taraxacum officinale* F.H.Wigg.) are edible, and their leaves and roots may be used for food or herbal medicine. Greater Burdock (*Arctium lappais* L.) common weed over much of the world, and is sometimes used to make soup and other medicine in East Asia. These so-called "beneficial weeds" may have other beneficial effects, such as drawing away the attacks of crop-destroying insects, but often are breeding grounds for insects and pathogens that attack other plants. Dandelions are one of several species which break up hardpan in overly cultivated fields, helping crops grow deeper root systems. Some modern species of domesticated flower actually originated as weeds in cultivated fields and have been bred by people into garden plants for their flowers or foliage.

An example of a crop weed that is grown in gardens is the corncockle (*Agrostemma githago* L.) which was a common field weed exported from Europe along with wheat, but now sometimes grown as a garden plant (Baker, 1974). White clover (*Trifolium repens* L.) is considered by some to be a weed in lawns, but in many other situations is a desirable source of fodder, honey and soil nitrogen (Andre, 1988). "Many gardeners will agree that hand-weeding is not the terrible drudgery that it is often made out to be. Some people find in it a kind of soothing monotony. It leaves their minds free to develop the plot for their next novel or to perfect the brilliant repartee with which they should have encountered a relative's latest example of unreasonableness (Christopher, 2001). Weeds have been found to represent a very important component of indigenous pharmacopoeias. The consumption of weedy greens has often been perceived to have a medicinal character (Govindaraj et al., 2011). In ancient Indian literatures all plants were not considered as weeds and it is clearly mentioned that every plant on this earth is useful for human beings, animals and other plants. It is ignorance of human beings as they consider some plants are useful and others as unwanted. Studies conducted by department of Agronomy. (IGAU), Raipur has revealed that weeds are a boon for tile farmers and industries. Uses of weeds of many important agricultural crops have been reported (http://www.ethnologue.com).

Nilgiris

The Nilgiri hills located in Western Ghats, Tamilnadu State, India have a history going backfor many centuries. It is not known why they were called the Blue Mountains (Table 1). Several sources cite the reason as the smoky haze enveloping the area, while other sources say it is because of the kurunji flower, which blooms every twelve years giving the slopes a bluish tinge. It was originally tribal land and was occupied by the todas around what is now the Ooty area, and by the Kotas around what is now the Kotagiri (Kothar Keri) area. The Badagas are one of the major non tribal populations in the district who reside in the mountain. Although the Nilgiri hills are mentioned in the Ramayana of Valmiki (estimated by Western scholars to have been recorded in the second century BCE), they remained all but undiscovered by Europeans until 1602.

Geographical distribution of the Nilgiris district

The district has an area of 2,452.50 km^2. The district is basically a hilly region, situated at an elevation of 2000 to 2,600 masl. Almost the entire district lies in the Western ghats. Its latitudinal and longitudinal dimensions being 130 km (Latitude: 10 - 38 WP 11-49N) by 185 km (Longitude: 76° E to 77.15° E). The Nilgiris district is bounded by Mysore district of Karnataka and Wayanad district of Kerala in the North, Malappuram and Palakkad districts of Kerala in the West, Coimbatore district of Tamil Nadu in the South and Erode district of Tamil Nadu and Chamarajanagar district of Karnataka in the East. In Nilgiris district the topography is rolling and steep. About 60% of the cultivable land falls under the slopes ranging from 16 to 35%.The altitude of the Nilgiris results in a much cooler and wetter climate than the surrounding plains, so the area is popular as a retreat from the summer heat. The temperature remains to the maximum of 25°C and reaches a minimum of 0°C (Wang et al., 2011).

Tribal communities in Nilgiris

The Niligiris is gifted with richest flora in which lot of medicinally important plants are present. But many of these plants are considered as weeds or useless plants. But many of these weeds will grow wildly and in cultivated fields. Many of these weeds having ethno medicinal and pharmacodynamic importance but due to lack of proper guidance and scientific documentation, many of these weeds are under destruction due to their short term useless selfish benefits of mankind, but some tribal people like Todas, Kotas, Kurumbas, Paniyas and Kattunayakas are safeguarding this type of plants and using as tribal medicine to cure lot of diseases.

Tribes and weeds of Nilgiris

Todas: *Centella asiatica* (L) Urban (Apiaceae), locally known as "Vallarai". Plant juice is considered as refrigerant to the body, when given orally.

Kotas:

1. *Achyaranthes aspera* L. (Amaranthaceae), locally known as "Uthrunk". Leaf paste is applied on cuts, wounds and sores for quick healing.
2. *Lantana camara* L. (Verbenaceae), locally known as "Thusik". Leaf juice is applied to the gum to stop

Table 1. Pharmacodynamic and ethnomedicinal uses of weed speices in nilgiris.

Scientific name of the weed	Common name	Family	Major constituents	Pharmacodynamic uses	Ethno medicinal uses
Achyranthes bidentata Blume	Ox Knee	Amaranthaceae	Alkaloids, glycosides, triterpenoids, saponins, flavonoids and mucilage	Antihypertensive agent , Antioxidant	For mosquito bites and as diuretic
Acalypha indica L.	Indian Nettle	Euphorbiaceae	Flavonoids, steroids, terpenoids and pyridone glucosides	In treatment of Myocardial ischemia), as a analgesic and anti- inflammatory activity	Leaf juice use to cure ear problems, drowsiness, and digestive problems and root as tooth brush
Aegeratum conyzoides L.	Goat weed	Asteraceae	Sterols, coumarins, alkaloids and essential oils .	Wound healing , antispasmodic and larvicidal activity	Whole plant as wound healing and mosquito repellent externally
Amaranthus spinosus L.	Spiny amaranth	Amaranthaceae	Betacyanins, phenolic compounds and terpenes	Antitumor activity , anti oxidant antimalarial, antidiabetic and immunomodulatory	Root and leaf juice promotes digestion
Anthoxnthum odoratum L.	Vernal grass	*Poaceae*	Coumarins	Nephroprotective and antioxidant activity	As a tincture to act as provocative incase of hay fever and hay asthma
Argemone mexicana L.	Mexican poppy	Papavaraceae	Alkaloids	Analgesic, antimicrobial, antidiabetic, atiarthritic and wound healing.	Flower decoction as a external medical agent in case of eye infections
Artemisia nilagirica Clarke.	Indian worm wood	Asteraceae	Flavanoid, steroids, terpenoids, saponins, tannins, proteins and essential oil.	Antileishmanial activity, antimalarial, anthelmintic, antiseptic, expectorant, astringent, and anti-inflammatory	Leaf and root decoction intrernally to treat fever and externally as mosquito repellent
Artemisia parviflora Roxb.	Japanese worm wood	Asteraceae	Sterols/ triterpenoids, flavonoids, phenols, saponin, alkaloids, tannins, carbohydrates, coumarins and lignins	Antihypertension, antihelmenthic, antidiabetic and antiviral	Leaf decoction as vermifuge internally and externally as wound healing agent in case of minor cuts and wounds
Asclepias curassavica L.	Scarlet Milkweed	Asclepiadaceae	Glycosides and saponins	Antitumor and anticancer	Leaf paste as a wound healing agent
Bidens biternata (Lour.) Merr.	Spanish needles	Asteraceae	Glycosides, flavonoid, alkaloid, tannin, seroid, terpenoid, coumarin, saponin, athraquinone, phlobatannin and iridoids	Anti-inflammatory, wound healing and as appetizer	Leaf juice in treatment of sores and root paste in case of tooth ache

Table 1. Contd.

Scientific name	Common name	Family	Phytochemical constituents	Pharmacological activities	Traditional use
Borreria latifolia Aubl.	Broad leaf button weed	Rubiaceae	Alkaloids, Flavanoid, Terpenoids and Iridoids.	Antifungal	Whole plant aqueous decoction to cure intestinal disorders
Brassica juncea (L.) Czern.	Mustard plant	Brassicaceae	Fatty acids, glucosinolate and allyl glucosinolates	Hepatoprotective	Leaf or seed aqueous decoction filtrate as eye drops externally to cure eye diseases (white patches in pupil)
Capsella bursapastoris L.	Shepherds Purse	Brassicaceae	Flavanoids, resins, saponins, amino acids, glucosinolates and glycosides	Anti-hemorrhagic, in treatment of menorrhagia	Leaf decoction to treat Genitor-urinary infections
Centella asiatica (L.) Urb.	Indian pennywort	Umbelliferae	Triterpenoids glycosides, flavanoids, tannins and phytosterols.	Wound and ulcer healing activities	Whole decoction internally as treatment for body swelling and menstrual pain
Cardiospermum halicacabum L.	Ballon wine	Sapindaceae	Saponins, traces of alkaloids, flavanoid, apigenin and phytosterol	Diaphoretic, emetic, laxative and emmenagogue.	Leaf decoction internally to cure cold, cough, fever, head ache and other minor disease
Crassocephalum crepidioides (Benth.) S.Moore.	Fire weed	Asteraceae	Phenols and flavanoid	Antioxidant, laxative, anti cancer and anti-inflammatory	Leaf extracts to treat wounds and inflammations
Chenopodium ambrosioides L.var.	Worm seed	Chenopodiaceae	Terpenoids and saponins	Amebicide, analgesic and vermifuge	Leaf decoction internally to treat stomach and intenstinal disorders
Chromolaena odorata (L.) R. king and Rabinson.	Devil weed	Asteraceae	Essential oils, terpenoids, sterols flavanoid and alkaloids	Antispasmodic, antiprotozoal, antitrypanosomal, antibacterial and atihypersensitive, anti-inflammatory and hepatotropic activities	Leaf juice used to treat wounds and insect bites
Cirsium wallichii DC.	Wallichs Thisthle	Asteraceae	Flavonoids, sterols, titerpenes, alkaloids	Antitumour activity.	Leaf decoction to treat gastric problem
Commelina benghalensis L.	Tropical Spider wort	Commelinaceae	Phlobatannins, carbohydrates, tannins, glycosides, volatile oils, resins	Cardio active, wound healing and anticancer	Plant decoction to cure worm infections
Dodonea viscose (L.) Jacq.	Hopbush	Sapindaceae	Tannins, flavonoids, steroids and triterpenes.	Analgesic, anti-inflammatory, antiviral, spasmolytic, laxative, antimicrobial and hypotensive agents	In treatment of bone fractures in animals

Table 1. Contd.

Scientific name	Common name	Family	Phytochemical constituents	Pharmacological activities	Traditional uses
Kalanchoe pinnata (Lam.) Pers.	Miracle leaf	Crassulaceae	Steroidal glycosides, flavonoids, triterpenoids and polyphenols	Antihelmentic, immunosuppressive, wound healing, hepatoprotective and antinociceptive	Raw leaves to treat stomach disorders
Lantana camara L.	Sleeper weed	Verbenaceae	Sterols, glycosides, saponins, carbohydrates, alkaloids, flavanoid,	Antitumor, antibacterial, and antihypertensive agent, roots for the treatment of malaria, rheumatism.	Leaf decoction as deworming agent
Leucas aspera (Wild.) Link.	Common Leucas	Lamiaceae	Triterpenoids, sterols, glucosides, and phenolic compounds.	Antifungal, antioxidant, antimicrobial, antinociceptive and cytotoxic activity.	The leaf decoction used as to cure cold, cough, and skin disorders
Lobelia nicotianaefolia Roth E & S.	Wild tobacco	Lobeliaceae	Alkaloids	Respiratory stimulant, smoking cessation and antiepileptic	Leaf paste is used to treat and cure foul smelling wounds
Melilotus indica L.	Indian sweet clover	Fabaceae	Flavanoids, coumarin glycosides, triterpenes and fatty acids	Emollient, carminative and digestive.	Leaf paste as good emollient, to treat diarrhoea and bowel complaints.
Mirabilis jalapa L	4° clock flower	Nyctaginaceae	Alkaloids, steroids, flavanoid, saponins, phenol compounds and tannins	Antinociceptive, antibacterial and antioxidant	Leaf paste, externally to cure wounds
Mollugo pentaphylla L.	Carpet weed	Molluginaceae	Flavanoids, glycosides and saponins .	Anti- Microbial, anti-Inflammatory, anticancer, hepatoprotective and antipyretic	Whole plant decoction can be used as mild laxative and emmenagogue
Nicandra physaloides (L.) Gaertn.	Apple of peru	Solanaceae	Flavanoids, saponins, carbohydrates, terpenoids and alkaloids	Diuretic and anticancer activities	Leaf paste is used externally to treat wounds.
Oenothera rosea L´ Hér. ex Ait.	Rosy evening Primrose	Onagraceae	Carbohydrates, steroids, Glycosides and Tannins	In the treatment of skin diseases, renal and inflammatory diseases, hepatic pain, liver and skin problems as well as anti-diarrheic effect	Aqueous infusion of the leaves has been used in hepatic pains and kidney problems.
Opuntia stricta (Ker-Gawler) Haw.	Noppales/ cactus	Cactaceae	Alkaloids, steroids, saponins and flavonoids	Antiulcer, anti-inflammatory, antiviral and anticancer	For menorrhagia and metrorrhagia, ten ml fruit juice

Table 1. Contd.

Plant	Common name	Family	Phytochemicals	Activity	Uses
Oxalis corniculata L.	Sleeping beauty	Oxalidaceae	Carbohydrates and glycosides, phytosterols, phenolic annins, flavonoids, proteins, amino acids and volatile oils	Antibacterial and antiulcer activity	mixed with 10 ml Of rice washed water and a spoonful of sugar is administered twice a day for 3 days till to cure
Persicaria nepalensis (Meissn.) H.	Nepalese Smart weed	Polygonaceae	Alkaloids, tannins, saponins and flavonoids.	Antibacterial, antifungal and insecticidal activities.	The fresh juice is beneficial for the treatment of anemia, tympanitis, dysentery and piles
Plantago erosa ex Roxb.	Plantain	Plantaginaceae	Tannins, diterpenoids and steroids.	Astringent, antitoxic, antimicrobial, demulcent, expectorant, diuretic, anti-inflammatory and analgesic	Root decoction in case of vomiting and fever
Plectranthus barbatus Andr.	Indian coleus	Lamiaceae	Flavonoid glucronide and diterpenoids	Antioxidant activity, anticonvulsant, spasmolytic and antihypertensive	The leaves as vegetable to treat constipation and also to improve digestion
Polygonum chinense L.	Tear thumb	Polygonaceae	Glycosides and flavonoids.	Antihelmenthic and antibacterial	Root decoction used as a general tonic for well being
Prinsepia utilis Royle.	Himalayan Cherry	Rosaceae	Hemiterpenoids, fatty acids, hydrocyanic acid and flavanoids	Anti rheumatic and anti diabetic.	Plant juice is used to treat paralysis, giddiness and quenching thrist
Ricinus communis L.	Castor oil plant	Euphorbiaceae	Tannins, saponins, alkaloids, carbohydrates, phenols, flavonoids, sterols and resins (Mary et al., 2011).	Antibacterial, purgative, anti-inflammatory, hepatoprotective, hypoglycemic and insecticidal (Zahir et al., 2010; Mary et al., 2011).	Seed oil warmed and massaged twice a day in arthritic pain. The paste of root is applied for healing of cuts. wounds and
Rubus ellipticus Smith.	Himalayan Raspberry	Solanaceae	Flavonoids, phenolic compounds and Tannins.	Diabetes, diarrhea, gastralgia, wound healing, dysentery, antifertility, al, analgesic, and epilepsy.	In Nepal, the root juice is used to treat pain and constipation (Ripu et al., 2010)
Rumex nepalensis Sprengel.	Dock weed	Polygonaceae	Anthraquinones, flavonoids and glycosides.	Psychopharmacological, antioxidant, antimicrobial, purgative, antidiarrhoeal and antiviral	Decoction of this plant as an abortifacient
					Root and leaf paste to treat jaundice

Table 1. Contd.

Scientific name	Common name	Family	Phytochemicals	Activities	Uses
Sarothamnus scoparius L.	Scotch broom	Fabaceae	Flavanoids, poly phenolics and alkaloids	Antioxidant and antibacterial	Leaf decoction as good diuretic.
Siegesbeckia orientalis L.	Holy herb	Asreraceae	Phenols, tannins, lignans, flavonoids, sterols and phenolic compounds, glycosides and Triterpenoids	Larvicidal, anti-inflammatory and analgesic	Plant decoction to treat various skin diseases
Silybum marianum L.	Mil thistle	Asteraceae	Alkaloids, amino acids, flavonoids, carbohydrates, phenolics, steroids and tannin.	Anti-inflammatory, anticancer, antioxidant and hepatoprotective	Leaf or seed decoction to treat burning pain in anus and in case of jaundice
Stellaria media L.	Chick weed	Caryophyllaceae	Saponin, flavonoids, steroids, triterpenoids, glycosides, and anthocynidine	Antiobesity, anti-inflammatory and antihepatitis B.	Plant juice is used in treatment of skin diseases, bronchitis, rheumatic pains and dysmenorrhoea,
Taraxacum officinale F.H. Wigg.	Dandelion	Asteraceae	Saponins, triterpenes, sterols, phenolics and tannins	Antibacterial, hepatoprotective and mild laxative	Root and leaves powder in treatment of migraine, cardiac complaints, Jaundice, abdominal complaints, as blood purifier and root, used as an antiseptic
Tephrosia purpurea (L.) Pers.	Wild indigo	Fabaceae	Flavooids, tannins, phenols and anthocyanins	Antioxidant, antiulcer, hepatoprotective and wound healing	The tribal uses the boiled extract of plant is a vermifuge and used to kill the intestinal worms
Tithonia diversifolia (Hemsl). A.Gray.	Marigold	Asteraceae	Alkaloids, flavanoids, Phlobatanins, Terpenoids and Saponins	Antidiarroheal, antiplasmodial, anti-inflammatory, analgesic and antimicrobial	Plant decoction to treat sore throat and malaria
Trifolium repens L.	White clover	Fabaceae	Phenolics, saponins, flavanoids and cyanogenic glucosides	Antirheumatic, antiscrophulatic, depurative, leucorrhoea and anticestodal	Plant decoction as deworming agent
Urtica parviflora Roxb.	Nettle	Utricaceae	Akaloids, flavanoid, terpenoids, glycosides, saponins and tannins	Wound healing, hepatoprotective, Antioxidant and hypoglycemic activity	The leaves and fresh roots are used for the treatment of fracture, bones, boils, and febrifuge

Table 1. Contd.

Verbascum thapsus L.	Common Mullein	Scrophulariaceae	Flavanoids, saponins, tannins, terpenoids, glycosides, proteins	Antihelmenthic, Antiviral, antiangiogenic and antiproliferative activities.	Leaf paste externally as antiseptic.

bleeding and to reduce tooth-ache.

3. *Rubia cordifolia* L. (Rubiaceae), locally known as "Sappli Koth". Decoction of stem is orally administered as a restorative tonic. Root juice is given orally to cure jaundice.

Kurumbas:

1. *A. aspera* L. (Amaranthaceae), locally known as "Nayurvi Geeda". Decoction of whole plant with root is orally given for ease child birth and to mitigate labour pain.

2. *Ageratum conyzoides* L. (Asteraceae), locally known as "Nasar soppu". Leaf juice is orally given as a cure for cough and cold.

D) Paniyas: *Oxalis corniculata* L. (Oxalidaceae), locally known as "Pulichen segae. The whole plant extract in water is orally given for piles and also used as a febrifuge.

Kattunayakas:

1. *A. aspera* L. (Amaranthaceae), locally known as "Cherukadalai". The whole plant with water is made into paste and applied on body to relive sprain ached in the Joints.

2. *Centella asiatica* (L) Urban (Apiaceae), locally known as "Gottala". Plant extract is orally given to allay toothache.

As a trail, in this review, we are trying to expose the important weeds and their pharmacodynamic importance and to educate the society to prevent the destruction of these important weeds and can be made them as economically important plants (Rajan et al., 2002).

DISCUSSION

Important weeds in Nilgiris

Achyranthes bidentata Blume (Amaranthaceae)

Description: A perennial herb grows up to height of 60 to 90 cm. It is hardy to zone 8. It is in flower from August to September, and the seeds ripen from September to October. The flowers are hermaphrodite (have both male and female organs). Achyranthes is an erect perennial with slender rambling branches, elliptical leaves, and greenish white flowers on terminal spikes grows up to 1 m tall. It is commonly known as Ox knee and locally known as Nayuruvi, Shiru-kadaladi.

Major constituents: Alkaloids, glycosides, triterpenoids, saponins, flavonoids and mucilage.

Pharmacodynamic uses: Antihypertensive agent, antioxidant (Babu et al., 2011).

Ethnomedicinal uses: In Mizoram (India) the tribes uses leaf paste externally in treatment of leach bites and mosquito bites and the root decoction is used internally as a diuretic (Sharma et al., 2001).

Acalypha indica L. (*Euphorbiaceae*)

Description: An annual herb to about 80 cm high, having catkin type of inflorescence. It is widespread throughout India, Srilanka and African countries. It has numerous medicinal uses in India and is official in Indian Pharmacopoeia as an expectorant. It is commonly known as Indian nettle, locally known as Poonamayakki, Kuppaimeni.

Major constituents: Flavonoids, steroids, terpenoids (Mouli et al., 2012) and pyridone glucosides (Hungeling et al., 2009).

Pharmacodynamic uses: In treatment of Myocardial ischemia), analgesic and anti-inflammatory activity (Mouli et al., 2012).

Ethnomedicinal uses: In Niligiris the tribal people uses leaf juice to cure ear problems, drowsiness, and digestive problems. The root of this plant is main source of tooth brush (Oudhia, 1999).

Aegeratum conyzoides L. (*Asteraceae*)

Description: Goat weed is a common tropical annual herbaceous weed. It is an erect softly hairy annual plant which grows up to a height of 2.5 feet. Oppositely arranged leaves are ovate to lance-like, coarsely rounded, and have toothed margin. Numerous pale blue or whitish flower heads are 6 mm across. It is commonly known as goat weed and locally known as Pumppillu, Appakkoti.

Major constituents: Sterols, coumarins, alkaloids and essential oils (Bhanu, 2011).

Pharmacodynamic uses: Wound healing, antinematocidal, anti-inflammatory, anticoagulant, smooth muscle relaxant, haemostatic, analgesic, antifungal, antibacterial and hypothermic activities (Bhanu, 2011).

Ethnomedicinal uses: The people of Pandalur village, Nilgiris uses the leaf paste externally used as wound healing activity and insect repellent (Govind raj et al., 2011).

Amaranthus spinosus L. (Amaranthaceae)

Description: It is an annual, erect monoecious herb up to 100(-130) cm tall and it is much branched. The stem is terete or obtusely angular, smooth or slightly hairy, and green or variably suffused with purple, having dense cluster of inflorescences. It is commonly known as spiny amaranth, locally known as mullukkeerai.

Major constituents: Alkaloid, terpene, glycoside, sugar, flavanoid and phenol compounds (Jhade et al., 2011).

Pharmacodynamic uses: Antioxidant, antimalarial, antidiabetic and immunomodulatory (Jhade et al., 2011).

Ethnomedicinal uses: The badaga community of Nilgiris uses the root and leaf decoction of this plant to promote digestion (Sathyavathi and Janardhanan, 2007).

Anthoxanthum odoratum L. (Poaceae)

Description: It is a grass grows in tufts and short lived. It can grow up to 100 cm. The stems are 25 to 40 cm (9.8-16 in) tall, with short but broad green leaves 3 to 5 mm (0.12-0.20 in) wide, which are slightly hairy. It flowers from April until June that is quite early in the season, with flower spikes of 4 to 6 cm (1.6-2.4 in) long and distributed in India, China and South Africa. It is commonly known as vernal grass, and locally known as vanilla grass.

Major constituents: Coumarins (Aziz and Islam, 2012).

Pharmacodynamic uses: Nephroprotective and antioxidant activity (Dheeraj et al., 2010).

Ethnomedicinal uses: In west Bengal (India), traditional practitioners uses its tincture as a provocative to hay fever and hay asthma, a medicinal tincture from it is sniffed well into the nose and throat for immediate relief

from congestion (Dheeraj et al., 2010).

Argemone mexicana L. (Papavaraceae)

Description: It is a prickly annual having strong branch grows 60 to 90 cm in height with yellow latex; leaves are simple, sessile and spiny; flowers are bright yellow. It is commonly known as Mexican poppy, and locally known as Mullmothakka.

Major constituents: Alkaloids (Singh et al., 2010).

Pharmacodynamic uses: Antimicrobial, antidiabetic, atiarthritic and wound healing (Charles and Kokati, 2012).

Ethnomedicinal uses: The badaga community in Nilgiris uses the flower decoction externally to treat eye infections (Sathyavathi and Janardhanan, 2007).

Artemisia nilagirica Clarke. (Asteraceae)

Description: It is an aromatic shrub, 1 to 2 m high, yellow or dark red small flowers, grows throughout India in hills up to 2400 m elevation. This medicinal herb is erect, hairy, often half-woody. The stems are leafy and branched. The leaves are pinnately lobed, 5 to 14 cm long, gray beneath. Mugwort blossoms with reddish brown or yellow flowers. The flowers are freely small and stand in long narrow clusters at the top of the stem. It is commonly known as Indian Wormwood, and locally known as Makkipu.

Major constituents: Flavonoids, steroids, terpenoids, saponins, tannins, proteins and essential oil (Devmurari and Jivani, 2010).

Pharmacodynamic uses: Antileishmanial activity, antimalarial, anthelmintic, antiseptic, expectorant, astringent, and anti-inflammatory (Devmurari and Jivani 2010).

Ethnomedicinal uses: The badaga community in Nilgiris uses the leaf and root decoction both externally and internally as mosquito repellent and in treatment of fever (Sathyavathi and Janardhanan, 2007).

Artemisia parviflora Roxb. (Asteraceae)

Description: It is a Perennial shrub growing to 1 m (3ft 3 in). It is hardy to zone 8. It is in flower from August to October, and the seeds ripen from September to October. The flowers are hermaphrodite (have both male and female organs) and are pollinated by Insects. It is commonly known as Japanese worm wood, and locally known as maccipatri.

Major constituents: Sterols/triterpenoids, flavonoids, phenols, saponin, alkaloids, tannins, carbohydrates, coumarins and lignins (Jitin et al., 2011).

Pharmacodynamic uses: Antihypertension, antihelmenthic, antidiabetic and antiviral (Jitin et al., 2011).

Ethnomedicinal uses: Tribal people in Nilgiris used the Decoction of leaves as vermifuge and leaves juice is used externally on cuts, wounds and skin infections (Srivastava and Nyishi Community, 2010).

Asclepias curassavica L. (Asclepiadaceae)

Description: It is an evergreen perennial sub shrub that grows up to 1 m (3.3 ft) tall and have pale gray stems. The leaves are arranged oppositely on the stems and are lanceolate or oblong-lanceolate shaped ending in acuminate or acute tips. Like other members of the genus, the sap is milky. The flowers are in cymes with 10 to 20 flowers each. They have purple or red corollas and corona lobes that are yellow or orange. Flowering occurs nearly year round. It is commonly known as Scarlet milkweed and locally known as kammalchedi.

Major constituents: Glycosides and saponins (Hemadri et al., 2012).

Pharmacodynamic uses: Antitumor and anticancer (Hemadri et al., 2012).

Ethnomedicinal uses: In Nilgiris, the toda tribes use the leaf paste externally to treat wounds (Rajan et al., 2005).

Bidens biternata (Lour.) Merr (Asteraceae)

Description: It is erect annual herb, up to 1 m. It can be distinguished by the leaves, which are usually 5 to 7 foliolate, the basal pair sometimes further divided. The flowers are yellow, including the ray-florets. It is commonly known as Spanish needles and locally known as mukkuthi.

Major constituents: Glycosides, Flavonoids, Alkaloids, Tannins, Seroids, Terpenoids, Coumarins, Saponins, Athraquinones, Phlobatannins and Iridoids (Sukumaran et al., 2012).

Pharmacodynamic uses: Antinflammatory, wound healing and as appetizer (Panda, 2000).

Ethnomedicinal uses: In Harayana and Uttarpradesh states of India, people use leaf juice externally for the treatment of sores and ulcers. Root is used for the treatment of tooth ache (Panda, 2000).

Borreria latifolia Aubl. (Rubiaceae)

Description: A branched herb, prostrate, ascendent or erect, usually branched from the base, stems fleshy, 4-winged, about 75 cm tall; leaves opposite, elliptical, broadest above the middle, tip broadly and shortly pointed, base tapered, variable in size about 2.5 to 5.0 cm long and 2.5 cm wide, thick, hairy on both sides, short leafstalk; leaf base joined with cup-shaped stipules with bristles on edges. Inflorescence in leaf axils, 0.6 to 1.2 cm across, off white, each flower with hairy calyx of four sepals; stamens 4 and stigma forked; flowers throughout the year; fruit hairy, splitting into two pairs to release seeds. It is commonly known as broad leaf button weed, and locally known as Kudalirakki.

Major constituents: Alkaloids, flavonoids, terpenoids and iridoids (Lucia et al., 2012).

Pharmacodynamic uses: Antifungal (Fezan et al., 2007).

Ethnomedicinal uses: In Pandalur village, Nilgiris. The people use the aqueous whole plant extract to cure the intestinal and appendages problems (Govind et al., 2011).

Brassica juncea (L.) Czern. (Brassicaceae)

Description: It is a Perennial herb, usually grown as an annual or biennial, up to 1 m or more tall; branches long, erect or patent; lower leaves petioled, green, sometimes with a whitish bloom, ovate to obovate, variously lobed with toothed, scalloped or frilled edges, lyrate-with 1 to 2 lobes or leaflets on each side; upper leaves sub entire, short petiole, 30 to 60 mm long, 2 to 3.5 mm wide, constricted at intervals, sessile, attenuate into a tapering, seedless, short beak 5 to 10 mm long. Rooting depth 90 to 120 cm. Seeds about 5,660 to 6,000 per 0.01 kg (1/3 oz). It is commonly known as mustard plant and locally known as Kadugu.

Major constituents: Fatty acids, glucosinolate and allyl glucosinolates (Anu et al., 2011).

Pharmacodynamic uses: Hepatoprotective (Anu et al., 2011).

Ethnomedicinal uses: The badagas of Nilgiris uses the leaf or seed aqueous decoction externally to treat eye diseases (White patches in pupil) (Sathyavathi and

Janardhanan, 2007).

Capsella bursapastoris L. (Brassicaceae)

Description: It grows from a rosette of lobed leaves at the base. From the base emerges a stem about 0.2 to 0.5 m tall, which bears a few pointed leaves which partly grasp the stem. The flowers are white and small, in loose racemes, and produce seed pods which are heart-shaped. It is commonly known as shepherds purse and locally known as Mumiri.

Major constituents: Flavanoid, resins, saponins, amino acids, glucosinolates and glycosides (Committee on Herbal Medicinal Products, 2010).

Pharmacodynamic uses: Anti-hemorrhagic, in treatment of menorrhagia (Committee on Herbal Medicinal Products, 2010).

Ethnomedicinal uses: In Indian medicine and Irula tribes of Nilgiris uses leaf juice externally and internally to treat menorrhagia and hemorrhages from renal and genito-urinary tract (Selva et al., 2009).

Centella asiatica (L.) Urb. (Umbelliferae)

Description: It grows in tropical swampy areas. The stems are slender, creeping stolons, green to reddish-green in color, connecting plants to each other. It has long-stalked, green, reniform leaves with rounded apices which have smooth texture with palmately netted veins. The leaves are borne on pericladial petioles, around 2 cm. The rootstock consists of rhizomes, growing vertically down. They are creamish in color and covered with root hairs. The flowers are pinkish to red in color, born in small, rounded bunches (umbels) near the surface of the soil. It is commonly known as Indian pennywort and locally known as Kuthirai kokku or vallrai.

Major constituents: Triterpenoids glycosides, flavanoid, tannins and phytosterols (Thangavel et al., 2011).

Pharmacodynamic uses: Wound healing, mental disorders, antibacterial, antioxidant and anticancer purposes (Thangavel et al., 2011).

Ethnomedicinal uses: The badagas of Nilgiris uses the aqueous whole plant decoction internally incase of body swelling, menstrual pain (Sathyavathi and Janardhanan, 2007).

Cardiospermum halicacabum L. (Sapindaceae)

Description: It is a woody perennial vine distributed

almost globally in the tropics. It is a fast growing vine up to 10 feet. Leaves are trifoliate, up to 4 inches long, with highly lobed leaflets. The plant climb with tendrils and needs some form of support. The small white flowers bloom from summer through fall, flowers are not very showy. The fruit is more interesting, from which the plant gets its common name. It is a brown, thin-shelled, inflated angled capsule up to 3 cm across, containing 3 black seeds each, with a white heart-shaped scar. It is commonly known as balloon wine and locally known as Kottavan.

Major constituents: Steroids, triterpenoids, sugars, alkaloids, phenols, saponins, amino acids, tannins, flavonoids and anthracene glycosides (Maluventhan and Sangu, 2010).

Pharmacodynamic uses: Diuretic, diaphoretic, emetic, laxative and emmenagogue (Maluventhan and Sangu, 2010).

Ethnomedicinal uses: In Pandalur village, Nilgiris. The people use the leaf decoction internally to cure cold, cough, fever, head ache and other minor diseases (Govind et al., 2011).

Crassocephalum crepidioides (Benth.) S. Moore. (Asteraceae)

Description: It is an erect annual slightly succulent herb growing up to 180 cm tall. Its use is widespread in many tropical and subtropical regions, but is especially prominent in tropical Africa. Its fleshy, mucilaginous leaves and stems are eaten as a vegetable, and many parts of the plant have medical uses. Commonly known as fire weed and locally known as Muyalkadhu.

Major constituents: Phenols and flavonoids (Sumi et al., 2011).

Pharmacodynamic uses: Antioxidant, laxative, anti cancer and anti-inflammatory (Odukoya et al., 2007; Chia Chung et al., 2007).

Ethnomedicinal uses: In Pandalur village, Nilgiris. The people use the broad leaf extracts are used to cure cut wounds and other inflammation (Govindraj et al., 2011).

Chenopodium ambrosioides L.var. (Chenopodiaceae)

Description: It is an annual or short-lived perennial plant, growing to 1.2 m tall, irregularly branched, with oblong-lanceolate leaves up to 12 cm long. The flowers are small and green, produced in a branched panicle at the apex of the stem. Commonly known as worm seed and locally known as Kadu soppu.

Major constituents: Saponins, tannins and alkaloids (Adejumo et al., 2011).

Pharmacodynamic uses: Anti-sickling agent (Adejumo et al., 2011).

Ethnomedicinal uses: The badagas of Nilgiris uses leaf decoction internally to treat intestinal worms and stomach disorders (Sathyavathi and Janardhanan, 2007).

Chromolaena odorata (L.) R. king and Rabinson. (Asteraceae)

Description: It is a rapidly growing perennial herb. It is a multi-stemmed shrub to 2.5 m tall in open areas. It has soft stems but the base of the shrub is woody. In shady areas it becomes petiolated and behaves as a creeper, growing on other vegetation. It can then become up to 10 m tall. The plant is hairy and glandular and the leaves give off a pungent, aromatic odour when crushed. The leaves are opposite, triangular to elliptical with serrated edges. Leaves are 4 to 10 cm long by 1 to 5 cm wide. Leaf petioles are 1 to 4 cm long. The white to pale pink tubular flowers are in panicles of 10 to 35 flowers that form at the ends of branches. Commonly known as devil weed and locally known as communist.

Major constituents: Alkaloids, flavanoid, tannins, cardiac glycosides, fats and oils (Larguita et al., 2008).

Pharmacodynamic uses: Antimicrobial and coagulant (Larguita et al., 2008).

Ethnomedicinal uses: In Pandalur village, Nilgiris. The people use Mature leaves used to cure wound healing, leaf extracts used to control mosquito bite and prevent insect bite (Govindraj et al., 2011).

Cirsium wallichii DC. (Asteraceae)

Description: It is an extremely variable plant, 4 to 10 ft tall, with spreading branches. White to purplish-white flower-heads, clustered or solitary, are borne on leafless stalks, or are stalkless. They are 2 to 3.8 cm across, with florets about 1.6 cm. There are lance shaped bracts ending in erect or recurved spines. Stalkless leaves are pinnately lobed, with margins having very long, stout spines. Leaves are hairless above and cottony beneath. Stems are hairy and leafy. Commonly known as Wallichs Thistle and locally known as Bungsee.

Major constituents: Flavanoid, sterols, titerpenes, alkaloids, polyacetylenes, aetylenes, hydrocarbons, sesquiterpene lactones, phenolic acids and lignans (Ingrid et al., 2003).

Pharmacodynamic uses: Antitumour activity (Vijaya et al., 2009).

Ethnomedicinal uses: The leaf decoction is used by chotta bhangal community in western Himalaya to treat gastric problems (Sanjay et al., 2006).

Commelina benghalensis L.

Description: It is an annual or perennial herb. Leaves are ovate to lancolate, 2.5 to 7.5 cm long, 1.5 to 4 cm wide, with parallel veination, entire leaf margins, and pubescence on top and bottom. The leaf sheath is covered in red and sometimes white hairs at the apex which is a primary identification factor for this species. Stems can be erect or crawling along the ground rooting at the nodes or climbing if supported, 10 to 30 cm in height, 20 to 90 cm in length, covered in a fine pubescence and dichotomously branched. Flowers are produced in spathes often found in clusters, funnel shaped, fused by two sides, 10 to 20 mm long, 10 to 15 mm wide, on peduncles 1 to 3.5 mm in length. Aerial flowers are staminate, perfect, and chasmogamous with 3 petals 3 to 4 mm long. The upper two flower petals are blue to lilac in color, with the lower petal lighter in color or white and much less prominent. Seeds are rectangular, 1.6 to 3 mm in length, 1.3 to 1.8 mm wide, brown to black in color, and have a netted appearance. Commonly known as tropical spider wort and locally known as Aduthinnathalai.

Major constituents: Phlobatannins, carbohydrates, tannins, glycosides, volatile oils, resins, balsams, flavonoids and saponins (Ibrahim et al., 2010).

Pharmacodynamic uses: Cardio active wound healing and Anticancer (Ibrahim et al., 2010).

Ethnomedicinal uses: The badagas of Nilgiris uses plant decoction internally to cure worm infections (Sathyavathi and Janardhanan, 2007).

Dodonea viscose (L.) Jacq. (Sapindaceae)

Description: It is a shrub growing to 1 to 3 m (3.3 to 9.8 ft) tall, rarely a small tree to 9 m (30 ft) tall. The leaves are simple elliptical, 4 to 7.5 cm (1.6 to 3.0 in) long and 1 to 1.5 cm (0.39 to 0.59 in) broad, alternate in arrangement, and secrete a resinous substance. The flowers are yellow to orange-red and produced in panicles about 2.5 cm (0.98 in) in length. The fruit is a capsule 1.5 cm (0.59 in) broad, red ripening brown, with two to four wings. Commonly known as hopbush and locally known as morantha.

Major constituents: Tannins, flavanoid, steroids and triterpenes (Venkatesh et al., 2008).

Pharmacodynamic uses: Anti-diarrheal, local anesthetic, smooth muscle relaxant, antidiabetic, anti-ulcer, anti-inflammatory and anti-microbial activities (Venkatesh et al., 2008).

Ethnomedicinal uses: The badagas of Nilgiris uses the stems and leaves externally to treat bone fractures in animals (Sathyavathi and Janardhanan, 2007).

Kalanchoe pinnata (Lam.) Pers. (Crassulaceae)

Description: It is a shrub that grows up to tall of 1.8 m. The pendent flowers are on short, lateral branches on tall, upright, chandelier-like flower stalks. The individual flowers are tubular, 1 inch (2.5 cm) long, enclosed in papery, inflated, green to reddish pink sepals, and have 4 red, narrowly triangular lobes. The flowers dry on the plant and gradually turn a light papery brown color. The leaves have scalloped, dark maroon margins and are green, succulent, opposite, and mostly pinnately compound with 3 to 5 elliptic leaflets. New baby plants can form along the edges of the leaves. Commonly known as miracle leaf and locally known as Runakkalli or Raga kanni.

Major constituents: Steroidal glycosides, flavanoid, triterpenoids and polyphenols (Zhang and Zhang, 2007).

Pharmacodynamic uses: Immunosuppressive, wound healing, hepatoprotective, antinociceptive, anti-inflammatory and antidiabetic, nephroprotective, antioxidant activity, antimicrobial activity, analgesic, and antipyretic (Zhang and Zhang, 2007).

Ethnomedicinal uses: The irular tribes of red hills, Tamilnadu, India eat raw leaves daily in empty stomach to treat stomach ulcers (Francisca and Rajendran, 2012).

Lantana camara L. (Verbenaceae)

Description: It is a low, erect or subscandent, vigorous shrub which can grow to 2 to 4 m in height. The leaf is ovate or ovate oblong, 2 to 10 cm long and 2 to 6 cm wide, arranged in opposite pairs. Leaves are bright green, rough, finely hairy, with serrate margins and emit a pungent odour when crushed. The stem in cultivated varieties is often non- thorny and in weedy varieties with recurved prickles. It is woody, square in cross section, hairy when young, cylindrical and up to 15 cm thick as it grows older. Lantana is able to climb to 15 m with the support of other vegetation. Flower heads contain 20 to 40 flowers, usually 2.5 cm across; the colour varies from white, cream or yellow to orange pink, purple and red. Commonly known as sleeper weed or wild sage and locally known as karadikke.

Major constituents: Sterols, glycosides, saponins, carbohydrates, alkaloids, flavanoid, tannins, proteins and ti-terpenoids (Jitendra et al., 2011).

Pharmacodynamic uses: Antiseptic, antispasmodic, carminative, diaphoretic, antinflammatory, antipyretic and analgesic (Jitendra et al., 2011).

Ethnomedicinal uses: The badagas of Nilgiris uses the leaf decoction internally to expel worms and externally to treat cuts and wounds (Sathyavathi and Janardhanan, 2007).

Leucas aspera (Wild.) Link. (Lamiaceae)

Description: It is an annual plant that can reach heights of 15 to 60 cm. The leaves can be obtuse, linear or linearly lanceolate or petiolate. The stem quadrangular and contains a wide stele. The epidermis of the stem is covered in a thick waxy cuticle and contains few traversed stomata. The roots contain epidermal cells which are very narrow and closely packed together. It is commonly known as common leucas and locally known as thumbaigidu.

Major constituents: Triterpenoids, sterols, glucosides, and phenol compounds (Prajapati et al., 2010).

Pharmacodynamic uses: Antifungal, antioxidant, antimicrobial, antinociceptive and cytotoxic Activity (Prajapati et al., 2010).

Ethnomedicinal uses: In Pandalur village, Nilgiris, the people uses leaf decoction used as to cure cold, cough, and skin disorders. Healthy leaves are used for the curry preparation (Govindaraj et al., 2011).

Lobelia nicotianaefolia Roth E & S. (Lobeliaceae)

Description: It is a tall, erect, much branched, somewhat hairy herb, which grows to 1.5 to 3 m in height. The leaves, resembling those of tobacco, are narrowly obovate-lanceolate, the lower ones being 30×5 cm, while the upper ones gradually become smaller. The flowers are large, white, and borne in terminal racemes 30 to 50 cm long. Flowers are 3 to 4 cm long, two-lipped. The sepal is smooth or hairy, narrow, about 1.2 cm long. The capsules are 2-celled, somewhat rounded, and about 1.5 cm in diameter. The seeds are numerous very small, ellipsoids and compressed. It is commonly known as wild tobacco and locally known as upperichedi.

Major constituents: Alkaloids (Abrar et al., 2012).

Pharmacodynamic uses: Respiratory stimulant, smoking cessation and antiepileptic (Arbar et al., 2012).

Ethnomedicinal uses: The kurichiar tribes of wayanad district, Kerala, uses Leaf paste mixed with a pinch of lime is applied to foul smelling wounds for speedy healing (Udayan et al., 2008).

Melilotus indica L. (Fabaceae)

Description: It is an annual herb, sprawling in the absence of a crop, but erect in crops such as wheat. The stem is thin and wiry when mature. A characteristic feature is the appearance of anthocyanin coloring in the form of a red stripe along the midrib of the leaflets which disappears at the time of flowering. The inflorescence is a dense raceme arising from leaf axils, 2 to 3 cm long with 15 to 50 flowers on a short peduncle 1.5 to 2 cm long. Flowers 2 to 2.5 mm long, yellow or pale-yellow, of typical leguminous structure. It is commonly known as Indian sweet clover and locally known as vana methika.

Major constituents: Flavanoid, coumarin glycosides, triterpenes and fatty acids (Suhail et al., 2012).

Pharmacodynamic uses: Emmolient, carminative and digestive (Suhail et al., 2012).

Ethnomedicinal uses: In Margallah hills (Pakistan), people uses plant paste traditionally leaf as Emollient, in treating swelling, and internally to treat diarrhoea, bowl complaints (Asma et al., 2009).

Mirabilis jalapa L. (Nyctaginaceae)

Description: It is a long-lived (perennial) herb growing up to 2 m high, with a tuberous root. Its leaves are egg-shaped in outline with broad end at base (ovate), oblong, or triangular, measuring to 9 cm long.; the leaf tip is acute, base cordate. The leaf stalk (petiole) is 4 cm long. Flowers of occur in groups of 3 to 7; flower stalks more or less absent; flowers are fragrant and open in the afternoon; flowers are tubular, white, pink or red in color, up to 6.5 long by 3.5 wide with 5 to 6 stamens. The fruit is a small, one-seeded capsule (anthocarp). It is commonly known as Four 0 clock flower and locally known as tolahoo.

Major constituents: Alkaloids, steroids, flavonoids, saponins, phenol compounds and tannins (Subin et al., 2012).

Pharmacodynamic uses: Antinociceptive, antibacterial and antioxidant activities (Subin et al., 2012).

Ethnomedicinal uses: The badagas of Nilgiris uses the leaf paste externally to cure wounds (Sathyavathi and Janardhanan, 2007).

Mollugo pentaphylla L. (Molluginaceae)

Description: It is Annual herb up to 24 cm tall. Stem and branches slender, glabrous. Leaves cauline and ramal, pseudo-verticillate, 1.2 to 4.0 cm long, 1.5 mm broad, elliptic lanceolate. Sepals 1.5 to 2 mm long, elliptic ovate. Stamens 5, antisepalous, c. 1.2 mm long; filaments dilated at the base, anthers less than 1 mm long, basifixed, dehiscing longitudinally. Ovary sub-globose, c. 1.5 mm broad, Fruit c. 9 mm broad, membranous. Seed less than 1 mm long, granulate dark brown. It is commonly known as carpet weed and locally known as kuttuttiray.

Major constituents: Flavanoid, glycosides and saponins (Valarmathi et al., 2010).

Pharmacodynamic uses: Anti-microbial, anti-inflammatory, anti-cancer, hepatoprotective and antipyretic (Valarmathi et al., 2010).

Ethnomedicinal uses: In India, the whole plant is used as a mild laxative medicine, stomachic, antiseptic and emmenagogue, while a decoction of the roots is used to treat eye diseases (Valarmathi et al., 2010).

Nicandra physaloides (L.) Gaertn. (Solanaceae)

Description: Plants grow to 1 m tall and are vigorous with spreading branches and ovate, mid-green, toothed and waved leaves. The flowers are bell-shaped and 5 cm or more across, pale violet with white throats. The flower becomes lantern-like towards the end of its bloom. It is commonly known as apple of peru and locally known as Ummathakkai.

Major constituents: Flavanoid, saponins, carbohydrates, terpenoids and alkaloids (Devi et al., 2010).

Pharmacodynamic uses: Diuretic and anticancer activities (Devi et al., 2010).

Ethnomedicinal uses: The badagas of Nilgiris uses the leaf paste externally to cure wounds (Sathyavathi and Janardhanan, 2007).

Oenothera rosea L´ Hér. ex Ait. (Onagraceae)

Description: It is an annual herb grows up to 2 ft (60 cm) tall. The flowers open at sunrise, are less than 1 inch (2.5 cm) wide, and have cream-colored anthers and 4 egg-shaped petals with conspicuous darker veins. The flowers fade to a dark pink color and are followed by club-shaped seed capsules that are widest near the tip. The leaves are both in a basal rosette and on the stems, and they are green, alternate, hairless to sparsely hairy, variably wavy-toothed, and elliptic or oblanceolate in shape. The lower leaves are sometimes pinnatifid at the base. The multiple stems are green, hairy, branched or not, and erect to ascending. It is commonly known as rosy evening – primrose and locally known as Mexican rose.

Major constituents: Carbohydrates, steroids, glycosides and tannins (Sumitra et al., 2012).

Pharmacodynamic uses: In treatment of hepatic pain, liver and skin problems and anti-diarrheic effect (Andrade-Cetto, 2009).

Ethnomedicinal uses: In the Himalayas. traditionally, the aqueous infusion of the leaves has been used in hepatic pains and kidney problems (Sumitra et al., 2012).

Opuntia stricta (Ker-Gawler) Haw (Cactaceae)

Description: It is a shrub or tree up to 5 m tall, forming sturdy trunk with age. Joints flattened, narrowly elliptic to ovate, varying in size, 30 to 60 cm long and 6 to 12 cm broad, attenuate below, often acute above, fairly thick, glaucous-green; areoles small to large, raised and woolly, with 3 to 6 radiating, unequally long, greyish white spines up to 3 (-10) cm long, straight or occasionally slightly curved, or spineless (in older plants and some cultivars). Leaves, if developed, are minute, subulate and early deciduous. Flowers about 7 cm long; hypanthium broadly cylindrical, contracted below, with numerous raised areoles spirally arranged, densely wooly and filled with glochidia, occasionally also bearing small spines and minute leaves; petaloid segments yellow or orange. Fruits ellipsoid, about 7 cm long, reddish, succulent, edible. Seeds are about 5 mm long. It is commonly known as noppales and locally known as nagakalli.

Major constituents: Alkaloids, steroids, saponins and flavanoid (Naod and Tsige, 2012).

Pharmacodynamic uses: Antiulcer, anti-inflammatory, antiviral and anticancer (Naod and Tsige, 2012).

Ethnomedicinal uses: The tribals of Sudi Konda Forest, East Godavari District, Andhra Pradesh used to treat for menorrhagia and metrorrhagia, 10 ml fruit juice mixed with 10 ml 0f rice washed water and a spoonful of sugar is administered twice a day for 3 days till to cure (Aniel et al., 2012).

Oxalis corniculata L. (Oxalidaceae)

Description: It is a annual/perennial growing to 0.1 m (0ft 4 in) by 0.3 m (1 ft). It is a somewhat delicate-appearing, low-growing, herbaceous plant. It has a narrow, creeping stem that readily roots at the nodes. The trifoliate leaves are subdivided into three rounded leaflets and resemble a clover in shape. Some varieties have green leaves, while others, have purple. The leaves have inconspicuous stipules at the base of each petiole.

The fruit is a narrow, cylindrical capsule, 1 to 2 cm long and noteworthy for its explosive discharge of the contained, 1 mm long seeds. It is commonly known as sleeping beauty and locally known as Kunnaullumajigai.

Major constituents: Carbohydrates glycosides, phytosterols, phenol compounds, tannins, flavanoid, proteins, amino acids and volatile oils (Raghvendra et al., 2009).

Pharmacodynamic uses: Antibacterial and antiulcer activity (Raghvendra et al., 2009).

Ethnomedicinal uses: In Jarkhand, India the people uses fresh juice internally to treat anemia, tympanitis, dysentery and piles (Hari et al., 2012).

Persicaria nepalensis (Meissn.) H. (Polygonaceae)

Description: A slender spreading or procumbent annual herb, rooting at the lower nodes. Stem pale green, greenish-brown or red to bright red, ascending up to 50 cm, glabrous or with scattered gland-tipped hairs, these more numerous below the ocrea. Leaf lamina with the upper part 0.5-6 × 0.5-3 cm, ovate to ovate-deltate, acute at the apex, tapering or abruptly contracted and decurrent below for up to 1.5 cm forming a false petiole with auricles at the base, glabrous or with scattered hairs, gland-dotted beneath. Ocrea brown, membranous, 6 to 8 mm long, entire at the apex. Flowers in small, pedunculate heads 6 to 9 mm in diameter; heads solitary or paired, c. 12-flowered, subtended by a sessile involucral leaf; peduncles up to 5 cm long, with deflexing glandular hairs below the inflorescence; bracts hyaline, 3.5-5 × 1.5-2.5 mm, broadly lanceolate to ovate. It is commonly known as Nepalese smartweed or knotweed and locally known as Actalaree.

Major constituents: Alkaloids, tannins, saponins and flavanoid (Faraz et al., 2003).

Pharmacodynamic uses: Antibacterial, antifungal and insecticidal activities (Farrukh et al., 2010).

Ethnomedicinal uses: In Nilgiris, the badaga community uses the root decoction internally to treat fever and vomiting as a home remedy (Sathyavathi and Janardhanan, 2007).

Plantago erosa ex Roxb. (Plantaginaceae)

Description: It is a sub shrub growing to 60 cm (23.5 in) tall. The leaves are sessile, but have a narrow part near the stem which is a pseudo-petiole. They have three or five parallel veins that diverge in the wider part of the leaf. Leaves are broad or narrow, depending on the species. The inflorescences are borne on stalks typically 5 to 40 cm (2.25 to 15.75 in) tall, and can be a short cone or a long spike, with numerous tiny wind-pollinated flowers. Commonly known as plantain and locally known as nelavarikke.

Major constituents: Tannins, diterpenoids and steroids (Baural et al., 2011).

Pharmacodynamic uses: Astringent, antitoxic, antimicrobial, demulcent, expectorant, diuretic, anti-inflammatory and analgesic (Baural et al., 2011).

Ethnomedicinal uses: In Nilgiris, the badaga community uses the whole plant juice uses externally and internally to treat muscle pains (Sathyavathi and Janardhanan, 2007) and the Adi tribe of Arunachal Pradesh, India uses the leaves as vegetable to treat constipation and also to improve digestion (Srivastava and Adi community, 2009).

Plectranthus barbatus Andr. (Lamiaceae)

Description: It is a densely hairy perennial herb, with pale blue flowers arranged in whorls, forming long leafless interrupted spikes. Flowers are up to 2 cm long, tube bent abruptly downward, longer than the sepal cup. Flowers are 2-lipped, the upper lip short, turned back, 3-lobed, the lower much longer, boat-shaped, pointed. Sepal cup is hairy, bell-shaped, with lance shaped, prickly-tipped sepals. Bracts are broadly ovate, pointed, overlapping in bud, soon falling. Leaves are ovate to oblong, blunt, rounded-toothed, short-stalked, 5 to 8 cm long. They are arranged in opposite pair's perpendicular to each other, along a 1 to 3 ft tall stem. It is commonly known as Indian coleus and locally known as karpuravalli.

Major constituents: Flavonoid glucronide and diterpenoids (Porfírio et al., 2010).

Pharmacodynamic uses: Antioxidant activity, acetyl cholinesterase Inhibition, anticonvulsant, spasmolytic and antihypertensive (Porfírio et al., 2010).

Ethnomedicinal uses: Kothas, the native tribes of

Trichigadi in Nilgiris, South India consider the decoction of tuberous roots as tonic for well being (Rakshapal et al., 2011).

Polygonum chinense L. (Polygonaceae)

Description: The genus primarily grows in northern temperate regions. They vary widely from prostrate herbaceous annual plants under 5 cm high, others erect herbaceous perennial plants growing from 3 to 4 m tall and yet others perennial woody vines growing from 20 to 30 m high in trees. Several are aquatic, growing as floating plants in ponds. The smooth-edged leaves range from 1 to 30 cm long, and vary in shape between species from narrow lanceolate to oval, broad triangular, heart-shaped, or arrowhead forms. The stems are often reddish or red-speckled. The small flowers are, pink, white, or greenish, forming in summer in dense clusters from the leaf joints or stem apices. It is commonly known as tear thumb and locally known as Kappu annu gidu.

Major constituents: Flavones and flavanones, phenol compounds and tannins (Shu et al., 2012).

Pharmacodynamic uses: Antihelmenthic (Shu et al., 2012).

Ethnomedicinal uses: In Nilgiris, the badaga community uses the whole plant juice internally to treat paralysis, giddiness, quenching thirst (Sathyavathi and Janardhanan, 2007).

Prinsepia utilis Royle. (Rosaceae)

Description: It is a spiny shrub, growing up to 1 to 5 m tall. Branches are grayish green, robust, branchlets green to grayish green, angled, brown velvety to hairless. Spines are up to 3.5 cm long. Winter buds are purplish red, ovoid to oblong, hairless. Leaf stalks are about 5 mm long. Leaf blade is oblong to ovate-lance shaped, 3.5 to 9 cm long, 1.5 to 3 cm wide, base broadly wedge-shaped to rounded, margin toothed, tip pointed to long pointed. Flowers are borne in racemes in leaf axils, or on short branchlets, 3 to 6 cm, many-flowered. Flowers are 1 cm in diameter. Flower- stalks are 4 to 8 mm, up to 1 cm in fruit. Flower base is cup-shaped, outside brown velvety. Commonly known as Himalayan cherry and locally known as cherara.

Major constituents: Hemiterpenoids, fatty acids, hydrocyanic acid and flavanoid (Jun et al., 2006).

Pharmacodynamic uses: Anti rheumatic and anti diabetic (Umar et al., 2008).

Ethnomedicinal uses: In Garhwal, Himalayas, seed oil

warmed and massaged twice a day in arthritic pain. The paste of root is applied for healing of cuts, wounds and boils.

Ricinus communis L. (Euphorbiaceae)

Description: It is a fast-growing, suckering perennial shrub that can reach the size of a small tree (around 12 m or 39 feet), but it is not cold hardy.The glossy leaves are 15 to 45 cm (5.9 to 18 in) long, long-stalked, alternate and palmate with 5 to 12 deep lobes with coarsely toothed segments. In some varieties they start off dark reddish purple or bronze when young, gradually changing to a dark green, sometimes with a reddish tinge, as they mature. The leaves of some other varieties are green practically from the start, whereas in yet others a pigment masks the green colour of all the chlorophyll-bearing parts, leaves, stems and young fruit, so that they remain a dramatic purple-to-reddish-brown throughout the life of the plant. The flowers are borne in terminal panicle-like inflorescences of green or, in some varieties, shades of red monoecious flowers without petals. The male flowers are yellowish-green with prominent creamy stamens and are carried in ovoid spikes up to 15 cm (5.9 in) long; the female flowers, born at the tips of the spikes, have prominent red stigmas. It is commonly known as castor oil plant and locally known as Aamanakku.

Major constituents: Tannins, saponins, alkaloids, carbohydrates, phenols, flavonoids, sterols and resins (Mary et al., 2011).

Pharmacodynamic uses: Antibacterial, purgative, anti-inflammatory, hepatoprotective, hypoglycemic and insecticidal (Mary et al., 2011).

Ethnomedicinal uses: In Nilgiris, the tribal people use the seed decoction in treatment of stomach pain and dysentery and externally the seed oil is applied over skin to remove black scars (Sathyavathi and Janardhanan, 2007).

Rubus ellipticus Smith. (Rubiaceae)

Description: It is a stout evergreen shrub with prickly stem that grows approximately 4.5 m tall. Its stems are covered with prickles and reddish hairs. Leaves are alternate and compound with three round to blunt leaflets of 5 to 10 cm long. The underside of the leaves is lighter than the upper surface and covered with downy hairs. The flowers are small and white with five petals. The fruit is a round yellow cluster of droplets which is easily detached from the receptacle. It is commonly known as yellow Himalayan raspberry and locally known as Tuppa mulli.

Major constituents: Flavonoids, phenolic compounds and tannins (Vadivelan et al., 2009).

Pharmacodynamic uses: Diabetes, diarrhea, gastralgia, wound healing, dysentery, antifertility, analgesic, and epilepsy (Vadivelan et al., 2009).

Ethnomedicinal uses: The tribal of Nilgiris uses the decoction of this plant as an abortifacient (Vadivelan et al., 2000).

Rumex nepalensis Sprengel. (Polygonaceae)

Description: A stout, perennial herb up to 1.8 m tall. Stems green to greenish-brown, hollow, glabrous, striate. Leaf lamina 25–45 × 7–9 cm, oblong-lanceolate to linear-lanceolate, obtuse or rounded at the apex, cuneate at the base, entire or crisped on the margin, glabrous or with scattered papillae on the undersurface, the upper leaves smaller; petiole of basal leaves 13 to 30 cm long. Flowers hermaphrodite, pedicellate, pendulous, in whorls borne in terminal racemose panicles; the basal whorls in the axils of foliaceous bracts; pedicels filiform, articulated near the base. Commonly known as dock weed and locally known as gongu.

Major constituents: Anthraquinones, flavonoids and glycosides (Surjeet et al., 2010).

Pharmacodynamic uses: Psychopharmacological, antioxidant, antimicrobial, antidiarrhoeal and Muscle relaxant (Surjeet et al., 2011).

Ethnomedicinal uses: In Nilgiris, the root and leaves paste is taken internally to treat jaundice (Sathyavathi and Janardhanan, 2007).

Sarothamnus scoparius L. (Fabaceae)

Description: It typically grow to 1 to 3 m (3–9 ft) tall, rarely to 4 m (13 ft), with main stems up to 5 cm (2 in) thick, rarely 10 cm (4 in). The shrub have green shoots with small deciduous trifoliate leaves 5 to 15 mm long, and in spring and summer is covered in profuse golden yellow flowers 20 to 30 mm from top to bottom and 15 to 20 mm wide. Flowering occurs after 50 to 80 growing degree days. In late summer, its legumes (seed pods) mature black, 2 to 3 cm long, 8 mm broad and 2 to 3 mm thick; they burst open, often with an audible crack, forcibly throwing seed from the parent plant. It is commonly known as Scotch broom and locally known as kothikeerai.

Major constituents: Phenolics, tannins and flavanoid (Jayabalan et al., 2008).

Pharmacodynamic uses: Antioxidant and anti-stress (Jayabalan et al., 2008).

Ethnomedicinal uses: In Nilgiris, the tribes use the leaf decoction internally as a good diuretic (Sathyavathi and Janardhanan, 2007).

Siegesbeckia orientalis L. (Asteraceae)

Description: It is an annual herb growing to 1.2 m (4 ft). The flowers are hermaphrodite (have both male and female organs. Leaves ovate, florets in solitary, axillary or terminal, heterogamous head, fruit achene and obovoid. It is commonly known as Holy herb and locally known as kadambu.

Major constituents: Phenols, tannins, lignans, flavonoids, sterols, phenolic compounds, glycosides and Triterpenoids (Geetha and Gopal, 2011).

Pharmacodynamic uses: Larvicidal, anti-inflammatory and analgesic (Geetha and Gopal, 2011).

Ethnomedicinal uses: In Nilgiris, the tribal people use the plant extract externally to treat various skin infections (Sasikumar et al., 2007).

Silybum marianum (L.) Gaertn (Asteraceae)

Description: It grows 30 to 200 cm tall, having an overall conical shape with a approx. 160 cm max. diameter base. The stem is grooved and more or less cottony, and with the largest specimens the 'trunk' is hollow. The leaves are oblong to lanceolate. They are either lobate or pinnate, with spiny edges. They are hairless, shiny green, with milk-white veins. The flower heads are 4 to 12 cm long and wide, of red-purple colour. They flower from June to August in the North or December to February in the Southern Hemisphere (Summer through Autumn). The bracts are hairless, with triangular, spine-edged appendages, tipped with a stout yellow spine. The achenes are black, with a simple long white pappus, surrounded by a yellow basal ring. Commonly known as milk thistle and locally known as dudh patra.

Major constituents: Alkaloids, amino acids, flavonoids, carbohydrates, phenolics, steroids and tannin (Bilani et al., 2006).

Pharmacodynamic uses: Anti-inflammatory, anticancer, antioxidant and hepatoprotective (Bilani et al., 2006).

Ethnomedicinal uses: In Nilgiris, Kattunayakas tribes use the leaves and seed decoction internally to treat burning pain in anus and jaundice pain (Selvaraj et al., 2009).

Stellaria media L. (Caryophyllaceae)

Description: It is an annual herb growing to 0.1 m (0ft 4in) by 0.5 m (1ft 8in). Cotyledons are ovate, 1-12 mm long by 0.25-2 mm wide, with a slender reddish hypocotyl that is sparsely hairy, stems are usually running prostrate along the ground, rooting at the nodes, with the upper portion erect or ascending and freely branching. Stems are light green in color and with hairs in vertical rows. Fruits are oval, one-celled capsule, whitish in color, containing numerous seeds and flowers are alone or in small clusters at the ends of stems. Flowers are small (3-6 mm wide) and consist of 5 white petals that are deeply lobed, giving the appearance of 10 petals. It is commonly and locally known as chick weed.

Major constituents: Saponin, flavonoids, steroids, triterpenoids, glycosides, and anthocynidine (Chidrawar et al., 2011).

Pharmacodynamic uses: Antiobesity, anti-inflammatory and antihepatitis B (Chidrawar et al., 2011).

Ethnomedicinal uses: In Dibrugarh District (Assam, India). The whole plant paste and decoction is used externally and internally in treatment of skin diseases, bronchitis, rheumatic pains and dysmenorrhoea (Dilip et al., 2005).

Taraxacum officinale F.H. Wigg. (Asteraceae)

Description: It grows from generally unbranched taproots and produces one to more than ten stems that are typically 5 to 40 cm tall but sometimes up to 70 cm tall. The stems can be tintled purplish and produce flower heads that are held as taller than the foliage; the stems can be glabrous or are sparsely covered with short hairs. It is commonly known as Dandelion and locally known as Kanphul.

Major constituents: Saponins, triterpenes, sterols, phenolics and tannins (Oseni and Yussif, 2012).

Pharmacodynamic uses: Antibacterial, hepatoprotective and mild laxative (Oseni and Yussif, 2012).

Ethnomedicinal uses: The Irula and Kurumba tribes of Niligiris use the fresh plant juice internally against liver diseases and intermittent fevers (Selvaraj et al., 2009).

Tephrosia purpurea (L.) Pers. (Fabaceae)

Description: A small spreading perennial herb grows up

to 50 cm in height. Leaves compound, imparipinnate, with 11 to 21 leaflets, oblong-lanceolate, small; flowers red or purple, in axillary racemes. Fruits pod, containing 4 to 8 seeds. It is commonly known as wild indigo and locally known as kavali.

Major constituents: Flavanoid, tannins, phenols and anthocyanins (Shivaraj and Khobragade, 2011).

Pharmacodynamic uses: Antioxidant, antiulcer, hepatoprotective and wound healing (Shivaraj and Khobragade, 2011).

Ethnomedicinal uses: In Jharkhand, India, the tribal uses the boiled extract of plant is a vermifuge and used to kill the intestinal worms (Hari et al., 2012).

Tithonia diversifolia (Hemsl.) A.Gray. (Asteraceae)

Description: It is 2 to 3 m (6.6–9.8 ft) in height with upright and sometimes ligneous stalks in the form of woody shrubs. The large, showy flowers are yellow to orange colored and 5 to 15 cm wide and 10 to 30 cm long. Leaves are sub-ovate, serrate, acute, 10 to 40 cm long, simply or mostly 3 to 7 lobed, somewhat glandular, and slightly grayish beneath. The seeds are achenes, 4-angled, and 5 mm long seeds are spread by wind. It is commonly known as tree marigold and locally known as Kattu suryakanthi.

Major constituents: Alkaloids, flavanoids, phlobatanins, terpenoids and saponins (Ezeonwumelu et al., 2012).

Pharmacodynamic uses: Antidiarroheal, antiplasmodial, anti-inflammatory, analgesic and antimicrobial (Ogundare, 2007).

Ethnomedicinal uses: In Nigeria, the decoctions of its various parts are used for the treatment of malaria, diabetes mellitus, sore throat, liver and menstrual pains (Ogundare, 2007).

Trifolium repens L. (Fabaceae)

Description: This perennial plant is about 6" tall, branching from the base. Initially, it produces several compound leaves from a short stem that grows only a little, after which this stem rapidly elongates and becomes up to 1' long. These elongated stems sprawl along the ground and have the capacity to root at the nodes. They are hairless and light green. The alternate compound leaves are trifoliate and hairless. They occur at intervals along the elongated stems and have long hairless petioles. The leaflets are obovate or ovate. Their margins are finely serrate. Across the upper surface of each leaflet are white markings in the form of a chevron

(an upside down "V"), although for this species these markings are often degenerate, irregular, or absent. Each leaflet is about ¾" long and about half as wide. At the base of each petiole there are a pair of small lanceolate stipules that are light green and membranous; sometimes they wrap around the elongated stems. Each stipule is less than ½" in length. It is commonly and locally known as white clover.

Major constituents: Phenolics, saponins, flavanoids and cyanogenic glucosides (Agnieszka and Maria, 2011).

Pharmacodynamic uses: Antirheumatic, antiscrophulatic, depurative, leucorrhoea and anticestodal (Agnieszka and Maria, 2011).

Ethnomedicinal uses: The naga tribes of India use the whole plant decoction as deworming agent (Hornoy et al., 2012).

Urtica parviflora Roxb.(Utricaceae)

Description: An erect biennial herb grows up to 60 cm in height. Leaves are solitary, alternate, dentate, chordate base, with minute hairs, which produce intense itching when touched. Flowers are minute, greenish yellow, seen in long axillary panicles, with numerous minute seeds. It is commonly known as Nettle and locally known as aanathumba.

Major constituents: Akaloids, flavanoids, terpenoids, glycosides, saponins and tannins (Prasanna et al., 2009).

Pharmacodynamic uses: Wound healing, hepatoprotective, antioxidant ad hypoglycemic activity (Prasanna et al., 2009).

Ethnomedicinal uses: In Sikkim, the leaves and fresh roots are used for the treatment of fracture, dislocation of bones, boils, and decoction of herb is used as a febrifuge (Srivastava, 1993).

Verbascum thapsus L. (Scrophulariaceae)

Description: It is a dicotyledonous plant that produces a rosette of leaves in its first year of growth. The leaves are large, up to 50 cm long. The second year plants normally produce a single unbranched stem usually 1 to 2 m tall. In the East of its range in China, it is, however, only reported to grow up to 1.5 m tall. The tall pole-like stems end in a dense spike of flowers that can occupy up to half the stem length. All parts of the plants are covered with star-shaped trichomes.This cover is particularly thick on the leaves, giving them a silvery appearance. It is commonly known as Common Mullein and locally known as kadu gidu.

Major constituents: Flavanoid, saponins, tannins, terpenoids, glycosides, carbohydrates and proteins (Ali et al., 2012).

Pharmacodynamic uses: Antihelmenthic, Antiviral, antiangiogenic and antiproliferative activities (Ali et al., 2012).

Ethnomedicinal uses: In Nilgiris, the badaga tribal people uses the leaf paste externally in treatment of cuts and wounds as an antiseptic (Sathyavathi and Janardhanan, 2007).

Conclusion

This review article has a paramount importance creating awareness for the public regarding the medicinal importance of weeds removing the misunderstanding from their minds that they consider it as useless. Moreover, it helps to motivate the public to safe guard these medicinally important weeds from all in once destruction. It is a misconception in people minds to consider all weeds as useless or hurdles to public, as some of these weeds having good ethno medicinal values globally and is good sources for new drug discovery and grows naturally in bulk, no need of specialized good agricultural practices, easily available in all the seasons. It is our duty to safe guards these beautiful nature gifts.

Globally some of these weeds are used as ethnomedicinal aids in treatment of fevers, pains, inflammations, microbial infections, worm infestations, cancer, wounds etc. But very less scientific validation is available on this area so there is a great scope for the phytoscientists to work on this area in order to explore the phytochemical or pharmacological importance of weeds. It is the duty of phytoscientists establish the scientific validation for these medicinally important weeds, so that the misconception of weeds as useless or public hurdle will convert to weeds as a pharmacologically and economically valuables.

REFERENCES

Abrar M. Tamboli, Rukhsana A Rub, Pinaki Ghosh, Bodhankar SL (2012). Antiepileptic activity of lobeline isolated from the leaf of Lobelia nicotianaefolia and its effect on brain GABA level in mice. Asian Pac J. Trop. Biomed. 2:537-542.

Adejumo OE, Owa-Agbanah IS, Kolapo AL, Ayoola MD (2011). Phytochemical and antisickling activities of Entandrophragma utile, Chenopodium ambrosioides and Petiveria alliacea. J. Med. Plants Res. 5:1531-1535.

Agnieszka K, Maria Wolbiś (2011). Study on the phenol constituents of the flowers and leaves of Trifolium repens L. Nat. Prod. Res. 26(21):2050-2054.

Ali N, Ali Shah SW, Shah I, Ahmed G, Ghias M, Khan I, Ali W (2012). Anthelmintic and relaxant activities of Verbascum thapsus Mullein. BMC Complement Altern. Med. 30:12-29.

Asma RM,Rahmatullah Q,Raza B (2010). Ethnomedicinal uses of herbs from Northern part of nara desert, Pakistan,Pak. J. Bot., 42(2): 839-851.

Andre V (1988). Grass Productivity, Island Press, Washington, D.C, pp. 48-53.

Andrade-Cetto (2009). Ethnobotanical study of the medicinal plants from Tlanchinol, Hidalgo, México.,J. Ethnopharmacol. 122:163–171.

Aniel kumar O, Krishna Rao M, Raghava Rao TV (2012). Ethno medicinal plants used by the tribals of Sudi Konda Forest, East Godavari District, Andhra Pradesh to cure women problems. J. Phytol. 4(1):10-12.

Anu W, Rajat M, Satish S Vipin S, Sumeet G (2011). Hepatoprotective effects from the leaf extracts of Brassica juncea in CCl4 induced rat model. Der Pharmacia Sinica 2(4):274–285.

Aziz B, Islam K (2012). Antibacterial activity of coumarine derivatives synthesized from 8-amino-4, 7-ihydroxy-chromen-2-one and comparison with Standard drug. J. Chem. Pharm. Res. 4(5):2495-2500.

Babu, Niranjan M, Elango K (2011). Pharmacognostical, Phytochemical and Antioxidant studies of Achyranthes aspera Linn and Achyranthes bidentata Blume. J. Pharm. Res. 4:1050-1053.

Baker HG (1974) The Evolution of Weeds, Ann. Rev. Ecol. Systematics 5:1-24.

Barua CC, Buragohain B, Roy JD, Talukdar A, Barua AG, Borah P, Lahon LC (2011). Evaluation of analgesic activity of hydroethanol extract of Plantago erosa ex roxb. Pharmacologyonline, 2:86-95.

Bhanu PK (2011). Evaluation of wound healing activity of leaves of Ageratum conyzoides. Int. J. P'Prac. Drug. Res 1:8-13.

Biller A, Boppre M, Witte L, Hartmann T (1994). Pyrrolizidine alkaloids in Chromolaena odorata, Chemical and chemo ecological aspects. Phytochemistry 35:615-619.

Charles LP, Kokati Venkata BR (2012). Ethanobotanical and Current Ethanopharmacological Aspects of Argemone Mexicana Linn: An Overview. Int. J. Pharm. Pharm. Sci. 3:2143-2148.

Chidrawar VR, Patel KN, Sheth NR, Shiromwar SS, Trivedi P (2011). Antiobesity effect of Stellaria media against drug induced obesity in Swiss albino mice. Ayu. 32:576-584.

Chharba SC, Mahunnah RLA, Mshiu EN (1993). Plants used in traditional medicine in eastern Tanzania. J. Ethnopharmacol. 39:83-103.

Christopher L (2001). The Well-Tempered Garden (Revised), Charles Elliot, Cassell & Co, London, pp. 25-27.

Devi PM, Muthumani R, Ratnaji CP, Vijayakumar T, Duddu VD, Murthy K, Jeyasundari (2010) Evaluation of alcoholic and aqueous Extracts of Nicandra Physalodes Leaves for Diuretic Activity. Int. J. Pharmaceut. Biological Arch. 1:331-334.

Devmurari VP, Jivani NP (2010). Annals of Biological Research, 1(1):10-14.

Dheeraj VL, Srikar Reddy A, Subramanyam S, Raj S (2010). Evaluation of Nephroprotective and antioxidant activity of Anthoxanthum odoratum on acetaminophen induced toxicity in rat. Int. J. Pharm. Res. Dev. 2:1-5.

Dilip K, Manashi D, Nazim FI (2005). Few plants and animals based folk medicines from Dibrugarh District, Assam. Indian J. Trad. Knowl. 4:81-85.

Ezeonwumelu JO, Omolo RG, Ajayi AM, Agwu E, Tanayen JK, Adiukwu CP (2012). Studies of phytochemical screening, acute toxicity and anti-diarrhoeal effect of aqueous extract of kenyan Tithonia diversifolia leaves in rats. Br. J. Pharmacol. 3:127-134.

Francisca GB, Rajendran A (2012). Ethnobotany of irular tribes in redhills, tamilnadu, India. Asian Pac. J. Trop. Biomed. 1:S874-S877.

Faraz M, Mohammad K, Naysaneh G, Hamid R, Vahidipour (2003). Phytochemical Screening of Some Species of Iranian Plants. Iran. J. Pharm. Res. 2:77-82.

Geetha R, Gopal GV (2011). Phytochemical screening of Siegesbeckia orientalis L., a medicinal plant of Asteraceae. J. Basic. Appl. Biol, 5:156-164.

Govindaraj R, Pandiarajan G, Balakumar, Makesh K, Sankarasivaraman K (2011). Beneficial usage of weeds in the tea fields of Pandalur Village, Nilgiris District, Tamilnadu. J. Biol. Res. 2:49-54.

Hari SL, Sanjay S, Kumari P (2012). Study of Ethno Medicinal Uses of Weeds in Rice Field Of Hazaribag District Of Jharkhand India. IJIIT 1:23-26.

Hemadri RS, Chakravarthi M, Chandrashekara, Naidu CV (2012). Phytochemical Screening and Antibacterial Studies on Leaf and Root Extracts of Asclepias Curassavica (L). IOSR-JPBS 2:39-44.

Hornoy B, Atlan A, Tarayre M, Dugravot S, Wink M (2012). Alkaloid concentration of the invasive plant species Ulex europaeus in relation to geographic origin and herbivory. Naturwissenschaften 99:883-892.

Hungeling M, Lechtenberg M, Fronczek FR, Nahrstedt A (2009). Cyanogenic and non-cyanogenic pyridine glucosides from Acalypha indica (Euphorbiaceae). Phytochemistry 70:270-277.

Ibrahim JA, Vivian C, Egharevba, HO (2010). Pharmacognostic and Phytochemical Analysis of Commelina benghalensis L. Ethnobotanical Leaflets 14:610-615.

Ingrid E, Jordon Thaden, Svata M Louda (2003). Chemistry of Cirsium and Carduus. A role in ecological risk assessment for biological control of weeds. Biochem. Sys. Ecol. 31:1353–1396.

Jayabalan N, Chidambaram SB Thanukrishnan H, Muthiah R (2008). Evaluation of behavioural and antioxidant activity of Cytisus scoparius Link in rats exposed to chronic unpredictable mild stress. BMC CAM 8:1-8.

Jhade D, Ahirwar D, Jain R, Sharma NK, Gupta S (2011). Pharmacognostic Standardization, Physico and Phytochemical Evaluation of Amaranthus Spinosus Linn. Root. J. Young Pharm. 3:221-225.

Jia RY, Yin ZQ, Wu XL, Liu DB, Du YH, Luo LY, Li C, Xiong Y (2008) Hypoglyceminc effect of flavonoids from Prinsepia utilis on alloxan-induced diabetic mice. Zhong Yao Cai. 31:399-403.

Jitendra Patel Kumar GS, Deviprasad SP, Deepika S, Shamim Qureshi Md (2011) Phytochemical and anthelmintic evaluation of Lantana camara (I.) var. aculeate leaves against Pheretima posthuma. JGTPS 2:11-20.

Jitin A, Suresh J, Deep A, Madhuri Pratyusha R (2011). Phytochemical Screening of Aerial Parts of Artemisia parviflora Roxb. A medicinal plant. Der Pharmacia Lettre 3:116-124.

Joanne P, Nynke B, David H, Jitendra G, Ronald H, Yaegl CE, Shoba R, Subramanyam V, Joanne J (2012). An ethnobotanical study of medicinal plants used by the Yaegl Aboriginal community in northern New South Wales, Australia. J. Ethnopharmacol. 139:244-55.

Jun Yi Hu, Wei Q, Yoshihisa T, Hong QD (2006). A New Hemiterpene Derivative from Prinsepia utilis. Chinese Chemical Lett. 17:198-200.

Larguita P, Reotutar ED, Gemma, AR, Supnet MST (2008). Phytochemical screening of wellawel (Chromolaena odorata) Leaves, its Antimicrobial and Coagulative Properties. JPAIR Multidiscipl. J. 1:111-1 22.

Lucia MC, Jesu C, Ferreira (2012). Borreria and Spermacoce species (Rubiaceae): A review of their ethno medicinal properties, chemical constituents, and biological activities. Pharmacognosy Rev. 6:46-55.

Maluventhan V, Sangu M (2010). Phytochemical analysis and antibacterial Activity of medicinal plant cardiospermum Halicacabum linn. J. Phytol. 2:68–77.

Mary, kensa V, syhed YS (2011). Phytochemical screening and antibacterial activity on Ricinus communis L. Plant Sci. Feed 1:167-173.

Mouli KC, Vijaya T, Dattatreya Rao S (2012). Effectiveness of flavonoid-rich leaf extract of Acalypha indica in reversing experimental myocardial ischemia: biochemical and histopathological evidence. Zhong Xi Yi Jie He Xue Bao. 10(7):784-792.

Naod G, Tsige GM (2012). Comparative Physico-Chemical Characterization of the Mucilages of Two Cactus Pears (Opuntia spp.) obtained from Mekelle, Northern Ethiopia. J. Bio. Nanobiotech. 3:79-86.

Odukoya OA, Inya-Agha SI, Segun FI, Sofidiya MO, Ilori OO (2007). Antioxidant activity of selected Nigerian green leafy vegetables. Am. J. Food Technol. 2:169–175.

Ogundare AO (2007). Antimicrobial effects of Tithonia diversifolia and Jathropa gosypifolia leaves extract collected from Ogbomoso, Oyo State. Nigeria. Adv. Nat. Appl. Sci. 4:31-54.

Okwu DE, Nnamdi FU (2011). Two novel flavonoids from Bryophyllum pinnatum and their antimicrobial Activity. J. Chem. Pharm. Res. 3:1-10.

Oseni LA, Yussif I (2012). Screening ethanolic and aqueous leaf extracts of Taraxacum offinale for in vitro bacteria growth inhibition. J. Biomed. Pharma. Sci. 20:1-4.

Panda H (2000). Medicinal Plants Cultivation and Their Uses. Asia Pacific Business Press Inc., Delhi, India. pp 503-05.

Prajapati MS, Patel JB, Modi K, Shah MB (2010). Leucas aspera: A review. Pharmacogn Rev. 4:85-87.

Porfírio S, Fale PL, Madeira PJA, Florencio MH, Ascensa L, Serralheiro MLM (2010). Antiacetylcholinesterase and antioxidant activities of Plectranthus barbatus tea, after in vitro gastrointestinal metabolism. Food Chem. 122:179-187.

Prasanna KK, Sutharson L, Lila KN, Bhagabat N (2009). Evaluation of wound-healing activity of leaves of Urtica parviflora Roxb and Callicarpa arborea Roxb in rats. Pharmacologyonline 1:1095-1103.

Rajamanickam V, Rajasekaran A, Anandarajagopal K, Sridharan D, Selvakumar K , Stephen Rathinaraj B (2010). Anti-diarrheal activity of dodonaea viscosa root extracts. Int. J. Pharma. Biol. Sci. 1:182-185.

Rajan S, Sethuraman M, Mukherjee Pulok K (2002) Ethnobiology of the Nilgiri Hills, India. Phytother. Res. 16:98-116.

Rajan S, Jayendran M, Sethuraman M (2005). Folk herbal practices among Toda tribe of the Nilgiri Hills in Tamil Nadu, India. J. Nat. Rem. 5:52-58.

Rakshapal S, Surendera PG, Deepmala S, Rachana S, Rakesh P, Alok K (2011). Medicinal plant Coleus forskohlii Briq. : Disease and management. Med. Plants 3:1-7.

Risky P (2008). Supplements to avoid. Consumer Rep. 73(1):46-47.

Samuel JL, Vipin CP, Senthil Kumar KL, Ram KS, Amit R (2010). Antitumor activity of the ethanol extract of Amaranthus spinosus leaves against EAC bearing Swiss albino mice. Der Pharmacia Lettre 2:10-15.

Sanjay KU, Singh KN, Brij L (2006).Traditional use of medicinal plants among the tribal communities of Chhota Bhangal, Western Himalaya.Pp.03-20.

Sasikumar JM, Thayumanavan Tha, Subashkumar R, Janardhanan K, Lakshmanaperumalsamy P (2007) Antibacterial activity of some ethnomedicinal plants from the Nilgiris, Tamil Nadu, India. NPR 6:34-39.

Sathyavathi R, Janardhanan KJ (2007). Folklore Medicinal Practices Of Badaga Community In Nilgiri Biosphere Reserve, Tamilnadu, India. IJPRD 3:50-63.

Selvaraj N, Mohandas B, Anita B, Murugesh KA (2009). Medicinal plants of Nilgiris- an organic prespective. Nandiseva printing works, Udhagamandlam- 643001, pp. 156-157.

Shivaraj H Nile, Kobragade, CN (2011). Phytochemical analysis, antioxidant and Xanthine oxidase inhibitory activity of Tephrosia purpurea Linn root extract. J. Nat. Prod. Res. 2:52-58.

Shu ML, Sudhahar D, Anandarajagopal K (2012). Evaluation of antibacterial and antifungal activities of Persicaria chinensis leaves. IJBPR 3:400-404.

Singh S, Singh TD, Singh VP, Pandey VB (2010). Quaternary Alkaloids of Argemone mexicana. Pharmaceutical Biol. 48:158-160.

Srivastava RC (1993). Medicinal plants of Sikkim Himalayan. J Res Ind Med, 121: 5-14.

Srivastava RC, Adi C (2009). Traditional Knowledge of Adi tribe of Arunachal Pradesh on plants. IJTK 8:146-153.

Srivastava RC, Nyishi C (2010). Traditional knowledge of Nyishi (Daffla) tribe of Arunachal Pradesh. IJTK 9:26-37.

Subin MZ, Aley kutty NA, Jayakar B, Vidya V, Halima OA (2012). Estimation of Antioxidant and Total flavanoid content of Mirabilis jalpa Linn using invitro models. Int. J. Pharm. 3:187-92.

Suhail AM, Mohtasheem M, Iqbal A, Ahmed SW, Bano H (2008). Chemical constituents from Melilotus officinalis. J. Basic. Appl. Sci. 4:89-94.

Sukumaran P, Nair AG, Chinmayee DM, Mini I, Sukumaran ST (2012). Phytochemical Investigation of Bidens biternata (Lour.) Merr. and Sheriff.-A Nutrient-Rich Leafy Vegetable from Western Ghats of India. Appl. Biochem. Biotechnol. 67:1795-1801

Sumi W, Ting KN, Khoo TJ, Wardah MD, Christophe W (2011). Antioxidant, Anti-Inflammatory, Cytotoxicity and Cytoprotection activities of Crassocephalum crepidioides (Benth.) S. Moore.extracts and its phytochemical Composition. Eur. J. Sci. Res. 157:157-165.

Sumitra S, Rupinder K, Surendra KR Sharma (2012). Pharmacognostical evaluation of Oenothera rosea L'Hér. ex Aiton. root. J. Pharm Res. 5:3269-3271.

Surjeet K, Lincy J, Mathew G, Lakhvir K, Vivek B (2011). Skeletal muscle relaxant activity of methanolic extract of Rumex nepalensis in albino rats. J. Chem. Pharm. Res. 3:725-728.

Thangavel A, Muniappan A, Yesudason J, Koil P, Thangavel S (2011). Phytochemical screening and antibacterial activity of leaf and callus extracts of Centella asiatica. Bangladesh J. Pharmacol. 6:55-60.

Udayan PS, Harinarayanan MK, Tushar KV (2008). Some common plants used by Kurichiar tribes of Tirunelli forest,Wayanad District, Kerala in medicine and other traditional uses. IJTK 4:250-255.

Umar RA, Hassan SW, Ladan MJ, Nma M, Jiya IK, Matazu MK, Abubakar U, Nata`ala, Abdullahi K (2008). Therapeutic Efficacy of Chloroquine for Uncomplicated Plasmodium falciparum Malaria in Nigerian Children at the Time of Transition to Artemisinin-Based Combination Therapy. Research Journal of Parasitology, 3:32-39.

Vadivelan R, Bhadra S, Ravi AV, Singh K, Shanish, Elango K (2009). Evaluation of anti inflammatory and membrane stabilizing property of ethanol root extract of Rubus ellipticus smith and albino rats. J. Nat. Remed. 9:74-78.

Vjanick CC Jules (1979). Horticultural Science. W.H.Freeman, San Francisco, CA, USA. P. 308

Valarmathi R, Rajendran A, .Akilandeswari S, Senthamarai R (2010). Study on Antipyretic Activity of a Mollugo pentaphylla Linn in Albino Mice. Int. J. Pharm. Tech. Res. 2:388-390.

Venkatesh S, Reddy YSR, Ramesh M, Swamy MM, Mahadevan N, Suresh B (2008). Pharmacognostical studies on Dodonaea viscose leaves. J. Pharm. Pharmacol. 2:83-88.

Vijaylakshmi S, Nanjan MJ, Suresh B (2009). In vitro anti-tumour studies on Cnicus wallichi DC. Anc Sci. Life 29:17-19.

Wang L, Zhu Y, Liao M (2011). Therapeutic effects of saponins from Achyranthes bidentata in SHRsp. 1. Zhongguo Zhong Yao Za Zhi, 36:1239-1241.

Zhang QC, Zhang H (2007). A review of researches on medicinal plants of Erigeron in Asteraceae. J. Tradit. Chin. Med. 27:68-69.

http://www.ethnologue.com.

http://www.fs.fed.us/ne/delaware/ilpin/ilpin.html
http://www.weeds.psu.edu

13

Spore density and diversity of Arbuscular mycorrhizal fungi in medicinal and seasoning plants

Regine Cristina Urcoviche[1], Murilo Castelli[1], Régio Márcio Toesca Gimenes[2] and Odair Alberton[1,2]

[1]Biotechnology Applied to Agriculture, Paranaense University – UNIPAR, Umuarama, Paraná, Brazil.
[2]Paranaense University – UNIPAR, Umuarama, Paraná, Brazil.

Arbuscular mycorrhizal fungi (AMF) set mutualistic symbiosis with most plants. Understanding this association and meet the diversity of AMF in both the medicinal and the seasoning herbs is very important, since these plants have increasingly contributed to improving the quality of human life. The aim of this study was to assess the spore density, taxonomic diversity, and root colonization by AMF in experimental beds of rosemary (*Rosmarinus officinalis* L.), nasturtiums (*Tropaeolum majus*), mint (*Mentha crispa* L.), boldo (*Peumus boldus*), oregano (*Origanum vulgare*) and chamomile (*Matricaria chamomilla*), all planted in the Medicinal Plant Nursery of the Paranaense University - UNIPAR, Umuarama – PR. Soil samples (0 to 10 cm depth) and plant roots were collected in two periods, June and November 2011. Colonization of plant roots by AMF ranged 17 to 48%. The rosemary treatment was highly responsive to the sampling periods, with only 17% of root colonization in June compared with 48% in November. The AMF spore density was higher in June than in November for all species of plants studied. Among the AMF identified within this study, the dominant genus was *Glomus sp.*, followed by *Acaulospora sp.* in all plants analyzed. Greater knowledge over diversity and density of AMF spores can strongly contribute to the sustainable management of nutrition for medicinal and seasoning plants, particularly on phosphorus supply.

Key words: Diversity of mycorrhizal fungi, symbiosis, mycorrhizae, medicinal and seasoning plants.

INTRODUCTION

Arbuscular mycorrhizal fungi (AMF) compose a key functional group of the soil biota that can substantially contribute to plant yields and ecosystem sustainability in crop production strategies. Presently, applications of beneficial microbial inoculants (biofertilizers) are increasingly attracting attention toward sustainable agriculture and life quality as a consequence of the need to solve health and environmental problems resulting from the excessive use of agrochemicals through conventional farming practices (Gianinazzi et al., 2010).

The AMF are commonly found in nature and very important as biofertilizers. They belong to Phylum *Glomeromycota*, Class *Glomeromycetes* and form a monophyletic group of fungi classified into four orders, thirteen families, and nineteen genera, with somewhat 215 species already described (Siqueira et al., 2010).

AMF form mutualistic symbiosis with the roots of most plants. Through this symbiosis, the fungus gets carbohydrates and other elements essential to their development from the host plant, forming new spores by sporulation processes. In contrast, the host plant obtains from the soil, with the help of the fungus, water and inorganic nutrients such as phosphorus (P), benefits by getting long and bulky roots, and acquires resistance to pathogens and abiotic stress, such as the presence of heavy metals and water shortage (Carrenho et al., 2007; Smith and Read, 2008).

Studies on mycorrhizal symbiosis with medicinal and seasoning plants are scarce. However, some of these few studies have shown that AMF can increase the production of secondary compounds containing medicinal active ingredients in plants under mycorrhizal symbiosis, in addition to promoting their growth (Faria et al., 2000; Russomano et al., 2008).

The aim of this study was to assess the content of soil organic matter (SOM), spore density and AMF root colonization in plants of rosemary (*Rosmarinus officinalis* L.), nasturtiums (*Tropaeolum majus*), mint (*Mentha crispa* L.), boldo (*Peumus boldus*), oregano (*Origanum vulgare*) and chamomile (*Matricaria chamomilla*) cultivated in experimental beds (plots) in the Medicinal Plant Nursery of the Paranaense University - UNIPAR, Umuarama – PR, in two periods, June and November, 2011.

MATERIALS AND METHODS

Experimental field: Soil and root sampling

Root and soil samples were collected at the Medicinal Plant Nursery of the Paranaense University - UNIPAR - Campus II, in the Umuarama city, northwestern Paraná State at coordinates S 23° 46' 11.34'' and WO 53° 16' 41.78''.

For each plant of rosemary, nasturtiums, mint, boldo, oregano and chamomile were assigned three experimental beds. Roots and rhizosphere soil were sampled in three points of each bed, giving a total of 9 replications per plant species in a completely randomized design. The beds received organic compost (coffee leaf straw transformed by fermentation process of composting) before being planted. Then, as with plants, the beds were irrigated daily by spraying when needed. Plants with the exception of boldo and rosemary, were at the phenological stage of pre-flowering.

The roots and soil sampling was performed at 0 to 10 cm, about 10 cm away from the stem of each plant, into two periods: June and November 2011. Sampling was done at the same point in each bed for the two periods. In each plot, three samples were collected for approximately 0.5 kg of soil, placed in plastic bags and stored in a refrigerator (4°C) until laboratory analyzes.

A soil sample was collected for chemical and granulometric analyses. One portion of that sample was utilized to determine the soil chemical characteristics at *Solo Fértil* Laboratory in the city of Umuarama, Paraná, Brazil. The characteristics determined were: pH in $CaCl_2$, Ca^{2+}, Mg^{2+} and Al^{3+} extracted in KCl (1 Mol L^{-1}), and P and K^+ extracted in Mehlich-1. All the analyses followed the CELA/PR standards to obtain a greater reliability of the results (Table 1). The other portion of the soil sample was intended for identification of the density and taxonomy of AMF spores. In both periods, the thinner roots of the plants were collected at three points of each bed (n = 9), washed in water, placed in flasks with preserving solution containing ethyl alcohol, acetic acid and formaldehyde (1:1:1) and stored in a refrigerator (4°C) (Souza, 2000) until laboratory analyzes for determining the percentage of AMF root colonization.

Spore density of arbuscular mycorrhizal fungal

The spores were extracted from 50 g of soil subsamples using the wet sieving method (Gerdemann and Nicolson, 1963). Each sample was suspended in 1 L of water and agitated in a beaker, kept at rest for 1 min so that the rougher particles of the soil were decanted, and then the content was poured on two juxtaposed sieves with 0.710 mm and 0.053 mm opening; the procedure repeated for four times. The material remained at the 0.053 mm sieve was transferred to 50 mL Falcon tubes, centrifuged in distilled water (3000 rpm, 3 min), and supernatant discarded. Next, saccharose solution (50%) was added into the tubes and they were agitated and centrifuged (2000 rpm, 2 min). The spores in the supernatant were transferred to the 0.053 mm sieve, washed to eliminate saccharose excess, transferred to Petri dishes and then counted under stereoscopic lens (40X).

Characterization and diversity of AMF

In Glomeromycota, taxonomy can be performed through morphological analysis of the formation, structure and germination of AMF spores. Spores were fixed on semi-permanent slides in two separate groups: one group with PVLG (polyvinyl alcohol and glycerol) resin and the other with PVLG resin + Melzer, and counted under a microscope (Morton et al., 1993). The sporocarps were carefully broken and the spores were counted.

Species taxa of AMFs were identified using Schenck and Pérez (1988) and INVAM - International Culture Collection of Arbuscular and Vesicular-Arbuscular Mycorrhizal Fungi (http://invam.caf.wvu.edu) in addition to other species descriptions. From the number of individuals of each genus, the indexes of dominance (Simpson) and diversity (Shannon-Wiener) were estimated according to Souza et al. (2010). They were calculated according to the equations:

$$C = -\Sigma \ (X_I/X_0) \times \log \ (X_I/X_0) \quad \text{Simpson,}$$

$$H' = -\Sigma \ (X_I/X_0)^2 \quad \text{Shannon-Wiener,}$$

where X_I is the spore density of each genus in 100 g of soil, X_0 is the total spore density of all AMF genera.

AMF root colonization

To determine AMF root colonization, six plants with roots were collected from each subplot (beds) and washed in running water. Plant roots were freshly cut at the length of ±1.5 cm so that they can represented as the whole radicular system. The lab procedure was done according to Phillips and Hayman (1970), where sample roots are placed in 10% KOH and closed in plastic Falcon Tubes. After heating the tubes with roots in water bath at 90°C for 1 h, the KOH solution was removed and the roots were washed in running water. A solution of 1% HCL was added in the tubes with roots and agitated for acidification for 5 min; next, the solution was removed. Then, roots were stained by adding 0.05% trypan blue to the tubes, which heated in water bath at 90°C for 30 min. At the end of the process, the roots were preserved in lactoglycerol. The root segments were examined in stereoscopic microscope (100x) for AMF structures and percentage root length colonization was

Table 1. Chemical properties of the experimental soil (0 – 10 cm) sampled in the experiment area during 1st and 2nd sampling period – June and November of 2011.

Plant	pH CaCl$_2$	P mg dm^{-3}	C g dm^{-3}	Al^{3+}	H$^+$+Al^{3+}	Ca^{2+}+Mg^{2+}	Ca^{2+}	Mg^{2+}	K$^+$	SB	CEC	V (%)
						--- Cmol$_c$ dm^{-3} ---						
					1st sampling – June						
Rosemary	4.63	52.40	7.60	0.0	4.96	3.50	2.00	1.50	0.15	3.65	8.61	42.42
Boldo	5.24	21.00	6.04	0.0	3.68	4.38	3.38	1.00	0.21	4.58	8.26	55.45
Chamomile	4.72	90.00	7.21	0.0	4.96	7.63	3.88	3.75	0.21	7.83	12.79	61.22
Nasturtiums	5.47	173.60	8.77	0.0	3.68	5.75	3.25	2.50	0.21	5.96	9.64	61.81
Mint	5.32	282.80	8.57	0.0	4.28	6.25	4.25	2.00	0.21	6.46	10.74	60.13
Oregano	5.20	57.40	8.57	0.0	4.28	5.50	3.25	2.25	0.21	5.71	9.99	57.14
					2nd sampling – November							
Rosemary	5.17	60.10	6.62	0.0	3.97	4.50	2.75	1.75	0.15	4.65	8.62	53.96
Boldo	5.30	23.70	7.21	0.0	3.97	6.88	4.00	2.88	0.10	6.98	10.95	63.74
Nasturtiums	5.69	248.20	10.91	0.0	3.42	6.25	3.50	2.75	0.26	6.51	9.93	65.55
Mint	5.46	207.20	7.60	0.0	3.97	5.25	4.25	1.00	0.15	5.4	9.37	57.65
Oregano	5.08	69.90	7.60	0.0	4.28	4.88	2.63	2.25	0.15	5.03	9.31	54.02

P – Phosphorus; C – Carbon; Al^{3+}– Aluminium; H$^+$+Al^{3+} – Potential Acidity; Ca^{2+} – Calcium; Mg^{2+} – Magnesium; K$^+$ – Potassium; SB – Sum of Bases; CEC –Cation Exchange Capacity; V – Bases saturation

estimated according slide method (Giovannetti and Mosse, 1980) for each replication of each treatment.

Statistical analysis

Data was subjected to one-way ANOVA using general linear model with mixed-effects and balanced design, considering each plant species as one treatment, and compared with the Duncan's test ($p \leq 0.05$), by using SPSS version 16.0 for Windows (SPSS Inc., Chicago, IL, USA). To comply with ANOVA assumptions, the data was previously checked with the Levene´s test. In the two periods June and November, t-test was done with independent bilateral averages.

RESULTS AND DISCUSSION

Soil chemical analysis from the first sampling period (June, 2011) presented the highest P level in soil with mint (282.8 mg dm^{-3}), followed by soil with nasturtiums (173.6 mg dm^{-3}) and boldo (21 mg dm^{-3}) (Table 1).

Both the P fixation and the natural P levels of soils vary according to plant variety or cultivar. Changes also may occur in the AMF activity depending on the conditions of soil fertility (Siqueira et al., 2010). However, the amount of organic matter from composting added to the beds is not exactly known for evaluating fertility within this study.

In general, stable levels of P were observed in the analysis of soils collected from the second sampling (November, 2011), with an increase in P levels observed only in the soil cultivated with nasturtiums (Table 1).

The availability of nutrients is affected by the soil pH. In this study, results of soil pH (Table 1) are according to the literature, which indicates values of 6 to 7 as ideal to grow most of plant species (Corrêa Junior and Scheffer, 2009); remembering that there are plants that can tolerate lower pH. On the other hand, the low pH, as of the soil with rosemary (4.63 – Table 1), may affect the mycorrhizal association with plants. This is due to the variation of the solubility of elements such as Al, Fe, Mn and Cu, which at toxic levels may reduce the germination of spores and germ tubes, reducing the sporulation of AMF (Lambais and Cardoso, 1989). Studies show that soil pH regulates the mycorrhizal condition and controls the distribution of AMF species (Moreira et al., 2003).

The C levels in the soil differed between sampling periods (Table 1). For example, the bed with nasturtiums had 8.77 g dm^{-3} C in June and started having 10.91 g dm^{-3} C in November. According to Kaschuk et al. (2010; 2011), the C level in soils can also be used as indicator of their fertility and quality as it supplies biological activity, maintains environmental quality and promotes the health of plants and animals.

The roots of all plants analyzed were colonized by AMF. The AMF root colonization in June was significantly lower than in November (Table 2), with averages of 25.72 and 35.36%, respectively. Boldo, mint, oregano and nasturtiums had no significant differences in AMF root colonization (Table 2). Rosemary had a significant increase in AMF root colonization in November compared to June. It indicates that AMF root colonization was maximum in winter season and lowest in early summer season. Kumar et al. (2010) observed similar results for *Spilantes acmella, Withania somnifera, Salvia officinalis, Mentha spicata* when AMF root colonization was significantly higher in November than in June.

Among all plants analyzed within this study, mint

Tabela 2. Means of AMF soil spore density (nº g⁻¹ of dry soil) and AMF root colonization (%) (± standard deviation, n = 9) in June and November, 2011.

Plant	AMF root colonization		AMF spore density	
	June	November	June	November
Rosemary	17.40±4.83 Bb	48.04±8.26 Aa	21.15±4.41 BCa	2.84±0.70 Ab
Boldo	30.74±4.92 Aa	45.67±8.82 Aa	16.80±3.11 Ca	3.01±1.09 Ab
Chamomile	20.80±4.82 B	ND	37.30±8.60 A	ND
Nasturtiums	21.88±4.31 Ba	18.35±4.45 Ba	16.52±2.37 Ca	3.05±0.47 Ab
Mint	42.37±7.47 Aa	45.35±7.01 Aa	19.96±2.65 BCa	3.84±0.77 Ab
Oregano	19.28±2.65 Ba	16.55±3.29 Ba	32.11±4.75 ABa	3.18±0.46 Ab
p value	> 0.001	> 0.002	> 0.001	0.902

ND = Not determined. Means followed by the same capital letter in the column are not significantly different by the Duncan test ($p \leq 0.05$) and by the same minor letter in the line did not differ by the t-test ($p \leq 0.05$).

showed the highest percentage of colonization, with 42.37% (June) and 45.35% (November), followed by boldo with 30.74% (June) and 45.67% (November). This result indicates that both the mint and the boldo are the species that are more depending on the AMF associations.

As AMF started establishing on the thinnest roots, temperature and humidity were likely to influence AMF root colonization as well as the nutrients intake by plants in this study (Smith and Read, 2008). According to Carrenho et al. (2007), the process of root colonization can be influenced by changes in seasonal periods. It was also observed in other studies in which the best root colonization was in the rainy season (Kumar et al., 2010). However, Radhika and Rodrigues (2010) states that mycorrhizal root colonization is present in all seasons, suggesting a plant dependence on AMF throughout the year.

A root can be colonized by more than one species of mycorrhizal fungus (Dood et al., 2000) and a fungus species can grow at different rates when associated to different species of plants (Smith and Read, 2008). Still, there may be several colonizing rates across genotypes of the same plant species (Grahman and Eissenstat, 1994).

Gupta et al. (2002) observed an increased percentage of root colonization in plants inoculated with AMF compared with respective controls of non-inoculated plants. The authors also noted a possible difference in response to mycorrhizal colonization across varieties of the same plant, as observed in the three cultivars of mint inoculated with G. fasciculatum in their study.

The major density of AMF spores (number of spores g⁻¹ of dry soil) was observed in the period of June, with emphasis on soils with oregano (32.11) and chamomile (37.29), which showed densities significantly higher than the soils of other plants studied (Table 2). However, AMF spores were present in November with minor density, but with no significant difference when compared to June (Table 2). It can be explained here by a wetter weather

(data not shown) affecting directly fungus sporulation. Kumar et al. (2010) observed similar results for soils with S. acmella and Mellisa officinalis, in which AMF spores density was significantly lower in November than in June. Radhika and Rodrigues (2010) reported a density of AMF spores varying in function of seasons with a higher number of spores in August than in January. On the other hand, Coppetta et al. (2006) developed a study suggesting that the density of spores in soils is most dependent on the extent of root colonization between the AMF and the plant.

Negative and significant correlation (p = 0.035) was observed between AMF root colonization and spore density. Similar results were found by Radhika and Rodrigues (2010) when they studied thirty-six medicinal plant species.

Studies demonstrate that the seasonality of mycorrhizal colonization is usually a function of environmental conditions as temperature, humidity, phenology and physiological condition of the plant (Mohammad et al., 1998; Brundrett, 2002) and in this way, significant differences in both the root colonization and the density of AMF spores were observed in this study (Table 2).

Phyla Glomeromycota is identified mainly from the analysis of the formation, structure and germination of spores. In this study, Glomus was the most predominant genus within the diversity of AMF in the two sampling periods (June and November - Tables 3 and 4, respectively). Acaulospora, Gigaspora, Scutellospora and Pacispora were others genera found in this study, but with lower frequency (Table 5). Similar results were found by Radhika and Rodrigues (2010) in samples of soils cultivated with thirty-six medicinal plant species. The great diversity of AMF found in this study indicates that the plants studied form a symbiotic-mandatory association with the AMF, regardless of the period analyzed.

The indexes of Shannon diversity and Simpson dominance (Table 5) revealed great spore diversity among species of plants regardless of the sampling

Table 3. Taxonomy and number of spores per species of mycorrhizal fungi (Phylum *Glomeromycota*) in the 1st sampling period (June, 2011) determined according to Siqueira et al. (2010).

Plant	Order	Family	Genus	Species	N°
Rosemary	Glomerales	Glomeraceae	*Glomus*	*Glomus aff. Lamellosum*	23
				Glomus mosseae	11
				Glomus microaggregatum	30
				Glomus aff. Tortuosum	3
				Gigaspora margarita	2
				Glomus claroideum	3
				Glomus etunicatum	7
	Diversispolares	Acaulosporaceae	*Acaulospora*	*Acaulospora sp.*	1
				Acaulospora koskei	2
Boldo	Glomerales	Glomeraceae	*Glomus*	*Glomus macrocarpum*	2
				Glomus aff. Lamellosum	27
				Glomus claroideum	1
				Glomus microaggregatum	5
				Glomus mosseae	1
				Glomus aff. Deserticola	9
				Glomus constrictum	1
				Glomus etunicatum	10
				Acaulospora delicata	5
	Diversispolares	Acaulosporaceae	*Acaulospora*	*Acaulospora koskei*	27
				Acaulospora morrowiae	13
				Acaulospora (Entrophospora) colombiana	3
				Acaulospora sp. (scro-reticulata)	1
		Gigasporaceae	*Gigaspora*	*Gigaspora margarita*	1
		Scutellosporaceae	*Scutellospora*	*Scutellospora aff. Verrucosa*	1
Chamomile	Glomerales	Glomeraceae	*Glomus*	*Glomus aff. Lamellosum*	7
				Glomus mosseae	7
				Glomus claroideum	10
	Diversispolares	Gigasporaceae	*Gigaspora*	*Gigaspora margarita*	1
		Scutellosporaceae	*Scutellospora*	*Scutellospora calospora*	1
Nasturtiums	Glomerales	Glomeraceae	*Glomus*	*Glomus mosseae*	4
				Glomus macrocarpum	13
				Glomus aff. Lamellosum	4
				Glomus tortuosum	31
				Glomus aff. Deserticola	5
				Glomus claroideum	2
				Glomus geosporum	1
				Glomus invermaium	5
				Glomus etunicatum	9
	Diversispolares	Acaulosporaceae	*Acaulospora*	*Acaulospora sp. (scro-reticulata*	2
				Acaulospora koskei	2
				Entrophospora infrequens	4
		Gigasporaceae	*Gigaspora*	*Gigaspora margarita*	1
Mint	Glomerales	Glomeraceae	*Glomus*	*Glomus claroideum*	3
				Glomus aff. lamellosum	3
				Glomus mosseae	1
				Glomus macrocarpum	1
				Glomus constrictum	1
				Glomus etunicatum	3
	Diversispolares	Acaulosporaceae	*Acaulospora*	*Acaulospora scrobiculata*	4
				Acaulospora delicata	1
				Acaulospora koskei	1
				Acaulospora delicata	1
		Gigasporaceae	*Gigaspora*	*Gigaspora margarita*	2
				Gigaspora decipiens	1
Oregano	Glomerales	Glomeraceae	*Glomus*	*Glomus macrocarpum*	4
				Glomus aff. lamellosum	1
				Glomus claroideum	1
	Diversispolares	Acaulosporaceae	*Acaulospora*	*Acaulospora koskei*	2
		Gigasporaceae	*Gigaspora*	*Gigaspora decipiens*	1

Table 4. Taxonomy and number of spores per species of mycorrhizal fungi (Phylum *Glomeromycota*) in the 2nd sampling period (November, 2011) determined according to Siqueira et al. (2010).

Plant	Order	Family	Genus	Species	N°
Rosemary	Glomerales	Glomeraceae	*Glomus*	*Glomus tortuosum*	5
				Glomus aff. lamellosum	5
				Glomus mosseae	2
				Glomus aff. luteum	3
				Glomus claroideum	1
	Diversispolares	Acaulosporaceae	*Acaulospora*	*Acaulospora (Entrophospora) colombiana*	2
		Gigasporaceae	*Gigaspora*	*Gigaspora decipiens*	4
Boldo	Glomerales	Glomeraceae	*Glomus*	*Glomus margarita*	1
				Glomus aff. lamellosum	5
				Glomus aff. luteum	1
				Glomus claroideum	3
	Diversispolares	Acaulosporaceae	*Acaulospora*	*Acaulospora koskei*	3
				Acaulospora (Entrophospora) colombiana	4
		Scutellosporaceae	*Scutellospora*	*Scutellospora rubra*	1
				Scutellospora heterogama	1
Nasturtiums	Glomerales	Glomeraceae	*Glomus*	*Glomus tortuosum*	6
				Glomus aff. lamellosum	7
				Glomus geosporum	1
				Glomus aff. luteum	1
				Glomus constrictum	1
				Glomus mosseae	2
				Glomus microaggregatum	30
				Gigaspora decipiens	1
Oregano	Glomerales	Glomeraceae	*Glomus*	*Glomus aff. lamellosum*	3
				Glomus mosseae	4
	Diversispolares	Acaulosporaceae	*Acaulospora*	*Acaulospora sp. (com espinhos)*	1
		Gigasporaceae	*Gigaspora*	*Gigaspora gigantea*	1
Mint	Glomerales	Glomeraceae	*Glomus*	*Glomus aff. lamellosum*	3
				Glomus geosporum	2
				Glomus macrocarpum	1
				Glomus etunicatum	1
				Glomus margarita	1
				Glomus aff. luteum	1
	Diversispolares	Gigasporaceae	*Gigaspora*	*Gigaspora ramisporophora*	2
		Acaulosporaceae	*Acaulospora*	*Acaulospora scrobiculata*	4
		Pacisporaceae	*Pacispora*	*Pacispora robiginia*	1

period. Shannon index for rosemary increased from 0.066 in June to 0.329 in November, indicating greater diversity of AMF in the period of June compared to November. However, the opposite was observed for nasturtiums (Table 5).

The plants of mint, oregano and bold had similar AMF indexes of Shannon diversity and Simpson dominance in the two sampling periods studied, thus these plants were efficient in their symbiosis in both seasons.

Among all AMF genera found, *Glomus sp.* was the most frequent in all species and in both periods analyzed in this study (Table 5). The dominance of *Glomus* sp. in soil cultivated with rosemary and sampled in June was 0.931 whereas in soil under nasturtiums and sampled in November was 0.96 (Table 5). Spores of *Glomus sp.* were observed in soils cultivated with all species studied. Their frequency was higher than 52%, reaching 98% in soils cultivated with nasturtiums.

Conclusions

All plants showed levels of AMF root colonization, and spore density of AMF decreased in June when compared

Table 5. AMF genera frequency, indexes of Shannon diversity and Simpson dominance (June and November, 2011).

Plantas	Glomus sp.	Acaulospora sp.	Gigaspora sp.	Scutellospora sp.	Pacispora sp.	Shannon	Simpson
	------------------ Relative frequency (%) ----------------------						
June							
Rosemary	96.47	3.53	0	0	0	0.066	0.931
Boldo	51.88	46.22	0.95	0.95	0	0.341	0.482
Chamomile	92.30	0	3.85	3.85	0	0.141	0.854
Nasturtiums	89.15	9.65	1.20	0	0	0.165	0.804
Mint	54.55	31.82	13.63	0	0	0.419	0.417
Oregano	66.67	22.22	11.11	0	0	0.368	0.506
November							
Rosemary	72.72	9.10	18.18	0	0	0.329	0.570
Boldo	52.63	36.84	0	10.53	0	0.409	0.423
Nasturtiums	97.96	0	2.04	0	0	0.043	0.960
Mint	56.25	25	12.50	0	6.25	0.403	0.394
Oregano	77.78	11.11	11.11	0	0	0.296	0.629

to November. The taxonomic diversity of AMF varied among species of medicinal and seasoning plants studied. Spores of *Glomus* sp. were observed in all species studied. Their frequency was higher than 52%, reaching 98% in soils cultivated with nasturtiums.

Conflict of Interests

The author(s) have not declared any conflict of interests.

ACKNOWLEDGEMENTS

The authors thank the Universidade Paranaense - UNIPAR by supporting research and Profa Dra Rosilaine Carrenho from Universidade Estadual de Maringá – UEM that identified AMF species. Regiane Cristina Urcoviche thanks for the scholarship PROSUP/CAPES.

REFERENCES

Brundrett MC (2002). Co-evolution of roots and mycorrhizas of land plants. New Phytol. 154:275–304. http://dx.doi.org/10.1046/j.1469-8137.2002.00397.x

Carrenho R, Marins JF, Lippert MAM, Stivanim SC (2007). Micorrizas arbuscular em plantas medicinais cultivadas na Universidade Estadual de Maringá. Rev. Bras. Bioci. 5:561–563.

Dood JC, Boddington CL, Rodriguez A, Gonzales-Chavez C, Mansur I (2000). Mycelium of arbuscular mycorrhizal fungi (AMF) from different genera: form, function and detection. Plant Soil 226:131–151. http://dx.doi.org/10.1023/A:1026574828169

Faria AYK, Matsuoka M, Oliveira EAS, Loureiro MF (2000). Fungos micorrízicos arbusculares em sucupirapreta: efeito de substratos sobre a colonização. Hortic. Bras. 18:913–914.

Gerdemann JW, Nicolson TH (1963). Spores of mycorrhizal endogone species extracted from soil by wet sieving and decanting. Trans. Br. Mycol. Soc. 46:235–246. http://dx.doi.org/10.1016/S0007-1536(63)80079-0

Giovannetti M, Mosse B (1980). An evaluation of techniques for measuring VA mycorrhizal infection in roots. New Phytol. 84:489–500. http://dx.doi.org/10.1111/j.1469-8137.1980.tb04556.x

Gupta ML, Prasad A, Ram M, Kumar S (2002). Effect of the vesicular-arbuscular mycorrhizal (VAM) fungus Glomus fasciculatum on the essential oil yield related characters and nutrient acquisition in the crops of different cultivars of menthol mint (Mentha arvensis) under field conditions. Bioresour. Technol. 81:77–79. http://dx.doi.org/10.1016/S0960-8524(01)00109-2

Grahman JH, Eissenstat DM (1994). Hot genotype and the formation of VA mycorrhizae. Plant Soil 159:179–185.

Gianinazzi S, Gollotte A, Binet M-N, van Tuinen D, Redecker D, Wipf D (2010). Agroecology: the key role of arbuscular mycorrhizas in ecosystem services. Mycorrhiza 20:519–530. http://dx.doi.org/10.1007/s00572-010-0333-3 PMid:20697748

Corrêa-Junior C, Scheffer MC (2009). Boas práticas agrícolas (BPA) de plantas medicinais, aromáticas e condimentares. Instituto Paranaense de Assistência Técnica e Extensão Rural – EMATER, Curitiba, Paraná.

Kumar A, Mangla C, Aggarwal A. Parkash V (2010). Arbuscular mycorrhizal fungal dynamics in the rhizospheric soil of five medicinal plant species. Middle East J. Sci. Res. 6:281-288.

Lambais MR, Cardoso EJBN (1989). Germinação de esporos e o crescimento do tubo germinativo de fungos micorrízicos vesículo-arbusculares em diferentes concentrações de alumínio. R. Bras. Ci. Solo 13:151-154.

Kaschuk G, Alberton O, Hungria M (2010). Three decades of soil microbial biomass studies in Brazilian ecosystems: lessons learned about soil quality and indications for improving sustainability. Soil Biol. Biochem. 42:1–13. http://dx.doi.org/10.1016/j.soilbio.2009.08.020

Kaschuk G, Alberton O, Hungria M (2011). Quantifying effects of different agricultural land uses on soil microbial biomass and activity in Brazilian biomes: inferences to improve soil quality. Plant Soil 338:467–481. http://dx.doi.org/10.1007/s11104-010-0559-z

Mohammad M, Pan WL, Kenndy AC (1998). Seasonal mycorrhizal colonization of winter wheat and its e effect on wheat growth under dryland field conditions. Mycorrhiza 8:139–144.

http://dx.doi.org/10.1007/s005720050226

Moreira SM, Trufem SFB, Gomes CSM, Cardoso EJBN (2003). Arbuscular mycorrhizal fungi associated with Araucaria angustifolia (Bert.) O. Ktze. Mycorrhiza 13:211–215.

Morton JB, Bentivenga SP, Wheeler WW (1993). Germplasm in the International Collection of Arbuscular and Vesicular-Arbuscular Mycorrhizal Fungi (INVAM) and procedures for culture development, documentation and storage. Mycotaxon 48:491–528.

Phillips JM, Hayman DS (1970). Improved procedures for clearing roots and staining parasitic and vesicular-arbuscular mycorrhizal fungi for rapid assessment of infection. Trans. Br. Mycol. Soc. 55:157–160. http://dx.doi.org/10.1016/S0007-1536(70)80110-3

Radhika KP, Rodrigues BF (2010). Arbuscular mycorrhizal fungal diversity in some commonly occurring medicinal plants of Western Ghats, Goa region. J. For. Res. 21:45–52. http://dx.doi.org/10.1007/s11676-010-0007-1

Russomano OMR, Kruppa PC, Minhoni MTA (2008). Influência de fungos microrrízicos arbusculares no desenvolvimento de plantas de alecrim e manjericão. Arq. Inst. Biol. 75:37–43.

Schenck NC, Pérez Y (1988). Manual for the identification of VA mycorrhizal fungi. 2nd edition. IFAS. Gainesville, University of Florida.

Siqueira JO, Souza FA, Cardoso EJBN, Tsai SM (2010). Micorrizas: 30 anos de pesquisa no Brasil. UFLA, Lavras, Minas Gerais.

Souza FA (2000). Banco ativo de Glomales da Embrapa Agrobiologia: Catalogação e introdução de novos isolados desde 1985. Seropédica: Embrapa Agrobiologia, Documentos P. 123.

Souza GIA, Caproni AL, Granha OJRD, Souchie EL, Berbara RLL (2010). Arbuscular mycorrhizal fungi in agricultural and Forest systems. Global Sci. Technol. 3:1–9.

Smith SE, Read DJ (2008). Mycorrhizal symbiosis, 3rd edition. Academic Press, New York.

14

The critical period of weed interference in upland rice in northern Guinea savanna: Field measurement and model prediction

Amadou Touré[1], Jean Mianikpo Sogbedji[2] and Yawovi Mawuena Dieudonné Gumedzoé[2]

[1]Africa Rice Center (AfricaRice), 01 BP 2031 Cotonou, Benin.
[2]University of Lomé, BP 1515, Lomé, Togo.

Luxuriant weed growth destroying rice crops is a major problem in tropical Africa. The objective of this study was to determine the critical period of weed infestation in upland rice varieties in order to enable the development of more precise weed management recommendations for farmers. The effects of 10 differing periods of weed management on upland rice yield were studied in experiments with five rice varieties (three interspecific NERICA: NERICA1, NERICA2, and NERICA4) and the parents (*Oryza sativa* WAB 56-104 and *Oryza glaberrima* CG 14) during the 2004 and 2005 rainy seasons at Farako (Mali). INTERCOM model was used to explore the relationship between duration and timing of weed competition and rice crop yield loss, and the applicability of the model in rice cropping based weed management. The critical period of weed infestation determined from the field experiment was similar for the three New Rice for Africa (NERICA) varieties and the *O. sativa* parent (WAB 56-104), and was between 14 and 42 days after seeding (DAS). For the *O. glaberrima* parent (CG 14), the critical period was between 28 and 42 DAS. Weed competition either before or after these critical periods had negligible effects on crop yield. During the 2 years, yields of NERICA varieties and WAB 56-104 averaged 2700 and 400 kg ha^{-1} under weed-free plots and no weed control plots, respectively, indicating a yield loss of 85%. For GG 14, yields averaged 900 and 300 kg ha^{-1} under weed-free plots and no weed control plots, respectively, resulting in a 66% yield loss. The occurrence and composition of weeds during the two years were similar with a mean of 40% broadleaves, 35% grasses and 25% sedges. The most important weeds were *Imperata cylindrica*, *Cyperus sphacelatus* and *Digitaria longiflora*. During both calibration and testing efforts, the INTERCOM model satisfactorily simulated rice NERICA1 LAI, shoot dry weight and yields (r^2 ranging from 0.71 to 0.87). There appears to be room for improvement in the model with regard to the assumption that nutrients are not limiting to crop growth, but the use of the model for simulating the interactions between rice crop yield losses, weed density, and duration of weed competition appears promising. Results of this study can serve as a guide for optimum timing of weed control to maximize upland rice yield in West Africa.

Key words: Critical period, northern Guinea savanna, upland rice, New Rice for Africa (NERICA), INTERCOM.

INTRODUCTION

Weed control is one of the main upland rice yield limiting factors in West Africa. Therefore, weeds should be controlled and eliminated before competing with rice plants for light, water and nutrients. Intensive manual weeding is often practiced by farmers because the use of herbicides is associated with high costs and the failure of the distribution market, and the low rate of literacy among farmers in Sub-Saharan Africa (SSA) limits still more the use of herbicides (Rodenburg and Johnson, 2009). But manual weeding faces the following constraints: (i) manual weeding becomes acute with increased weed infestation if the field is cultivated for many years without a period of sufficient fallow, (ii) the weeding becomes difficult when the labor is insufficient mainly at the beginning of the rainy season when land preparation, planting and weeding all compete for the farmer's limited labor and (iii) manual weeding is done very often too late, when the weeds have outcompeted the crops leading to crop loss. In West Africa, is there a crucial period during which weed infestation is particularly harmful to upland rice? Thus for a sound integrated weed management in rice cropping, it is necessary to determine when rice plants will be the most and least harmed by weeds. The concept of critical periods of weed competition, during which weeds have the greatest effect on crop growth, was verified by Nieto et al. (1968). It is a specific minimum period of time during which the crop must be free of weeds in order to prevent loss in yield and represents the overlap of two separate components (Weaver and Tan, 1983). The first component is the length of time weeds can remain in a crop before interference begins. The second component is the length of time that weed emergence must be prevented so that subsequent weed growth does not reduce crop yield. The critical period is the prime period most suitable for conducting weeding operations taking into account the following factors: the environment (climate and soil), the period of weed infestation in the field, the weed species, the cultural practices including crop rotation, fertilization, density and methods of seeding (broadcast, hill seeding or transplanting), and the relative growth rates of the crop and its associated weeds. For example, according to Le Bourgeois and Marnotte (2002), the critical period is generally located between 15 and 60 days after seeding (DAS) for short-cycle annual crops (cotton, corn, sorghum, rice, etc.) and between 30 and 90 DAS for long-cycle crops (yams, cassava, sugarcane, etc.). In rainfed rice in southern Togo, weed competition is more harmful between 21 and 30 DAS (Boyoda, 1991). In the areas of northern Guinea savanna characterizing southern Mali, farmers generally weed their fields one or twice (extension services often recommend two weedings), but the weeding operations are often late. Inevitably in rainfed rice cultivation in these areas, the concept of critical period or critical threshold leads to severe competition between weeds and rice plants.

The use of the term critical threshold in integrated weed management to predict when weeds must be controlled to prevent yield loss was proposed by Dawson (1986). The economic threshold could be also calculated to indicate the length of time during which a crop could tolerate the competition of weeds before yield losses exceeded the costs of control (Weaver et al., 1992). This would lead to the early-season threshold that signals the beginning of the critical period, and the late-season threshold the end. Van Heemst (1985) has shown that the end of the critical period is related to the competitive ability of the crop. Thus, a crop with a high competitive ability has a critical period that ends early. Critical period of weed control has commonly been reported as day after seeding (DAS), but due to differences in planting dates and environment, this may generate different results among sites, seasons, and varieties (Anwar et al., 2012). Studies have also reported critical period as growing degree days because it is a biologically meaningful measure of time required for plant growth and development, and therefore, it would be applicable for comparing critical period across different agro-climatic conditions (Evans et al., 2003; Anwar et al., 2012). The critical period is usually determined through empirical mean comparison and multiple regression statistical tests. Cousens (1988, 1991) suggested using fitted responses curves to determine these critical thresholds. This allows a more accurate estimation of yield losses but still suffers from problems associated with empirical relationships. Because these parameters of response curves can vary depending on factors such as the crop and the associated weed species, the weed density, and especially the environmental conditions (Weaver et al., 1992). A dynamic simulation tool such as INTERCOM can be used to examine in detail the effect of these factors on the length of the critical period for upland rice crops. The model has been used for crop-weed competitions including crops such as corn (Lindquist and Mortensen, 1997; Cavero et al., 2000), leek and celery (Baumann, 2001), and rice (Kropff and van Laar, 1993; Akanvou, 2001). This model is basically a growth model of two or more species that are linked through additional routines that govern distribution of resources such as light and water over the competing species (Bouman et al., 1996; Akanvou, 2001). Precise technical guidelines to identify the critical period of weed competition of upland rice and mainly for one group of interspecific rice varieties known as NERICA developed by the Africa Rice Center (AfricaRice) and partners, are still scant (Wopereis et al., 2008).

The objectives of this study were to identify the weed flora, to assess the in-field critical period for weed competition with upland rice, and to evaluate the performance of the INTERCOM, a dynamic and process-based simulation model that can be used to realistically address the effects of the duration of weed competition on upland rice crop yields.

MATERIALS AND METHODS

Experimental site

The experiment was conducted from 2004 to 2005 under rain fed conditions in northern Guinea savanna agroecology in southern Mali at the agricultural research station of the IER (*Institut*

Table 1. Air temperature and rainfall data during cropping seasons (June-October) 2004-2005.

Month	2004			2005		
	T_{min} (°C)[a]	T_{max} (°C)	Rainfall (mm)	T_{min} (°C)	T_{max} (°C)	Rainfall (mm)
June	23.1	33.6	78.6	22.2	32.2	292.1
July	21.9	30.2	325.5	22.0	30.4	197.7
August	21.6	30.3	285.3	21.5	29.6	275.8
September	21.5	31.3	141.5	21.7	31.2	174.0
October	22.5	34.2	52.7	22.0	33.6	52.3

[a]T_{min} (°C), Minimum air temperature in degree Celsius; T_{max} (°C), Maximum air temperature in degree Celsius

d'Economie Rurale) of Farako (Sikasso) (11° 12' 48.9"N, 5° 27' 16.7"W, 400 m above sea level). The climate falls within the open woodland savanna agroecological zone with a monomodal rainfall pattern averaging annually 1130 mm. The rainfall pattern is characterized by one single pick, increasing in amount and frequency, reaching a maximum in July/August/September. The average daily temperature is 28°C with a range between 22 and 34°C. The air temperature and rainfall data during the cropping season (June to October) were collected during the experiment and are presented in Table 1.

According to the analytical procedures of the International Institute of Tropical Agriculture (1989), the average chemical analysis of topsoil 0 to 20 cm showed soil pH in water 1:1 = 5.8, organic carbon content of 0.46%, organic matter 0.79%, nitrogen 0.30‰. The textural class of the soil is sandy loam with sand, silt and clay content of 84, 11 and 5%, respectively. At field capacity on wet basis, soil water retention was 19% and wilting point was 9%, and the soil is classified as acidic Acrisol (FAO, 1998). The study area has been previously sown to sorghum (*Sorghum bicolor* L.) for 2 years, and left to a short fallow of 1 year of *Imperata cylindrica* (L.) Raeuschel and *Digitaria longiflora* (Retz.) before the experiment. These two weed species accounted for more than 80% of the weed population found on the site at the onset of the experiment.

Field experiment

Experimental design

A split-plot design was used, with ten weeding regimes on the plot level and five upland rice cultivars on the sub-plot level, in four replicates. Ten weeding regimes treatments (WD14–WFharv) were devised to examine the effects of differing periods of weed control and interference, and were similar to those of Nieto et al. (1968) and Johnson et al. (2004). The treatments were:

(1) WD14: Weedy until 14 DAS,
(2) WD28: Weedy until 28 DAS,
(3) WD42: Weedy until 42 DAS,
(4) WD56: Weedy until 56 DAS,
(5) WDharv: Weedy from seeding to maturity,
6) WF14: Weed-free until 14 DAS,
(7) WF28: Weed-free until 28 DAS,
(8) WF42: Weed-free until 42 DAS,
(9) WF56: Weed-free until 56 DAS,
(10) WFharv: Weed-free from seeding to maturity.

Weed growth was controlled in the required periods for each of the above treatments, and hand weeding was weekly undertaken as needed. The term "weed-free" in the treatments therefore indicates the period during which weeds were removed at weekly intervals.

The five varieties for the sub-plots were three NERICAS: NERICA1, NERICA2, and NERICA4, and the two parents: *O. sativa* L. WAB 56-104 and *O. glaberrima* Steud, GC 14. Sub-plot size was 4 by 3 m with rice hill distances of 0.2 by 0.25 m.

Soil and crop management

The land was disc-ploughed and harrowed once before the plots were laid out. Seeding dates were 26 June in 2004 and 23 June in 2005. Rice was dibble-seeded at a rate of five to six seeds per hill, and then seedlings were thinned to four plants per hill at 14 to18 DAS in order to have a population density of 800000 plants ha⁻¹. Fertilizer at a rate of 10N-18P₂O₅-18 K₂O kg ha⁻¹ was uniformly broadcast on the tilled fields and incorporated into the soil prior to rice seeding. In addition, 33 kg N ha⁻¹ (as urea) of fertilizer was broadcast at 21 and 42 DAS. These fertilizer rates are recommended by Africa Rice (Sahrawat et al., 2001). The plants were protected against nematodes and termites by applications of carbofuran at the rate of 2.5 kg ai ha⁻¹ at seeding.

Data collection and statistical analysis

Weeds and rice were sampled from two 0.5 m² quadrats taken in each plot at 14, 28, 42, 56, and at harvest, with plants being cut at ground level. The weeds were separated from the rice into different species and all biomass was weighed. Two 500 g sub-samples of each species were oven-dried and weighed to allow correction of the fresh weight data. The numbers of days to maturity for rice varieties were recorded. At harvest, yield components were observed by taking samples from a 0.25 m² quadrat per plot. Yield components recorded were: the number of panicles, tiller number per square meter, percent of full grains, number of spikelets per panicle, and 1000-grain weight. Grain yield was recorded from 6 m² quadrats and corrected for 14% moisture content. The relative frequency of major weeds was determined as the percentage of plots in which the species were present.

Statistical analyses were performed using the mixed model with maximum likelihood (REML) for the estimation of the variance over the years (SAS Institute, 2004). Fixed effects were the year, weeding regimes and varieties, while replicates and their interactions with weeding regimes accounted for random effects. Mean separation was performed using the SAS LSMEANS test (pair-wise comparisons) at P ≤ 0.05.

Simulating the effects of the duration of weed competition on crop yields

The model INTERCOM (Kropff and van Laar, 1993) was used to assess the relationship between duration and timing of weed competition and rice crop yield losses. Indeed, the influence of length of time that weeds were present in the crop and the associated crop yield was assessed in the context of the critical period of weed competition. The critical periods were determined by changing the dates of weed emergence (DOYEM parameter) and the dates of weed removal (KILDAY parameter) in the model in order to determine associated yield by simulating crop growth. Overall, the INTERCOM model was evaluated by first calibrating the model based on the sensitivity analysis using measured data of LAI and shoot dry weight and its performance was tested against measured yield data.

Model overview

The structure of the simulation model INTERCOM was described by Kropff and van Laar (1993). The model simulates growth of the crop and weeds from emergence through crop maturity as a function of solar radiation, temperature, water availability, and species characteristics on a daily time step basis. Interactions are simulated by distributing the growth-limiting resources of light and water over the competing species and assuming that neighboring species (rice crop and weeds) mutually reduce their growth only by modifying the environment and changing water and light availability. The amount of resources acquired by a species determines its growth rate. Nitrogen and other nutrients are assumed to be available in sufficient amount and the effects of insects and diseases on crop and weed growth are neglected (Kropff et al., 1992). Input data for the model are daily weather information (maximum and minimum temperatures, total global radiation, and rainfall), weed densities and dates of crop and weed emergence.

In a potential production system, where light, temperature and physiological and morphological characteristics determine the growth of a plant community, plants only compete for the resource light. In agricultural systems where other factors like nitrogen or water limit crop production weeds compete with the crop for light as well as for the other resources (Kropff and van Laar, 1993). Key complex interrelationship processes are described in the INTERCOM model (Kropff and van Laar, 1993) as follows:

Light interception by the canopy: The photosynthetically active radiation (PAR) supplies the plants with energy for CO_2 assimilation. The assimilated CO_2 is converted into carbohydrates (CH_2O). The overall simplified chemical reaction of the process is:

$$CO_2 + H_2O \xrightarrow{\text{Light}} CH_2O + O_2 \qquad (1)$$

The incoming radiation is partly reflected by the canopy. The reflection coefficient (ρ) of a green leaf canopy with a random spherical leaf angle distribution, which indicates the fraction of the downward radiation flux that is reflected by the whole canopy can be approximated by the following equation (Goudriaan, 1986):

$$\rho = [(1 - \sqrt{(1 - \sigma)})/(1 + \sqrt{(1 - \sigma)})] \bullet [2/(1 + 1.6 \sin \beta)] \qquad (2)$$

in which σ represents the scattering coefficient of single leaves for visible radiation and the $\sin\beta$ is the sinus of the solar elevation (β).

Then the radiation fluxes decrease within the canopy with the cumulative leaf area index (LAI), counted from the top downwards (Kropff and van Laar, 1993):

$$I_L = (1 - \rho) I_0 \exp(-k \times LAI) \qquad (3)$$

Where, I_L is the net PAR flux at depth L in the canopy (MJ m^{-2}ground s^{-1}), I_0 is the flux of visible radiation at the top of the canopy (MJ m^{-2}ground s^{-1}), LAI is the cumulative leaf area index from top

downwards of the canopy (m^2 leaf m^{-2} ground), ρ is the reflection coefficient of the canopy (-), and k is the extinction coefficient for PAR (-).

The light absorbed by species (I_{abs}, MJ m^{-2} s^{-1}) is obtained by taking the first derivative of Equation 3 with respect to LAI:

$$I_{abs} = -dI_L/dL = k (1 - \rho) I_0 \exp(-k \times LAI) \qquad (4)$$

Biomass production: Gross canopy photosynthesis of the species is calculated based on the photosynthesis light-response of individual leaves which is characterized by the initial light use efficiency of leaf CO_2 assimilation (ε, kg CO_2 ha^{-1} leaf h^{-1}/J m^{-2} leaf s^{-1}) and the light saturated of CO_2 assimilation (Amax, kg CO_2 ha^{-1} h^{-1}, Spitters et al., 1989; Akanvou, 2001).

Leaf area: The expansion of leaf area determines the amount of intercepted light by the canopy, and is simulated as an exponential function of accumulated degree-days (Kropff and van Laar, 1993):

$$LAI (tsum) = LA0 \times N \times \exp(RGRL \times tsum) \qquad (5)$$

Where, LA0 is the leaf area index at seedling emergence (m^2 leaf plant^{-1}); tsum, the accumulated degree-days since emergence (°Cd); RGRL, the relative leaf area growth rate (°Cd)$^{-1}$; and N the number of plants (m^{-2}).

Model calibration

Sogbedji et al. (2001) defined the calibration as being the process of adjustment of the model parameters within an expected range of published values to minimize the difference between observed and simulated data. Based on sensitivity analyses (Kropff et al., 1994), we calibrated the model by performing multiple runs and sequentially adjusting the following input parameter: (1) the maximum assimilation rate of individual leaves, AMAX, (2) the specific leaf area, SLA and (3) the leaf area index at seedling emergence, LA0 to optimize the fit between simulated and measured data of shoot dry matter and leaf area index. All model parameters input parameter values (Table 2), except those adjusted in the calibration procedure, were selected from Kropff et al. (1994) and Akanvou (2001). Simulations covered the crop growth period of the two years (2004 and 2005) during which shoot dry matter and leaf area index data were collected at specific dates including emergence and 14, 28 and 56 DAS. The calibration consisted of slight increases or decreases of each parameter within a range of published values (Table 2) during each run, and was completed when adjustments to the specific parameter no longer reduced the difference between measured mean and simulated values of shoot dry matter and leaf area index. We followed the methods of Addiscott and Whitmore (1987), using a positive, highly significant correlation coefficient, and a reduced mean difference between simulated and measured data as criteria for goodness of fit of model predictions. To assess the accuracy of simulations, we used graphical and statistical methods (Loague and Green, 1991; Willmott, 1981). Simulated values were plotted against the corresponding measured values on a 1:1 scale to examine trends. We assumed a linear relationship between measured and simulated data, and used PROC REG of the SAS software package (SAS Institute, 2004) to conduct least squares regression analysis. The root mean square error (RMSE) was compared to the mean measured value (normalized root mean square error, NRMSE) to determine the prediction error. The statistical methods also included calculation of Willmott's index of agreement (d). The value of d reflects the degree to which the simulated variation accurately estimates the measured variation, and its value is 1.0 when there is a perfect agreement between simulated and measured values.

Table 2. INTERCOM parameter input values used in the simulations for NERICA1. Functions in the table are related to thermal time (°Cd).

Function description	Abbreviations	Units	Values
Development rate during vegetative phase	DVRV	$(°Cd)^{-1}$	0.000845
Development rate during reproductive phase	DVRR	$(°Cd)^{-1}$	0.00152
Light extinction coefficient for leaves	KDF		0.6
Photosynthetic rate	AMAX	$kg\ CO_2\ ha^{-1} leaf\ h^{-1}$	0,51;1000,40; 1200,27; 2200,5
Dry matter distribution pattern above ground (leaves-stems-panicles)	RGRL	$(°Cd)^{-1}$	0.0075
Initial leaf area	LA0	$m^2\ plant^{-1}$	0.0000682
Relative death rate of the leaves	RDRLV		1162,0 ; 1222,0.0071; 614,0.0028; 2200,0.029
Specific leaf area	SLA	$m^2\ kg^{-1}$	0,20.4; 141,21.4; 15,24; 418,25.2; 656,17.6 ; 912,15; 230,22; 2200,23

Mean difference, RMSE, NRMSE, and d are defined as follows:

Mean Difference (MD) $= \sum (O_i - S_i)/n$

$$RMSE = \left[\sum_{i=1}^{n} (oi - si)^2 / n \right]^{0.5}$$

NRMSE = RMSE $/o$

$$d = 1 - \frac{\sum_{i=1}^{n} (o_i - s_i)^2}{\sum_{i=1}^{n} (|o'_i| + |s'_i|)^2}$$

Where $O'_i = O_i - o$ and $S'_i = S_i - o$. n is the number of observations, O_i is the value observed, and S_i is the corresponding simulated value, and o is the mean observed value.

Data collection for model simulations

In the two field experiments conducted in 2004 and 2005 as described above, focus was placed on the variety of rice NERICA1 and the calibration task was performed using data from field plots without weed infestation. Measurements started one week after rice emergence and during both years samplings were taken every two weeks. At each sampling date, the height of the plants was recorded. Destructive samplings were performed on a quadrat of 0.5 m². The above ground parts of plants were separated from the roots. The samples were further partitioned between the leaves, stems and the storage organs, and dried in the oven at 70°C for 72 h and the leaf area index was measured on a subsample of leaves using the LiCor LI-3000 (Lincoln, Nebraska). Phenological and physiological data for model parameterization for the rice variety NERICA1 and the weed species were derived from literature (Akanvou, 2001) and from the first experiment in 2004. Data on densities of rice plants and weeds, and the dates of emergence (50%) of rice plants and weeds were collected on the experiments conducted in 2004 and 2005. Weather data was collected at the

Sikasso airport located at around 20 km from the study area. Yield data were collected from the ten weed regimes (weed-free and weed-infested regimes) as described above and were used for model performance testing.

Model testing

The performance of the calibrated model was tested, using the yield data from both experimental plots with and without weed infestation. The calibrated model was executed for the 2004 and 2005 years without any changes to the values of the calibrated parameters (AMAX, SLA, and LA0) and simulated and observed yield data were compared. The graphical and statistical methods described in the calibration section were used for the comparisons.

RESULTS AND DISCUSSION

The relative frequencies of major weeds

During the 2 years of experimentation, 22 main weed species were identified in 2004, and 26 in 2005 (Table 3). There was the same number of species for broadleaf weeds and grasses, and sedges represented the lowest number. The vegetation was almost homogeneous with the grasses and sedges represented by *Imperata cylindrica* (grass) (66%), *Cyperus sphacelatus* (sedge) (51%) and *Digitaria longiflora* (grass) (37%) that were the dominant weed species. By grouping weeds according to their methods of reproduction and dispersal determining their life cycle, the following groups were distinguished (annual grasses, broadleaved species and sedges, and perennial grasses, broadleaved species and sedges) (Table 3). Thus, the annual weeds that complete their life cycle within one year or less were the most common group during the two \\years of study. The group of perennial weeds was the second group in terms of fre-

Table 3. Relative incidence (%) of main weeds at harvest, Farako, (2004-2005).

Species	2004	2005
Annual sedges species		
Cyperus sphacelatus	45	56
Mariscus squarrosus	9	3
Perennial sedges species		
Cyperus rotundus	6	9
Cyperus spp.	25	36
Cyperus tenuiculmis	15	23
Annual grasses species		
Dactyloctenium aegyptium	24	11
Digitaria horizontalis	5	15
Digitaria longiflora	35	38
Digitaria spp.	10	24
Eleusine indica	6	5
Paspalum scrobiculatum	1	3
Pennisetum spp.	-	3
Pennisetum polystachion	1	-
Rottboellia cochinchinensis	1	3
Setaria pumila	-	4
Perennial grasses species		
Imperata cylindrica	63	68
Annual broadleaved species		
Acanthospermum hispidum	18	21
Aeschynomene americana	-	6
Ageratum conyzoides	3	1
Borreria stachydea	1	-
Borreria verticillata	-	4
Indigofera hirsuta	5	-
Mitracarpus scaber	-	1
Oldenlandia herbacea	-	1
Spilanthes filicaulis	3	1
Tephrosia argentea	-	1
Vernonia pauciflora	15	23
Perennial broadleaved species		
Scoparia dulcis	6	1
Smilax krausiana	8	3

quency. Broadleaved weeds were numerous but they had lower frequency than grasses and sedges, and have also experienced the most significant interannual floristic changes (Table 3).

Effects of critical periods on grain yields

Rice grain yields from the different periods of weed competition during the raining seasons of 2004 and 2005 are shown in Table 4. Rice grain yields were calculated in relation to the control plot (weed-free) to harvest (WFharv). Average yields (kg ha^{-1}) for the two years (2004-2005) of each variety in weed-free plots were: NERICA1: 1735; NERICA2: 1706; NERICA4: 2698; WAB 56-104: 1648; GC 14: 965. In the unweeded plots throughout the cropping cycle of varieties, yields (kg ha^{-1}) were: NERICA1: 450; NERICA2: 410; NERICA4: 728; WAB 56-104: 459; GC 14: 252. The average relative yields of the unweeded plots compared to the weed-free

Table 4. Effects of the period of interference of weeds on the yield of rice, Farako, 2004 and 2005.

Weeding regimes	Variety	2004 rice yield (kg ha^{-1})	Rice grain yield (%) compared to weed-free	2005 rice yield (kg ha^{-1})	Rice grain yield (%) compared to weed-free
Early competition	NERICA1				
WD14		1771[a]	98	1599[a]	96
WD28		1601	88	1453	88
WD42		1520	84	1380	83
WD56		1123	62	842	51
WDharv (unweeded)		502	27	398	24
Late competition					
WF14		483	27	626	38
WF28		1376	76	985	59
WF42		1720[a]	95	1562[a]	94
WF56		1756[a]	97	1625[a]	98
WFharv (weed-free)		1811[a]	100	1658[a]	100
LSD (P < 0.05)		192		155	
Early competition	NERICA2				
WD14		1632[a]	92	1718[a]	105
WD28		1501	85	1503[a]	92
WD42		1452	82	1224	75
WD56		532	30	568	35
WDharv (unweeded)		434	25	386	24
Late competition					
WF14		917	52	863	53
WF28		1235	70	1110	68
WF42		1716[a]	97	1586[a]	97
WF56		1683[a]	95	1510[a]	92
WFharv (weed-free)		1771[a]	100	1642[a]	100
LSD (P < 0.05)		220		140	
Early competition	NERICA4				
WD14		2881[a]	94	2315[a]	99
WD28		2388	78	2176[ab]	93
WD42		2092	68	1910	82
WD56		1731	57	1546	66
WDharv (unweeded)		803	26	653	28
Late competition					
WF14		991	32	785	34
WF28		1894	62	1689	72
WF42		2907[a]	95	2010[ab]	86
WF56		2968[a]	97	2290[a]	98
WFharv (weed-free)		3059[a]	100	2337[a]	100
LSD (P < 0.05)		468		365	
Early competition	WAB 56-104				
WD14		1588[a]	96	1514	93
WD28		1326	80	1289	79
WD42		1237	75	1165	71
WD56		840	51	687	42
WDharv (unweeded)		485	29	431	26
Late competition					
WF14		839	51	769	47

Table 4. Contd.

WF28	1254	76	1032	63
WF42	1577[a]	95	1485[a]	91
WF56	1611[a]	97	1602[a]	98
WFharv (weed-free)	1660[a]	100	1636[a]	100
LSD (P < 0.05)	242		154	
Early competition CG 14				
WD14	983[a]	105	936[a]	94
WD28	997[a]	106	897[a]	90
WD42	706	75	676	68
WD56	682	73	666	67
WDharv (unweeded)	275	29	229	23
Late competition				
WF14	685	73	280	28
WF28	740	79	642	65
WF42	890[ab]	95	920[a]	93
WF56	909[ab]	97	972[a]	98
WFharv (weed-free)	937[a]	100	992[a]	100
LSD (P < 0.05)	192		228	

The averages in a column followed by the same lowercase letter are not significantly different from the control plot weed-free until the harvest at P < 0.05.

(reflecting relative yield losses) for the five upland rice varieties were 74, 76, 73, 72, and 74%, respectively for NERICA1, NERICA2, NERICA4, WAB 56-104 and CG 14, with an average of 74%. This figure lies in the range of yield loss due to uncontrolled weed growth in upland rice ecosystems in West Africa (Akobundu, 1980; Dzomeku et al., 2007). For the two scenarios in Mali (weed free and unweeded plots to harvest), NERICA4 variety had significantly higher yield (P<0.05) than the other varieties, implying a better weed competitiveness of this variety in this northern Guinea savanna environment. This character of NERICA4 may have played a predominant part in its dissemination and adoption in the southern Mali agroecology (AfricaRice, 2008). In the present study, NERICA4, with a height of 120 cm, was the tallest variety among the NERICAS tested, and the advantage of height was seen as a morphological advantage for competition with weeds (De Vida et al., 2006; Zhao et al., 2006; Moukoumbi et al., 2011).

Increasing periods of weed interference in the early stages of the rice plants (WD14-WDharv) caused a steady decrease in rice yields for the five varieties. For the 2 years combined, daily yield losses of 17, 27, 23 and 18 kg ha^{-1} of rice grain were found respectively for varieties NERICA1, NERICA2, NERICA4, and WAB 56-104, when weeding was delayed between 14 and 56 DAS. For CG 14, yields loss was less significant and was around 6 kg ha^{-1}. In the early competition group, mean rice yields for the three NERICAS and WAB 56-104 were equal to that of the weed-free control when the first weeding was performed at 14 DAS (Table 4). There were significant yield differences relative to the weed-free

control when this first weeding was done at 28 DAS or later. For the late competition group treatments, the results did not differ significantly from the weed-free control when weeding was stopped 42 DAS or later. Under the experimental conditions, only plots in which weeding was stopped at 14 and 28 DAS gave significantly lower yields than the weed-free control. The early weed competition threshold occurred at 14-28 DAS, and the late weed competition threshold was between 28-42 DAS. Thus concerning the NERICAS and their sativa parent WAB 56-104, the critical period for weed competition was estimated as the time interval between these two thresholds, that is, 14-42 DAS. For O. glaberrima CG 14, the critical period was between 28-42 DAS (Table 4). For CG 14, the yield loss between the start and the end of the critical period was less important, indicating the ability of this variety to better withstand weeds during this period, and also suggesting that the first weeding for this variety may be delayed up to 28 DAS. Thus the critical period for this variety would then extend from 28 to 42 DAS or 14 days instead of 28 days for the other varieties.

First weeding of CG 14 may be delayed up to 28 DAS without significant yield loss because during its vegetative phase, this variety produces more vigorous seedlings and many tillers to better compete with weeds (Koffi, 1980). Although CG 14 could be competitive with weeds, it had low yield potentials (Table 4). And for CG 14 and other O. glaberrima varieties, yield losses are mainly due to their lodging and grain shattering characteristics (Koffi, 1980).Figure 1 highlights the negative effect of the early competition with the most important rice yield loss

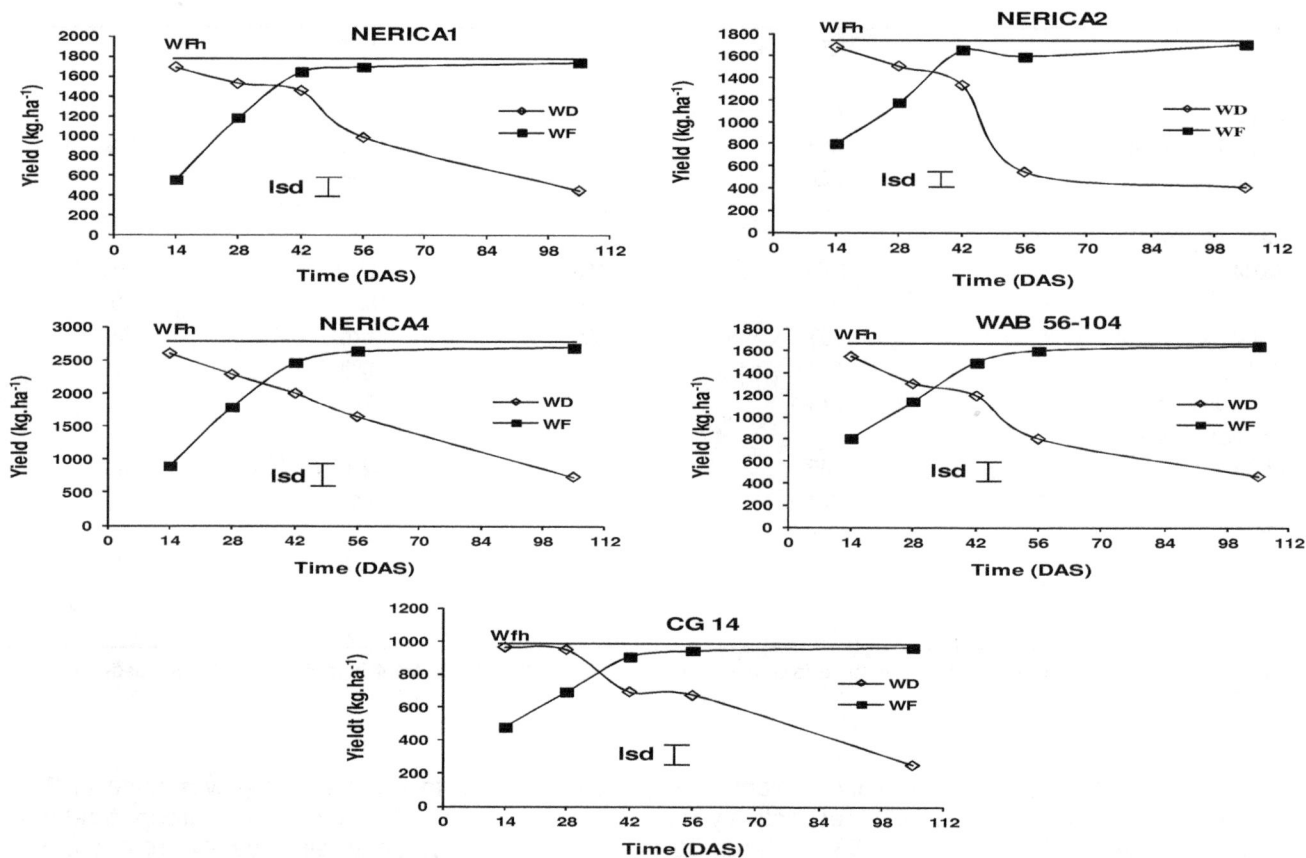

Figure 1. Rice yields in plots subjected to early competition (WD) and those subjected to late competition (WF). WFh: weed-free plots to the harvest. Farako (2004-2005).

due to early competition, and a lesser loss due to late competition. Then in Figure 1, there is a point of intersection (approximately 35 DAS) between the time during which weeds may remain in the plots and the period of time during which the plots should be weeded, suggesting that a single weeding at this time can prevent significant yield loss. But the present study did not include the effect of weeding on this specific date of 35 DAS. Nevertheless a previous study (Touré et al., 2011) were able to establish that a single weeding done at 31 DAS (close to 35 DAS) had a yield comparable to the double weeding done at 21 and 42 DAS. If a single weeding done on a specific date between 31 DAS and 35 DAS did not have a significantly lower yield than the wee-free control, then it would not be a critical period, but a critical date for weeding. Therefore, the critical period for the different rice varieties was from 14 to 42 DAS or 28 days, and during this period weeds should be theoretically removed. This critical period of weed control is in compliance with previous studies. Le Bourgeois and Marnotte (2002) located this critical period between 15 and 60 DAS for annual short-cycle crops such as rice and other cereals (maize and sorghum). In Ghana, Dzomeku et al. (2007) determined in rainfed condition that the critical period of two varieties of NERICA rice (NERICA1

and NERICA2) was between 21 and 42 DAS. For irrigated rice in the Sahel, this critical period was between 29 and 32 DAS during the rainy season and between 4 and 83 DAS during the dry season (Johnson et al., 2004). In rainfed rice in southern Togo, this critical period was a little shorter, and weed competition was much more harmful between 21 and 30 DAS (Boyoda, 1991).

For both groups of competition (early and late), average yields of the five rice varieties during the two years of experimentation were almost equal to the weed-free plots when plots were unweeded or weeded during the first two weeks (Figure 1). In this case, weeds germinating very early during the crop cycle did not significantly affect yields. In addition, during the early stage of the cropping cycle, weed flora is less developed; making weed controls easier with greater efficiency.

These early weeding controls avoid rhizomes and cuttings of some frequent perennial weeds (*Imperata cylindrica* and *Cyperus* spp) from growing on the experimental site. Weeds with higher relative frequency were annual grasses such as *Digitaria longiflora* and *Dactyloctenium aegyptium* (Table 3). For these annual species with short growth cycle, the early weedings prevent development, flowering, fruiting, and seed production which would increase the seed stock in the

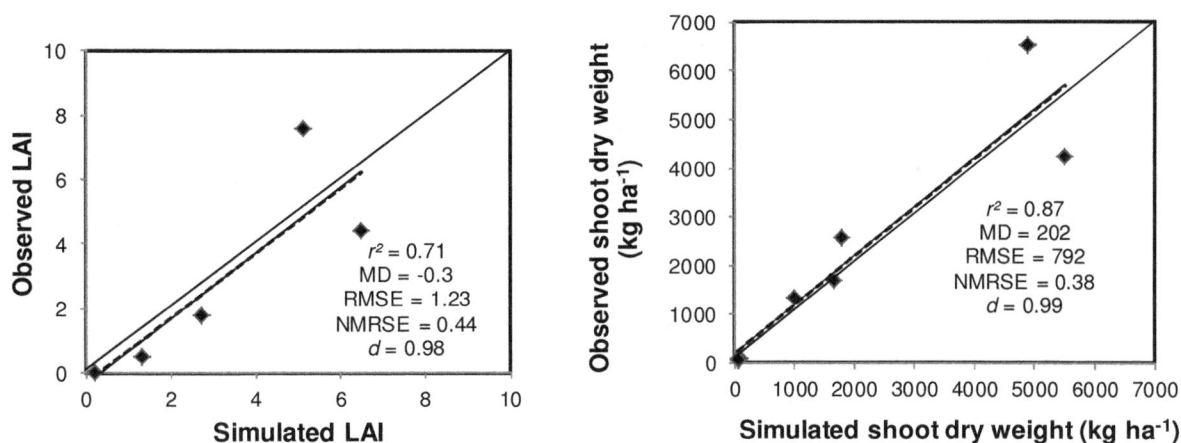

Figure 2. 1:1 scale plot, regression and statistics of observed and INTERCOM-simulated values of rice NERICA1 LAI and shoot dry weight during model calibration.

soil (Akobundu, 1987). The drawback of the early weeding resides in the close resemblance of those grass weeds with rice plants at their seedling and vegetative stage, and those weeds can be mistaken for the rice crop, and thus evade eradication during hand weeding (Akobundu, 1987).

Simulating the effects of the duration of weed competition on crop yields

Model calibration

The calibrated values of AMAX, SLA and SL0 are presented in Table 2. Upon calibration, the model simulations of LAI were reasonably satisfactory early in the crop growth cycle but discrepancies appeared noticeable later during the cycle. Measured and simulated LAI values were fairly well correlated ($r^2 = 0.87$) and the Willmott's index of agreement (d) was 0.98, indicating that simulated and measured data agreed well (Figure 2). However, the mean difference value was -0.3 and the RMSE was 1.23 resulting in a fairly high prediction error of 44% for the measured mean value. Storkey et al. (2003) used an INTERCOM-based eco-physiological model to simulate LAI for a winter wheat crop and reported measured and simulated data with r^2 of 0.69 and 0.72. Our r^2 value of 0.71 was well in the range of their values but they did not report any other statistics. Simulated and measured values of shoot dry weight matched reasonably well early in the crop growth period but were less satisfactory later during the growth period where the model either underestimated or overestimated shoot dry weight (Figure 2). Measured and simulated shoot dry weight values were highly correlated ($r^2 = 0.87$) and the Willmott's index of agreement (d) was 0.99, indicating a good match between simulated and measured data (Figure 2). The mean difference value

was 202 kg ha^{-1} and the RMSE was 792 kg ha^{-1} leading to a prediction error of 38% for the measured mean value. Weaver et al. (1992) used the model to simulate dry matter under a tomato crop and found that the model accurately simulated the increase in dry matter, but did not provide any statistics on the comparisons of measured and simulated data sets. Overall, our results during the calibration process showed a generally good match between simulated and observed LAI and shoot dry weight data (r^2 values of 0.71 and 0.87, and d values of 0.98 and 0.99), but the 44 and 38% prediction error values for the measured mean values of LAI and shoot dry weight, respectively, indicate that there may be room for improvement of the model simulations.

Model performance

When the calibrated model was tested against measured yield data of 2004, 2005 and pooled yield data from the two years from plots with and without weed infestation, it performed reasonably well. There was a good agreement between measured and simulated values and their trends did not display any noticeable deviations (Figure 3). The data sets were highly correlated with r^2 values ranging from 0.84 to 0.87 and Willmott's index of agreement values in all cases were 0.99. The RMSE values were low (typically between 200 and 240 kg ha^{-1}) with prediction errors ranging from 15 to 19% (Figure 3). The mean difference values were negative in all cases (typically ranging between -50 and -140 kg ha^{-1}) which indicates that in general the model tended, although slightly, to overestimate the yield data. This suggests that the model might underestimate yield losses especially under weed infestation conditions presumably because it assumes that nutrients are not limiting to crop growth. Our findings on the simulation of crop yields corroborate those reported by several other similar studies. Storkey et

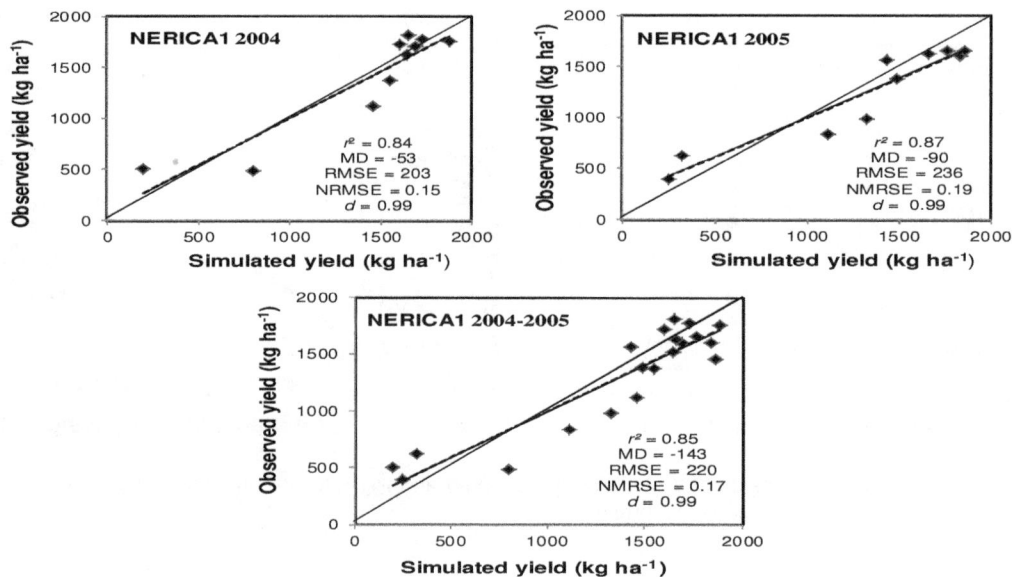

Figure 3. 1:1 scale plot, regression and statistics of observed and INTERCOM-simulated values of rice NERICA1 yield during model performance testing.

al. (2003) used a simple thermal time model and an INTERCOM-based eco-physiological model to simulate winter wheat yield loss and found that both approaches underestimated yield damage coefficient although the eco-physiological model performed better. The simple thermal model was only able to describe a maximum of 55% of the variation in yield loss. Weaver et al. (1992) used the INTERCOM model to simulate yield losses under sugar beets and tomato crops, and reported r^2 values ranging from 0.81 to 0.94 between simulated and observed data. They found that under both crops the model underestimated yield losses when weeds were allowed to compete with the crop for longer than 20 days after transplanting and 45 days after emergence, and argued that these trends in the simulations were linked to the assumption that nutrients were not limiting to crop growth in the model.

Conclusion

Weed flora at the experimental site was variable in composition but grasses were the most dominant flora, followed by sedges. Season-long weed infestation resulted in reduction in grain yields of about 74% in the varieties, suggesting the vulnerability of rice crop to weed infestation. In this study, the critical period of weed competition was approximated for upland rice varieties in southern Mali (14 to 42 DAS), and the harmful effects of early weed competition was demonstrated. Overall, the INTERCOM model proved to be capable of simulating the interactions between NERICA1 rice crop yield losses, weed density, and duration of weed competition under rainfed conditions in northern Guinea savanna agro-

ecology in southern Mali. A better and more realistic performance of the model requires further analysis of the assumption that nutrients are not limiting to crop growth, through quantitative field experiments at different levels of nutrients.

ACKNOWLEDGMENTS

Financial support of the African Rice Initiative (ARI), a network coordinated by the Africa Rice Center (AfricaRice) and funded by the African Development Bank is gratefully acknowledged.

REFERENCES

Africa Rice (2008). Africa Rice trends 2007. Cotonou, Benin: 5th Edition, (eds) Diagne A, Bamba I, Touré A A, Medagbé, A. P. 84.
Addiscott TM, Whitmore AP (1987). Computer simulation of changes in soil mineral nitrogen and crop nitrogen during autumn, winter and spring. 109(1):141-157.
Akanvou R (2001). Quantitative understanding of the performance of upland rice-cover legume cropping systems in West Africa. PhD thesis. Wageningen University. P. 149.
Akobundu IO (1980).Weed science research at the International Institute of Tropical Agriculture and research needs in Africa. Weed Sci. 28:439–445.
Akobundu IO (1987). Weed science in the tropics. Principles and practices. John Wiley & Sons, Ltd. P. 522.
Anwar MP, Juraimi AS, Samedani B, Puteh A, Man A (2012). Critical period of weed control in aerobic rice. The Scientific World J. doi:10.1100/2012/603043.
Baumann DT (2001). Competitive suppression of weed in a leek-celery intercropping system. An exploration of functional biodiversity. PhD thesis, Wageningen University. P. 190.
Bouman BAM, van Keulen H, Rabbinge R, van Laar HH (1996).The 'School of de Wit' crop growth simulation models: A pedigree and historical overview. Agric. Syst. 52:171–198.

Boyoda TBK (1991). Contribution à l'étude de la nuisibilité des mauvaises herbes en riziculture pluviale au Togo. Direction de la recherche agronomique de Lomé et Université du Bénin. Lomé, Togo.
Mémoire de fin d'études agronomiques. P. 130.

Cavero J, Zaragosa C, Suso ML, Pardo A (2000). Competition between maize and *Datura stramonium* in irrigated field under semi-arid conditions. Weed Res. 39:225-240.

Cousens R. (1988). Misinterpretations of results in weed research through inappropriate use of statistics. Weed Res. 28:281–289.

Cousens R (1991). Aspects of the design and interpretation of competition (interference) experiments. Weed Technol. 5:664–673.

Dawson JH (1986). The concept of period thresholds. In: Proceedings of the European Weed Research Society. Symposium on Economic Weed Control, Stuttgart-Hohenheim, Germany. pp. 327–331.

De Vida FBP, Laca EA, Mackill DJ, Grisel M, Fischer AJ (2006). Relating rice traits to weed competitiveness and yield; A path analysis. Weed Sci. 54:1122–1131.

Dzomeku IK, Dogbe W, Agawu ET (2007). Response of NERICA rice varieties to weed interference in the guinea savannah uplands. J. Agron. 6:262–269.

Evans SP, Knezevic SZ, Lindquist J L, Shapiro CA, Blankenship EE (2003). Nitrogen application influences the critical period for weed control in corn. Weed Sci. 51:408–417.

FAO (1998). World reference base for soil resources. World Soil Resources Report 84, Food and Agriculture Organization of the United Nations, Rome.

Goudriaan J (1986). A simple and fast numerical method for the computation of daily totals of canopy photosynthesis. Agric. Meteorol. 43:251-255.

Heemst HDJ van (1985). The influence of weed competition on crop yield. Agric. Syst. 18:81–93.

International Institute of Tropical Agriculture [IITA] (1989). Automated and semi-automated methods for soil and plant analysis. Manual series. IITA, Ibadan, Nigeria 7:33.

Johnson DE, Wopereis MCS, Mbodj D, Diallo S, Powers S, Haefele SM (2004). Timing of weed management and yield losses due to weeds in irrigated rice in the Sahel. Field Crops Res. 85:31–42.

Koffi G (1980). Collection and conservation of existing rice species and varieties of Africa. Agron. Trop. 34:228–237.

Kropff MJ, Weaver SE, Smits MA (1992). Use of ecophysiological models for crop-weed interference: relations amongst weed density, relative time of weed emergence, relative leaf area, and yield loss. Weed Sci. 40:296–301.

Kropff MJ, van Laar HH (eds) (1993). Modelling crop-weed interactions. CAB International, Wallingford, UK. P. 267.

Kropff MJ, van Laar HH, Matthews RB (eds) (1994). ORYZA1: An ecophysiological model for irrigated rice production. International Rice Research Institute, Manilla, Philippines. P. 110.

Le Bourgeois T, Marnotte P (2002). La lutte contre les mauvaises herbes. In: Mémento de l'Agronome. CIRAD-GRET, Ministère des Affaires Etrangères, Paris. pp. 663–684.

Lindquist JL, Mortensen DA (1997). A simulation approach to identifying the mechanism of maize tolerance to velvet leaf competition to light. In: The 1997 Brighton Crop Protection Conference-Weeds, Brighton, UK. pp. 503–508.

Loague K, Green RE (1991). Statistical and graphical methods for evaluating solute transport models: overview and application. J. Contam. Hydrol. 7:51–73.

Moukoumbi YD, Sie M, Vodouhe R, Bonou W, Toulou B, Ahanchede A (2011). Screening of rice varieties for their weed competitiveness. Afr. J. Agric. Res. 24:5446–5456.

Nieto JH, Brondo MA, Gonzalez JT (1968). Critical periods of the growth cycle for competition from weeds. PANS (C) 14:159–168.

Rodenburg J, Johnson DE (2009). Weed management in rice-based cropping systems in Africa. Adv. Agron. 103:149–218.

Sahrawat KL, Jones MP, Diatta S (2001). Response of upland rice to fertilizer phosphorous and its residual value in an Ultisol. Commun. Soil Sci. Plant Anal. 32:2457–2468.

SAS Institute Inc. (2004). SAS System Version 9.1. SAS Institute Inc., Cary, NC.

Sogbedji JM, van Es HM, Hutson JL, Geohring LD (2001). N rate and transport under variable cropping history and fertilizer rate on loamy sand and clay loam soils: II. Performance of LEACHMN using different calibration scenarios. Plant Soil 229:71–82.

Spitters CJT, van Keulen H, van Kraalingen DWG (1989). A Simple and Universal Crop Growth Simulator: SUCROS 87. In Rabbinge R, Ward SA, van Laar H H (eds): Simulation and systems management in cop protection. Simulation Monographs, Pudoc, Wageningen, The Netherlands. pp. 147-181.

Storkey J, Cussans JW, Lutmana PJW, Blairb AM (2003). The combination of a simulation and an empirical model of crop/weed competition to estimate yield loss from *Alopecurus myosuroides* in winter wheat. Field Crops Res. 84:291–301.

Touré A, Rodenburg J, Saito K, Oikeh S, Futakuchi K, Gumedzoé D, Huat J (2011). Cultivar and weeding effects on weeds and rice yields in a degraded upland environment of the coastal savanna. Weed Technol. 25:322–329.

Weaver SE, Tan CS (1983). Critical period of weed interference in transplanted tomatoes (*Lycopersicon esculentum*): growth analysis. Weed Sci. 31:476–481.

Weaver SE, Kropff MJ, Groeneveld RMW (1992). Use of ecophysiological models for crop-weed interference: The critical period of weed interference. Weed Sci. 40:302–307.

Wopereis MCS, Diagne A, Rodenburg J, Sié M, Somado EA (2008). Why NERICA is a successful innovation for African farmers: a response to Orr et al. from the Africa Rice Center. Outlook Agric. 37:169–176.

Zhao D, Atlin GN, Bastiaans L, Spiertz JHJ (2006). Cultivar weeds competitiveness in aerobic rice; heritability, correlated traits, and the potential for indirect selection in weed-free environments. Crop Sci. 46:372–380.

15

Effect of post-emergence dual herbicides on weeds and yield of maize (*Zea mays* L.) in order to decrease environmental biology pollution of Atrazine in semi-arid region of Khuzestan, Iran

Fateme Nuraky[1] and Hassan Rahmany[2]

[1]Shoushtar Branch, Islamic Azad university, Shoushtar, Iran.
[2]Payame Noor University of Khuzestan, Iran.

Many of the chemicals used in pesticides are persistent soil, groundwater and drinking water contaminants. Use of efficient methods of weed integrated management with as regards environmental sustainability and reduce pollution as well as increased crop yield and also weed resistance to herbicides is essential. An experiment was conducted in 2010 in the north of Khuzestan in Iran. Experiment treatments were compared in a split plot design by a randomized completely block design with 4 replication. The used variety of maize was S.C. 704. Main factors included 3 levels of cultivation, once, twice and without cultivation. Sub factors were weed control by application of indicated herbicide in 4 levels: Nicosulfuron, Foramsulfuron, Atrazin + Alachlor and no control. The results conducted that the highest of weed control followed the highest yield by ranged 15.47 ton per hectare related to Nicosulfuron + once cultivation treatment and lowest yield by ranged 10.56 ton per hectare related to Atrazin + Alachlor + once cultivation treatment. There were difference between treatments in yield and yield components during the whole growing season, the kind of index harvest in the level of probability 1% and all in the level of 5% significant.

Key words: Maize, dual herbicides, Atrazine, biology pollution, weed integrated management.

INTRODUCTION

Maize (*Zea mays* L.) is the world's third most important cereal grain after wheat and rice. It is grown primarily for grain and secondarily for fodder in Iran. It is grown on an area of 320,000 ha with a production of 2,560000 tons an average grain yield of 8,000 kg ha^{-1}. Among various factors responsible for low yield, weed infestation is of supreme importance. Worldwide maize production is hampered up to 40% by competition from weeds which are the most important pest group of this crop (Oerke and Dehne, 2004). Atrazine is an herbicide registered in the United States for the control of broadleaf weeds and some grassy weeds. It is currently used on corn, sorghum, sugarcane. Atrazine acts by inhibiting photosynthesis. Many Atrazine-tolerant mutations have begun to appear in weeds, and this tolerance is predominantly based on detoxifying Atrazine by binding it to glutathione, a mechanism in naturally Atrazine-tolerant corn. Efforts have been made to select or produce Atrazine-tolerant mutant's crops such as soybean that is otherwise difficult to rotate with Atrazine-treated corn or potato (Joe and Mae-Wan, 2011). The environmental impact of pesticides is often greater than what is intended by those who use them. Over 98% of sprayed insecticides and 95% of herbicides reach a destination other than their target species, including no target species, air, water, bottom sediments, and food (Miller, 2004). Pesticide residues have also been found in rain and groundwater (Kellogg et al., 2000). Studies by the UK government showed that pesticide concentrations exceeded those allowable for drinking water in some samples of river water and ground water (Bingham, 2007). Atrazine can be presented in parts per million in agricultural run-offs and can reach 40 parts per billion in precipitation. The global impact of Atrazine is staggering. Significant Atrazine pollution has been found in the Lio-He and Yangtse rivers of China, and a review of the atmospheric dispersion of Atrazine shows impacts of the herbicide even in isolated areas of the globe (Joe and Mae-Wan, 2011). Atrazine has a higher risk than Metolachlor in all soils because of its higher toxicity. Surface application of pesticide generally increases the chance of pesticide loss in runoff, which possess a greater risk to surface water, while soil incorporation may increase pesticide loss in percolation. There is a trade-off in managing practices to protect surface and groundwater quality. The objective of this research was to determine how well selected post emergence herbicides worked when applied at normal use rates for weed control instead of Atrazin towards environmental sustainability and reduce groundwater contamination as well as increased crop yield with in-row cultivation in maize.

MATERIALS AND METHODS

This study was carried out in north of Khuzestan in Iran during summer 2010. The experimental site had mean soil pH of 7.70 with 22.8, 55.7 and 21.5% clay, silt and sand, respectively. The experimental was split plot in randomized completely block design (RCBD) design with four replications. Maize variety (Single cross 704) was used in the study as this is the widely used variety used in the area. Soils were fertilized according to NMSU recommendation based on soil tests. The field were plowed, fertilized, and leveled before the field maize was planted. The size of each treatment was 6 × 5 m². There were 12 treatments in the experiment with row to row distance of 75 cm, each treatment having eight rows. Distance of seeds inter row was 17 cm. Experimental field was irrigated as and when needed. Main factor was cultivation in three levels and sub factor was herbicides in four levels. Herbicides included Atrazin (WP80, P80), Alachlor (EC48), Foramsulfuron (OD 22.5) and Nicosulfuron (SC4) by the balance (1 kgha^{-1}), (4 Lha^{-1}), (2.5 L/ha) and (2 Lha^{-1}) respectively. The experiment comprised of the following treatments:

1. Foramsulfuron + once cultivation
2. Nicosulfuron + once cultivation
3. Atrazin + Alachlor + once cultivation
4. Weedy + once cultivation
5. Foramsufuron + twice cultivation
6. Nicosulfuron + twice cultivation
7. Atrazin + Alachlor + twice cultivation
8. Weedy + twice cultivation

9. Foramsulfuron + non-cultivation
10. Nicosulfuron + non-cultivation
11. Atrazin + Alachlor + non-cultivation
12. Weedy + non-cultivation

During the course of experiment, the data were recorded on weed density m^{-2} 26 days after sowing, yield and particulars included 100 grains weight (g) number of grains per row – number of row in ear, biological yield (tha^{-1}) and economical yield (tha^{-1}).

Each time quadrate having size 0.5 × 0.5 m² was placed randomly four times in each treatment and the weeds inside the quadrate was counted. For recording the grain yield data, two central rows were harvested in each treatment bundled, sun dried and weighed. The data recorded were statistically analyzed using MSTAT-C software. The purpose of analysis of variance was to determine the significant effect of treatments on weeds and maize. Duncan's multiple range tests at 1% probability level was applied for mean separation of significant parameters.

RESULTS AND DISCUSSION

Dominant weed species in field were Cyperus and Chenopodium, respectively

Weed density (m²) 26 days after sowing (15 days after herbicides application). The data regarding weed population revealed that weed density at 26 days after sowing (DAS) was significantly affected by all weed control treatments (Table 1). The results indicated that maximum weed density 26 days after sowing was recorded in weedy (Table 1). Table 2 shows the control percentage for the treatments. The data (Table 2) reveal that maximum control percentages for Cyperus, Convolvulus, Chenopodium and Nicosulfuron are 55.63, 74.42 and 100 respectively. Maximum control of Amaran tus was for Foramsulfuron treatment. These results are in line with that of Jodie (2008) and Nurse et al. (2007).

100 - Grain weight

The highest 100 – grain weight (31 g) was recorded in Table 3 weedy + twice cultivation. Significantly minimum 100 – grain weight (25.75 g) was recorded in Atrazin + Alachlor + non cultivation. In those treatment where the weeds were controlled, 100 – grain weight were greater as compared to uncontrolled treatments as weeds share the resources with the crop plants. These results were in agreement with Khan et al. (2002) and EL- Bially (1995). They reported that weed infestation decreased the 100 – grain weight in maize.

Number of grain per row

Number of grain is an important yield contributing trait and can greatly affect the economic return. It could be inferred from the data that maximum (45.33) number of grains per row was obtained in Nicosulfuron + once cultivation. Minimum (33.41) grains per row were recorded

Table 1. Weed control percentage with herbicides.

Treatment	Cyperus (control %)	Convolvulus (control %)	Chenopodium (control %)	Amaranthus (control %)
Foramsulfuron	26.94	71.47	91.20	100
Nicosulfuron	55.63	74.42	100	93.80
Atrazin+lasso	27.92	36.70	84.88	56.01
Weedy	0	0	0	0

Table 2. Weed density at 26 days after sowing.

Treatment	Cyperus (number)	Convolvulus (number)	Chenopodium (number)	Amaranthus (number)
Foramsulfuron	1.56	1.28	0	0
Nicosulfuron	1.25	0	0	0
Atrazin+lasso	3	1.41	0	0
Weedy	9.48	2.31	9.25	6.91

Table 3. Results of mean comparisons between treatments.

Treatment	100 – grain weight (g)	Number of grain per row	Number of row in ear	Economical yield (tha⁻¹)	Biological yield (tha⁻¹)
1	28.25^{bcd}	38.42^{bced}	14.99^a	12.81^c	24.19^{bcd}
2	28.75^{abc}	45.33^a	15.16^a	15.47^a	29.55^a
3	25.7^{cde}	33.41^f	14.49^a	10.65^d	20.09^f
4	27.75^{cde}	36.33^{cdef}	14.49^a	11.46^d	21.81^{def}
5	30.25^{ab}	40^{bc}	14.83^a	14.10^b	26.44^{bc}
6	30.5^{ab}	39.33^{bcd}	14.66^a	13.82^{bc}	26.06^{bc}
7	26.25^{de}	37.67^{bcde}	14^a	10.87^d	19.91^f
8	31^a	40.17^b	14^a	13.73^{bc}	26.74^b
9	29.25^{abc}	39.41^{bcd}	14.33^a	12.96^{bc}	23.88^{cde}
10	28.75^{abc}	41.16^b	14.50^a	13.45^{bc}	25.56^{bc}
11	25.75^e	36.17^{def}	14.83^a	10.84^d	20.0^{5f}
12	29^{abc}	35.33^{ef}	14.16^a	11.31^d	21.28^{ef}

Means with similar letter(s) in each trait is not significantly different at 1% probability level according to Duncan's multiple range test.

in Atrazin + Alachlor + once cultivation. From these results it was observed that good weed control was effective to get higher number of grain per row and it was also observed that less grain per row in uncontrolled plot (Naveed et al., 2008)

Number of row in ear

The results revealed that, treatments have no significant variance by genetics but affected by environmental factors.

Economical yield (tha-^{1})

The highest grain yield (15.47 tha-^{1}) was recorded in Nicosulfuron + once cultivation. Higher grain yield was due to more number of grains per cob, grain weight per cob and 100 – grain weight as compared to uncontrolled treatments.

Efficiency of chemicals and other weed control practices in increasing grain yield had also been demonstrated by some scientists (Khan and Hag, 2004).

Biological yield (tha^1)

The data presented indicated that maximum biological yield (29.55 tha^1) was recorded in Nicosulfuron + once cultivation. As all vegetative parameters were significantly affected by different treatments, the biological yield was also significantly affected because leaf area, number of leaves plant, plant height and number of grains cob contributes in increasing the biological yield. Ullah et al. (2008) also reported similar results.

Conclusion

Results of study indicated that additional of grain yield, others affected by treatments. Single herbicide application cannot control weeds but integrated control was exceedingly weed control significantly. Thus for high yield and better control of weed with modern methods of weed integrated management In order to decrease of environmental biology pollution of Atrazine on base of use of integrated methods, the most qualify treatment of weed control that recommending is Nicosulfuron +once cultivation instead of Atrazin (conventional herbicide) in Maize field.

REFERENCES

Bingham S. (2007). Pesticides in rivers and groundwater. Environment Agency, UK. Retrieved on 2007-10-12.

EL-Bially ME (1995). Efficiency of Atrazine with other herbicides used alone, in sequence or as tank mix in maize Ann-Agri. Sci. 40(2):709-721.

Jodie K (2008). Annual weed control with balance flexx capreno, and Corvus in maize at Lamberton. South West Research and outreach center. University of Minnesota, Lamberton.

Joe C, Mae-Wan H (2011). Atrazine Poisoning Worse than Suspected, Institute of science in society

Kellogg RL, Nehring R, Grube A, Goss DW, Plotkin S (February 2000). Environmental indicators of pesticide leaching and runoff from farm fields. United States Department of Agriculture Natural Resources Conservation Service. Retrieved on 2007-10-03.

Khan MA, Marwat KB, Gul H, Naeem K (2002). Empact of weed management on maize (zea mays L.) planted at night. Pak. J. Weed Sci. Res. 8(1,2):57-62.

Khan M, Hag N (2004). Weed control in maize (Zea mays L.) with pre and post emergence herbicides Pak. J. Weed Sci. Res 10(1-2):39-46.

Miller GT (2004). Sustaining the Earth, 6th edition. Thompson Learning, Inc. Pacific Grove, California. Chapter 9:211-216.

Naveed M, Ahmad R, Nadeem MA, Nadeem SM, Shahzad K (2008). Effect of a new post emergence herbicide application in combination with urea on growth, Yield and weed control in maize (Zea mays L.). J. Agri. 46(2).

Nurse RE, Hamill AS, Swanton CJ, Tardif FJ, Sikkema PH (2007). Weed control and yield response to Foramsulfuron in maize. Weed Tech 21:453-458.

Oerke EC, Dehne HW (2004). Safeguarding production losses in major crops and the role of crop protection. Crop Port 23:275-285.

Ullah W, Khan MA, Arifullah SH, Sadiq M (2008). Evaluation of integrated weed management practices for maize. Pak. J. Weed Sci. Res. 14(1.2):19-32.

16

Distribution status and the impact of parthenium weed (*Parthenium hysterophorus* L.) at Gedeo Zone (Southern Ethiopia)

Talemos Seta*, Abreham Assefa, Fisseha Mesfin and Alemayehu Balcha

Department of Biology, Dilla University, Ethiopia.

A study was conducted in Gedeo Zone, Southern Ethiopia to determine the impact and distribution status of Parthenium weed, *Parthenium hysterophorus* L. in the area. To collect data related with the impact of parthenium, 14 Peasant Association (PAs) were purposefully selected along the high way. A total of 140 quadrats (1 m^2 area each) from 14 PAs were selected by using stratified random sampling for herbaceous vegetation data associated with parthenium. The plant species found in each quadrat were counted, recorded and identified. The data collected from farmers' perception on the impact of parthenium weed were analyzed by using descriptive statistics. Shannon Diversity Index, evenness, species richness and Jaccards Similarity Index to determine parthenium impact on species diversity were calculated from the vegetation data. To see correlation of vegetation variables among and between sample PAs, R.2.14.0 Package was used. This study revealed that high infestation level of parthenium weed is confined to Dilla administrative town as supported by 73% of the respondents. A total of 45 plant species under 20 families were recorded from this study. The sampled PAs in Dilla town showed high infestation level with lower diversity index. Thus, it is an urgent task to draw the attention of relevant responsible bodies and public in general for managing and preventing further introduction and dissemination of the weed in this study area.

Key words: Diversity Index, Farmers' Perception, herbaceous vegetation, *Parthenium hysterophorus* L.

INTRODUCTION

Weed and undesirable woody plants encroachment have been threatening the agricultural system and pastoral production system in the Horn of Africa, particularly Ethiopia (Amaha, 2003; Gemedo et al., 2006). Moreover, population pressure, over-stocking, overgrazing and deforestation have facilitated the disturbance of the Ethiopian ecosystem and enhanced the effect of weed invasion by threatening biodiversity of the country (EARO, 2003). Herbaceous weedy species like *Xanthium* like *Prosopis juliflora* (Sw.) DC. (Fabaceae), *Acacia mellifera* (Vahl) Benth. (Fabaceae), *Acacia nubica* (Fabaceae), *Lantana camara* L (Verbenaceae) and succulents such as *Opuntia* spp. (Cactaceae) are

nowadays increasing in different regions of the country. They are responsible for a significant reduction in production of the potential of the rangelands and arable lands (SERP, 1990). Among others, *Parthenium hysterophorus* L. (Asteraceae) is a Parthenium (both are Asteraceae), woody species aggressive invasive alien weed species (Kohli et al., 2006), native to the Americas but now widely spread in Asia, Africa and Australia (Evans, 1997).

Parthenium weed was first introduced accidentally into Ethiopia in the 1970s. It was first reported from Ethiopia in 1988 at Dire- Dawa and Harerge, Eastern Ethiopia (Seifu, 1990) and subsequently found near Desse, North-eastern Ethiopia as well. Both are major food-aid distribution centers and there is a strong assertion that parthenium weed seeds were imported from subtropical North America as a contaminant of grain food aid during

---

*Corresponding author. E-mail: talemos.seta@yahoo.com.

the 1980s famine and distributed with the grain (Tamado et al., 2002).

However, it has now emerged as one of the most aggressive weeds of both grazing land and cereal-based agriculture and crop lands (Tamado and Milberg, 2000). *P. hysterophorus* is considered as a noxious weed because of its prolific seed production and fast spreading ability (Haseler, 1976), allelopathic effect on other plants (Adkins, 1996), strong competitiveness with crops (Tamado et al., 2002) and health hazard to humans as well as animals (Chippendale and Panetta, 1994). Parthenium is so devastating that very little and sometimes no other plant species are seen in areas where it has gained dominance (Shabbir and Swhsana, 2005). In areas where the weed occurs, the productivity of forage is reduced by 90% and the weed make lands infertile and weakens the quality of grazing land, animal health, meat and milk products, agricultural production (Rezene et al., 2005). It also poses a serious threat to the environment and biodiversity owing to its high invasion and allelopathic effect which has the capacity to rapidly replace the native vegetation (Tamado and Milberg, 2000). Parthenium exerts strong allelopathic effect and reduces the growth and reproduction of associated crops. It does these by releasing phytotoxins from its decomposing biomass and root exudates in soil. Bioassay, pot culture and field studies have revealed that all plant parts (shoot, root, inflorescence and seed) are toxic to plants (Mulatu et al., 2009)

As mentioned earlier, *P. hysterophorus* was first observed in eastern part of Ethiopia, especially in Dire-Dawa and Hararghe, Eastern Ethiopia. Currently, the weed has been distributed to different regional states of ethiopia, eastern to southern including SNNPR (Southern Nations Nationalities and Peoples Region)

However, the local people have not yet noticed its effect on arable land, rangeland, animals and human health. The occurrence of *P. hysterophorus* is very frequent in urban, semi-urban and rural areas of Gedeo Zone, which is found in Southern Nations, Nationalities and peoples Region. However, no data of scientific studies have been documented regarding the diversity and abundance of plant species where there is *P. hysterophorus* infestation in both arable and non-arable lands. It is imperative to identify plant species that may have the ability to resist or overcome the challenges of the weed, which is increasingly reducing the quality and quantity of the composition and biomass of the herbaceous species. Thus, major objective of this present study was to determine the impact of *P. hysteroporus and* its distribution status in Gedeo Zone.

MATERIALS AND METHODS

Description of the study area

This study was conducted in Gedeo Zone which is found in the

Southern Nations, Nationalities and Peoples Region (SNNPR). Gedeo Zone extends south as a narrow strip of land along the eastern escarpment of the Ethiopian Highlands into the Oromia Region, which borders the zone on the east, south and west. Gedeo zone shares its northern boundary with Sidama zone. Dilla is the zonal administrative town.

The altitude of the zone ranges from 1268 m above sea level in the vicinity of Lake Abaya to an elevation of 2993 m at Haro Wolabu Pond (Tadesse, 2002). Among various zones of SNNPR, the Gedeo zone was given the major emphasis during this study (Figure 1). The Gedeo zone exhibits bimodal rainfall distribution, from March to May (Belg) and from July to October (Keremet) with an interval of 4 to 5 dry months. The source of rain from March to May is the monsoon wind from the Indian Ocean and from July to October from equator western region from the Atlantic Ocean. The mean annual rainfall ranges from 800 to 1400 mm with mean annual temperature of 11 to 29°C. The major proportion of the zone constituted from Weyena Dega (71%), Dega (21%) and Kola (8%). According to FAO classification, the soil type is farrsole and Nithosole; having greater depth. The traditional soil classification shows that the *zone* comprises brown soil (90%), red soil (5%) and black soil (5%).

Perception of local people on the impact of parthenium weeds

Data related to peoples' perception about the impact of this invasive weed from urban, semi-urban and rural areas (both arable and non-arable lands) were collected from *Gedeo Zone*. From three districts and Dilla administrative town of this zone, a total of 14 PAs were selected for this study. The researchers purposefully selected these PAs on the basis of the aggressive invasiveness of parthenium weed towards arable land following roadway. A single visit for an informal survey was conducted before the beginning of the actual research work. The survey was undertaken to observe the presence of weed, the impact of the weed on human health (if any) and to mark out the possible dispersal agents, cause of aggressiveness and the suitable seasons for the distribution of this weed in both arable and non arable lands. A total of 100 local people (farmers) and development agents at the age of above 30, unequal number of farmers from each PA were identified and selected by using purposive sampling procedure. This selection was based on the awareness of local farmers about the aggressive colonization of *P. hysterophorus* on arable land and non-arable land (road sides) and its impact on plant biodiversity, livestock and on themselves. In order to get adequate information on the overall impact of parthenium in the study area, semi-structured interview questions were prepared. Furthermore, observations, interview, and focus group discussions was made with development agents and agricultural experts of each district.

Sampling of herbaceous vegetation cover

The field study was undertaken between January and September 2011 to collect the vegetation data depending up on the ecology of *P. hysterophorus*. Where massive growth of the weed occurs, road transect survey method was employed (Wittenberg et al., 2004) in 50 m distance to lay a quadrat. Since the study area is well known in its agroforestry system other than grazing land, one hundred and forty (140) quadrats (10 from each kebele) around roadsides, gardens and the farmlands of 14 sampled kebeles/PAs each measuring 1 m × 1 m (1 m^2) were laid in order to collect herbaceous vegetation data and assess impact of parthenium on aboveground herbaceous vegetation cover in this study area. GPS readings to record altitude, latitude and longitude for each sample site was recorded using GPS reader in order to locate the global position of each quadrat as well as the study site.

The majority of the plant species collected from the quadrats was identified in the field. For species difficult to identify in the field, voucher specimen were collected, pressed and dried properly using plant presses and transported to the Addis Ababa University, National Herbarium for identification and proper naming. The nomenclature of the plant species followed the Flora of Ethiopia and Eriterea (Hedberg and Edwards, 1995). In order to investigate the relative abundance and composition of the herbaceous vegetation as impacted by parthenium, the proportion of individual species (cover and abundance of the plant species) encountered in each of the quadrats was recorded using the procedure documented by Wittenberg et al. (2004). This method involves a total estimate based on abundance and cover of the species where invasion is spatially patchy. The total estimate scale (abundance plus coverage) can be shown as follows. A plant species covers a very small area (+), cover small (1), less or equal to 5% area coverage (2), 6 to 25% area coverage (3), 26 to 50% area coverage (4), 51 to 75% area coverage (5) and 76 to 100% area coverage (6). Following the methods suggested by Chellamuthu et al. (2005), the sample sites were categorized into different groups based on parthenium infestation levels: None, very low (< 10%), low (11 to 25%), moderate (26 to 50%) and high (> 50%) of the total percent area coverage of parthenium weed.

Data analysis

Diversity of the species for the vegetation data from the sample sites in the study areas were compared using Shannon Diversity Index. This index accounts both for the abundance and the evenness of the species in natural environment as shown by the equation below (Shannon and Wiener, 1949). It is also used to assess the impact of parthenium on the diversity of herbaceous plant species. The higher value of index of diversity indicates the variability in the type of species and heterogeneity in the community where as the lesser values point to the homogeneity in the community.

$$H' = -\sum_{i=0}^{S} Pi \ln Pi$$

Where H' = Shannon diversity index; Pi = the importance value of the ith species; S = total number of species in the sample quadrat
The evenness of species will be calculated as proposed by (Hill, 1973):

$$E = \frac{H'}{\ln S}$$

Where E= evenness
This index explains how equally abundant each species would be in the plant community and high evenness is a sign of ecosystem health. This is because it does not have a single species dominating the ecosystem. The evenness or equitability assumes a value 0 and 1 with 1 being complete evenness and 0 a single species dominating the area. The similarity of the standing vegetation (herbaceous vegetation layer) among the sample sites in this study area was compared using Jaccard's coefficient of similarity (JCS) as shown by the equation below. This coefficient of similarity has been recognized robust and unbiased compared with other similarity indices, even with small sample size (Ludwing and Reyonlds, 1988).

$$JCS = \frac{a}{a+b+c}$$

Where, JSC = Jaccard's coefficient of similarity; a = species common to quadrat 1 and 2; b = species present in quadrant 1 but absent in quadrat 2; c= species present in quadrat 2 but absent in 1

The coefficient has a value from 0 to 1, where 1 reveals complete similarity and 0 complete dissimilarity. The data collected from the respondents on the impact of parthenium were arranged and analyzed by using 'Microsoft office excels' for descriptive statistics (frequency and percentage). Moreover, to analyze vegetation data obtained from the field on the impact of parthenium weed, R.2.14.0.package was used. The correlation of vegetation variables such as species composition, species richness, average parthenium density/sample, and average number of flower head/plant among sample sites were done to check if there is any association among and between sample sites.

RESULTS

Farmers' perceptions about parthenium weed

Information collected from the respondents with regards to their perception on the parthenium weed was analyzed and also interpreted as follows (Table 1).

Data obtained from the responses given by local people (Table 1), show that 73% (73 in numbers) of the respondents from semi-urban and rural areas heard about parthenium weed and its presence in the area and they do have some information about the impact of parthenium on their surroundings and on themselves. Only 27% (27 in number) of the respondents did not hear and know about the presence of the weed in their surrounding and they did not have any information about the effect of the weed on animal, human health, and bio-diversity. As further evidence, the researchers observed the people cleaning the floor and front yard of their house with matured dried parthenium weed. From the total number of respondents who heard and knew about the weed before, 44% were from Dilla administrative town, which is a major town of Gedeo Zone, while residents are in and closer to the town and they observed dense parthenium growth in their surroundings. The highest numbers of respondents (46%, 46 in number) for this study were deliberately selected from Dilla town as the highest distribution of the weed has been observed.

From the table it can be seen that 88% (64 in numbers) of the respondents that heard and knew about parthenium weed before this study were able to identify it from other weeds by its morphology or physical form. As indicated in the Table 1, 85% respondents believed that the first appearance of parthenium was observed particularly in specific localities of Dilla town at the beginning of 2001 where donated food grain is stored (warehouse for grain storage) and temporary station for grain carrying trucks. According to the respondents (Table 1, item number 5), it is apparent that the weed seeds may have arrived with introduced grain and vehicles that carry them.

For the extent of parthenium weed distribution status in this study area (in item no.4), sixty eight percent of the respondents indicated that the weed have been highly distributed in Gedeo Zone specifically Dilla administrative town since the beginning of 2001 regardless of low

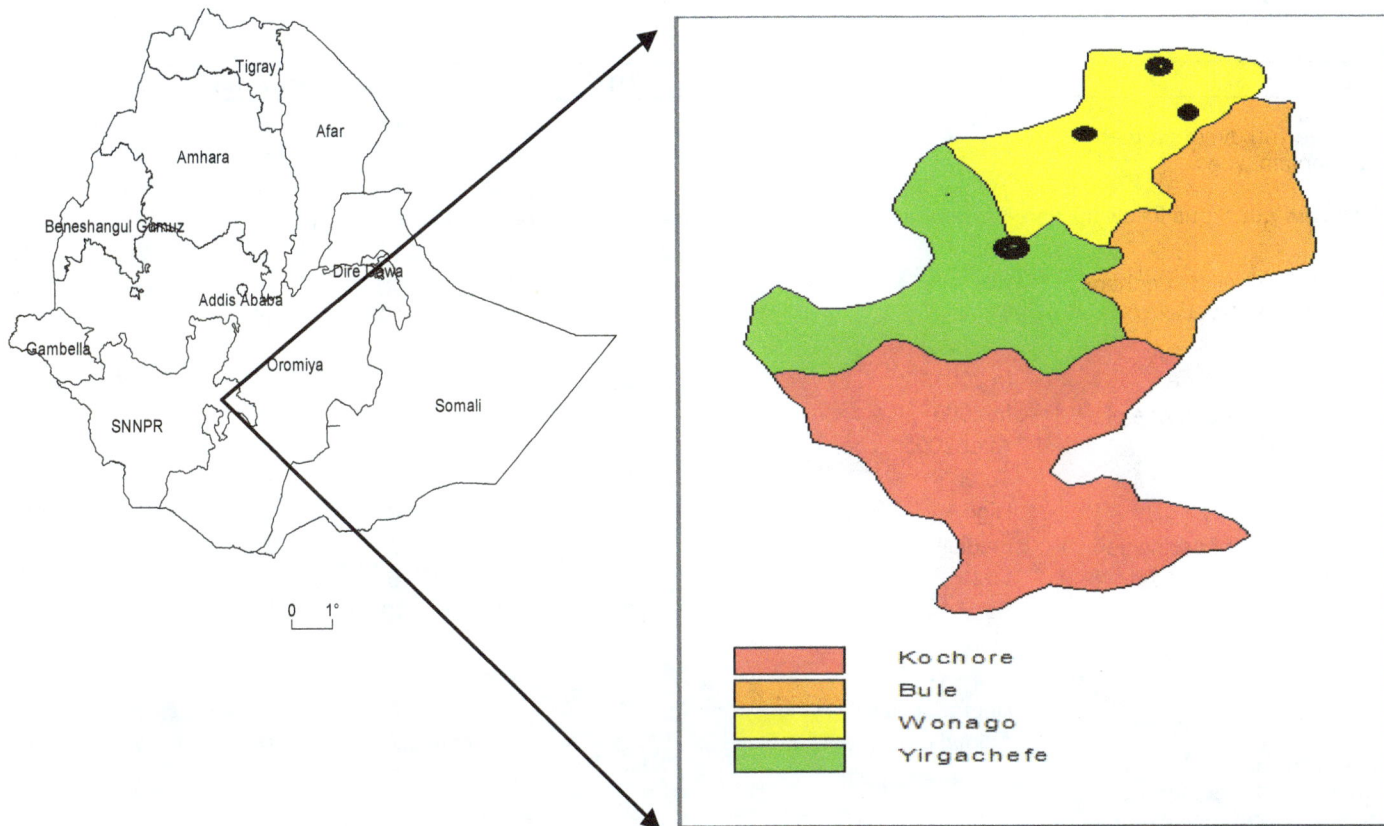

Figure 1. Map of the study area

distribution level of parthenium weed in *Yirga cheffe* and Wonago district.. As to the respondents in this study area, the agents for the wide spread dispersal of parthenium seed is mainly with the introduced grain donated for the food aid. This is supported by 72.6% of the respondents in this study area. The vehicle and farm implements as agents for the dispersal of the weed were supported by 17.8% of the respondents, flood (supported by 6.8%) and animal movement (supported by 2.7%).

However, 68.5% of the respondents explained that most parthenium invasion was observed along road sides of town (non-arable land) because of long distance dispersal of the seed by the vehicles and farm implements followed by crop field (16.4%) and wasteland (as supported by 15.0%). It was described by them that the road sides of Dilla administrative town are highly infested with parthenium weed unlike the other parts of Gedeo zone. Farmers perception on distribution and impact of parthenium weed From all the study sites, as mentioned by the respondents who knew parthenium weed (68.5%) ranked first because of its high spread and invasion on road sides and on the margin of farmland, *Lantana camara* ranked as first by 13.7% of the respondents, *Euphorbia pulcherrima* ranked as first by 8.2%, *Argemon mexicana* ranked as first by 5.5% of the respondents and *Cuscuta* spp. with 4.1% respondents

ranked first well known distributed weeds in their surroundings with negative effects on native plants and human health. Based on the above first rank frequency researchers gave the general ranking from 1st to fifth (Table 2).

In contrast, those respondents who were unaware of parthenium weed before the study (27% of the respondents) ranked *Euphorbia pulcherima* as the first, *Cuscuta spp* as the second, *A. mexicana* third and *L. camara* as the fourth well known and highly distributed weed in their surroundings. Those respondents who did not know parthenium weed and ranked the other weeds as the above mentioned one were mainly from the Peasant associations/kebeles of Dilla Zuria Woreda, *Wonago*, and Yirga Cheffe district. Moreover, the farmers from Wonago and Yirga Cheffe emphasized mainly on *Cuscusta* sp which is a parasitic liana of coffee plant in their homegarden. According to them, this parasitic liana on coffee reduces the annual production of coffee cherries. Therefore, a research should be conducted on the ecology, biology and management system of this parasitic plant on the area to prevent the coffee plant from being attacked by it. The respondents (100% of the respondents who were unaware of the weed) also explained that the parthenium weed has an impact on the growth and development of different plant species that

Table 1. Farmers perception on the weed

Items	Alternatives	DT	DZ	W'go	Y/Cheffe	Total	%
1. Have you ever heard about parthenium weed	Yes	32	15	10	16	73	73
	No	14	5	2	6	27	27
	Total	46	20	12	22	100	
Interview questions following were forwarded to only those respondents who heard about the weed (73 Respondents)								
2. Can you identify parthenium among other weed	Yes	32	10	10	12	64	88
	No	0	5	0	4	9	12
	Total	32	15	10	16	73	
3. The first appearance of weed in the area	In 2001/02	32	13	7	10	62	85
	In 2005	0	2	3	6	11	15
	After 2005	0	0	0	0	0	-
	Total	32	15	10	16	73	
4. The extent of parthenium distribution status in your area	High	32	10	0	8	50	68.5
	medium	0	5	5	2	12	16.5
	Low	0	0	5	6	11	15.0
	Total	32	15	10	16	73	
5. Which dispersal agents would you expect for high spread of parthenium?	Fodder	0	0	0	0	0	0
	Human activity	0	0	0	0	0	0
	Animal movement	0	0	2	0	2	2.7
	Introduced grain	25	10	8	10	53	72.6
	Wind	0	0	0	0	0	0
	Vehicle and farm implements	5	5	0	3	13	17.8
	Flood	2	0	0	3	5	6.8
	Total	32	15	10	16	73	
6. In which land use type is parthenium infestation highest	Roadside	24	9	7	10	50	68.5
	Crop field	4	3	2	3	12	16.4
	Wasteland	4	3	1	3	11	15.0
	Rangeland	0	0	0	0	0	0
	Others	0	0	0	0	0	0
Total		32	15	10	16	73	100

DT, Dilla Town; DZ, Dilla Zuria district; W'go, Wonago district; Y/Cheffe, Yirga Cheffe district.

Table 2. Weed mentioned by farmers as first rank (N=73).

Weed type mentioned	Respondents	%	Preference rank
Argemone mexicana	4	5.5	4
Cuscuta spp.	3	4.1	5
Euphorbia pulcherrima	6	8.2	3
Lantana camara	10	13.7	2
Parthenium hysterophorus	50	68.5	1

have either medicinal value or no negative effect on other crops or plants. As to their explanation, those parthenium impacted plant species are *Cyanodon dactylon, Amaranthus*spp., *Solanumincanum, Nicandraphysaloides, Oxalis corniculata* and *Solanum dasyphyllum* which are palatable and have various medicinal values. This was so because those plant species mentioned above have been decreasing from time to time in the highly infested area. According to the farmers and agricultural experts' observation, the plant species in their surrounding that outcompete parthenium weed for long period of time are mainly *Achyranthes aspera, Argemone mexicana Cynodon dactylon, Lantana camera, and Xanthium stumarium.*

Table 3. Sample sites of study area with infestation level and Land-use type.

Number	Kebele (PA)	Woreda	Number of sample	Infestation level *	Land use type
1	Aroresa	D.A. town	10	High	Cultivated land, garden, roadside
2	Ase Dela	D.A.town	10	High	Garden, road side
3	Buno	D.A.town	10	High	Roadside, wasteland, waterway
4	Haroke	D.A.town	10	High	Roadside, wasteland
5	Bareda	D.A.town	10	High	Roadside, wasteland, water way
6	Odaya	D.A.town	10	High	Garden, roadside, wasteland
7	Boyiti	D.A.town	10	High	Waterway, garden
8	Chicho	D.Zuria	10	Moderate	Road side, waterway
9	Andida	D.Zuria	10	Moderate	Roadside, waterway
10	Deko	Wonago	10	Low	Roadside, cultivated land
11	Tumata Chirecha	Wonago	10	Low	Roadside, closer to cropland
12	Bule Bukisa	Wonago	10	Low	Roadside, closer to cropland, wasteland
13	Konga	Cheffe	10	Low	Roadside
14	Cheffe town	Cheffe	10	Low	Roadside
Total	14		140		

D.A. town, Dilla Administrative town; D. Zuria, Dilla Zura woreda; *,Groups based on parthenium infestation levels: None, very low (< 10%); low (11-25%); moderate (26-50%) and high (> 50%) of the total percent area coverage of parthenium weed.

According to the respondents (100% who are aware of the weed), they did not recognize the impact of parthenium weed on the area but had heard about its impact on milk quality, cropland, human health and biodiversity. They described that if a cow feeds on grass mixed with parthenium weed the quality of milk would be reduced by saying that the "milk becomes bitter". They also heard that allergy on the skin of man can be caused by Parthenium weed contact. As long as the management is concerned, hundred percent of the respondents who knew the weed before this study only mentioned hand hoeing/hand pulling/uprooting and burning of parthenium weed in its early period before it sets flowers.

Vegetation data associated with parthenium weed

The vegetation data associated with parthenium weed with different invasion levels and the respective land-use type in 14 different sample sites/Kebeles are presented below in Table 3.

As can be seen from the table, 70/140 (50%) of the samples are taken from Dilla town due to the fact that there is high level of parthenium infestation in the area as compared to other sample sites. The basis for this classification was based on the parthenium weed cover in all sample sites of the kebeles in the town. The states of infestation are high in *Aroresa, Ase Dela, Buno, Haroke, Odaya, Bareda* and *Boyti* which are small kebeles in Dilla Administrative town. Except *Aroresa, Ase Dela and Odaya* where the parthenium weed are observed in both arable land and roadsides, the other parthenium infestation are only observed in non-arable land containing road side and wasteland.

Plant diversity, species composition and evenness

The herbaceous vegetation data associated with parthenium had shown 45 plant species recorded and classified under 20 families. Of 20 plant families, *Poaceae* accounts the largest (17.8%), Asteraceae accounts 15.6%, Amaranthaceae, Solanaceae and Euphorbiaceae accounts 8.9% each holding the third place among the plant species recorded in this study area.

The biodiversity impact of parthenium weed on highly infested areas was more visible than the moderate and low infested areas. The sampled sites in Dilla town starting from Aroresa to Boyti have lower diversity index (H) (highly infested area) as compared to other sites where the calculated Shannon Diversity Index is high (where invasion was low) (Table 4). Aroresa and Buno showed the lowest Shannon Diversity Index value of 1.6 and 1.37 respectively due to the fact that their species diversity was highly affected by the high infestation of parthenium. Comparatively, the Tumata Chirecha, Bule Bukisa, Andida Konga and Cheffe have a diversity Index of greater than 3.0 showing low parthenium infestation in the area and the impact of it on species diversity is low. Similarly, the evenness index was found to be higher in uninfested areas which indicated that the species are evenly distributed. This is true in this present study sites such as Chicho, Andida, Tumata Chirecha, Bule Bukissa, Konga and Cheffe where the evenness index were above

Table 4. Shannon Diversity index (H), Species richness (S) and evenness (E).

Study sites	H	S	E
Aroresa	1.6	32	0.46
Odaya	2.08	39	0.57
Ase Dela	1.66	40	0.45
Buno	1.37	33	0.39
Haroke	1.73	38	0.48
Bareda	1.76	37	0.49
Boyti	1.72	32	0.50
Tumata Chirecha	3.04	32	0.88
Bule Bukisa	3.13	34	0.89
Deko	2.98	29	0.88
Chicho	2.74	30	0.81
Andida	3.2	34	0.91
Konga	3.3	36	0.92
Cheffe town	3.41	39	0.93

0.80 which shows the sampled area were uninfested by the weed. In contrast, the fact that it was lesser in the weed-infested area indicated patchiness in distribution where a few species dominate the area.

This current study showed that the number of desirable species declined in the high parthenium infested sites of the study areas. Such a reduction could be attributed to the increasing abundance of the weed in the sites.

The result of this study goes in line with the study of Sakai et al. (2001) and Kohli et al. (2004) where invasive plants are known to exert significant impact on the natural communities as they cause their displacement and hence exert imbalance in the natural and agricultural ecosystem.

This imbalance causes the formation of large monoculture of invasive plants in the alien environment. The weed affects not only the species diversity of the native areas, but also their ecological integrity. Moreover, environmental degradation and disturbance favours the invader species such as parthenium weed. Under such circumstances the weeds easily establish themselves in the sites and start interfering with other native species by suppressing their potential growth and biomass production. Parthenium was known to suppress the associated species through the release of allelochemicals from decomposing biomass and root exudates into the soil environment (Pandey et al., 1993).

There is a high negative correlation between mean parthenium density and Shannon diversity index with R^2 = 0.89, p<0.001. The equation for the regression line is y = 398.6 - 126x indicated in the Figure 2. High negative correlation means that as the mean parthenium density increases, the Shannon Diversity Index decreases. This in turn shows the parthenium effect on the biodiversity of plants.

Jaccards similarity coefficients (JSC) of herbaceous vegetation

The standing herbaceous vegetation associated with parthenium among 14 sampled PAs were recorded and Jaccards Similarity coefficients were calculated (Table 5). The values in Table 5 are calculated by using Jaccards Coefficient of Similarity. This coefficient of similarity has values from 0 to 1, where 1 reveals complete similarity and 0 complete dissimilarity. Thus, the standing herbaceous vegetation from the sample sites of the Dilla town where there is high parthenium infestation level showed higher Jaccards Similarity coefficient (greater or equal to 0.683). This may be due to the complete dominance of parthenium weed over other palatable herbaceous plant species in both arable and non-arable lands of the study area. The high similarity value between parthenium invaded areas of Dilla town indicated that there is no radical change on species composition within the area. In contrast, other kebeles from Dilla zuria, Wonago, and Yirga Cheffe woreda showed relatively lower similarity coefficient than Kebeles from Dilla town revealing low impact of the weed in the area.

Comparison and correlation of mean parthenium density

For 14 PAs, the mean parthenium density and species richness were compared (Table 6). Aroresa Site in Dilla town has the first highest mean parthenium density (Table 6) which accounts $267/m^2$ but with relatively low species richness. Buno accounts the second highest mean parthenium density which is 235.5 with the species richness value of 33 and Ase Dela accounts the third highest with its mean parthenium density comprising species richness of the same value as of Buno. In general, the table shows the trend of decrease of species richness as the mean parthenium density is increasing in sites/kebeles. This is particularly true in Dilla town where the distribution of parthenium weed is highest as compared to other study sites. In contrast, species richness increases as mean parthenium density decreases as in the case of sample sites such as Chicho, Andida, Tumata Chirecha, Bule Bukisa, Deko, Konga and Chefe town. This definitely shows the impact of parthenium weed on species diversity and hence in species richness of the study area.

Mean of parthenium height, branches and flower head (capitula)

The mean of parthenium height, number of branches and number of flower heads counted in three different times (at the beginning of February, April, and June) in 14 sample sites are shown in the (Table 7).

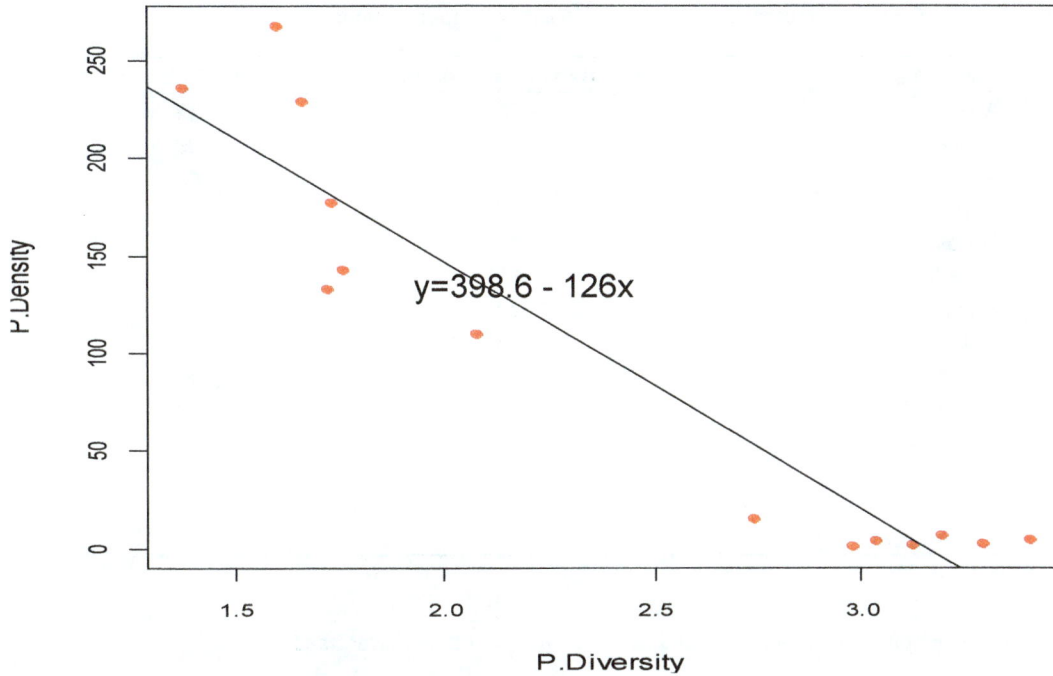

Figure 2. Correlation between parthenium density and Shannon Diversity index.

Table 5. The JSC of the standing herbaceous vegetation among 14 kebeles.

	Aroresa	Odaya	Ase Dela	Buno	Haroke	Bareda	Boyti	Tumata Chirecha	Bule Bukisa	Deko	Chicho	Andida	Konga	Yirga Cheffe
Aroresa														
Odaya	0.77													
Ase Dela	0.85	0.86												
Buno	0.73	0.76	0.76											
Haroke	0.88	0.83	0.93	0.73										
Bareda	0.85	0.81	0.90	0.71	0.88									
Boyti	0.68	0.73	0.73	0.97	0.71	0.68								
Tumata Chirecha	0.62	0.55	0.58	0.55	0.60	0.65	0.48							
Bule Bukisa	0.64	0.61	0.65	0.63	0.67	0.73	0.55	0.91						
Deko	0.57	0.55	0.58	0.51	0.60	0.61	0.45	0.81	0.85					
Chicho	0.52	0.60	0.57	0.58	0.55	0.60	0.52	0.60	0.68	0.74				
Andida	0.61	0.66	0.66	0.56	0.64	0.69	0.50	0.62	0.69	0.70	0.88			
Konga	0.66	0.71	0.71	0.61	0.68	0.70	0.48	0.57	0.64	0.67	0.74	0.84		
Yirga Cheffe	0.75	0.80	0.80	0.66	0.77	0.56	0.55	0.64	0.71	0.75	0.75	0.85	0.90	

From the table, the highest mean height (1.6 m) of parthenium was recorded in Aroresa kebele. The specific localities where this height recorded were along the way to Dilla University, main Campus, Dilla. The average number of branches counted in the same localities was 16 followed by Odaya where number of branches per plant was 11 with an average height of 1.2 m. Generally, the findings of this research shows a decreasing trend of mean height, number of branches, and flower head/plant from Aroresa to Cheffe town (that is, from sample sites along the way from Dilla town, Dilla Zuria wereda, wonago to Yirga Cheffe wereda).

Table 6. Comparison of mean parthenium density and species richness.

Sample sites	Mean parthenium density	Species Richness (R)
Aroresa	267	32
Odaya	110.2	39
Ase Dela	229	40
Buno	235.5	33
Haroke	177.2	38
Bareda	143	37
Boyti	132.8	32
Chicho	15.6	30
Andida	7.1	34
Tumata Chirecha	4	32
Bule Bukisa	2	34
Deko	1	29
Konga	3	36
Cheffe town	5	39

Table 7. Comparison of mean height, stem branches and number of flower heads.

Sites/Kebele	Mean height (cm)	Number basal branches/plant	Flower heads/plant
Aroresa	160	16	4751.0
Odya	130	11	4171.0
Ase Della	110	8	3663.3
Buno	100	9	3476.7
Haroke	80	12	2132.3
Bareda	82	10	1681.0
Boyti	78	7	1598.0
Chicho	50	5	114.0
Andida	30	6	112.0
Tumata Chirecha	35	5	56.0
Bule Bukisa	40	6	52.3
Deko	35	5	49.7
Konga	25	6	51.7
Cheffe	30	6	46.7

The mean number of flower head/plant in sample sites

The mean number of flower head counted per plant showed a considerable increase from February, April to June particularly in highly infested sites of Dilla town. The increased number of flower head/plant counted in June may be because of small rain that the area has got during the time of data collection. From this study, it is found that the average number of flower heads/capitulla counted in June was greater than that of the remaining two rounds for all the sites in the Dilla town where the level of infestation was higher. In comparison, Aroresa kebele has the highest average number of flower head/capitula counted (4751) in three rounds followed by Odaya (4171). The average numbers of flower heads counted for the sample sites other than Dilla Town are very low as

its level of infestation was also moderate and low.

The competitive ability of plant species

Even though the experiment on the competitive ability of plants was not done in this present study, it is feasible to list down plant species that frequently occurred in the sampled sites of infested areas tolerating the ill effects of the weed. Thus, the plant species that can tolerate the effect of parthenium weed have the highest relative frequency indicating its frequent occurrence in both highly infested and low infested areas (Table 8).

It is obvious that the occurrence of parthenium weed was in all 140 sampled sites (100% relative frequency) even though their abundance varies. Accordingly, *Cyanodon dactylon* was the second frequently occurred

Table 8. The relative frequency of ten plant species in the sample sites.

Plant species	Frequency	Relative frequency
Achranthus aspera L.	130	93
Ageratum conyzoides L.	132	94.3
Amaranthus caudatus	120	85.7
Cyanodon dactylon	135	96.4
Datura stramonium L.	56	40
Eragrostis spp	115	82.14
Euphorbia hirta	50	35.7
Lantana camara L.	78	55.7
Parthenium hysterophorus	140	100
Xanthium strumarium	100	71.4

plant species with the relative frequency of 96.45 in this study area and thirdly *Ageratum conyzoides* which occurred in 132 sampled sites out of 140 (94.3%). This is so because *P. hysterophorus, Ageratum conyzoides* and also *Lantana camara,* are tropical in origin and they possess similar growth strategies. They grow fast, have short life cycle and except *Lantana camara,* they have greater reproductive potential, competitive ability, and allelopathy that make them successful invaders of non native habitat (Grice, 2006). Due to its high growth rate, *Parthenium* becomes competitive and develops the ability to exclude the growth of other species.

DISCUSSION

Studies by Taye (2007), show heavy and widespread infestation mostly on roadsides, wastelands, towns, villages and gardens in the central farmlands of east Shewa: Dukem, Bishoftu, Modjo, and Koka areas. One can also see parthenium infestation on field borders and in some fields; parthenium grew in crop field during fallow period. In Ziway, Awassa and Wolkite, the weed has been observed only in the town along the road and near dwelling sites indicating its recent introduction into the area. This report is in agreement with the present finding which shows us that roadside of Dilla town and surrounddings are highly infested with the weed. Moreover, this was also observed by the researchers during the period of data collection and some authors (Haseler, 1976) who reported the initial occurrence of *P. hysterophorus* in a new area usually occurs along roadsides and it is from this foothold that it spreads extensively into agricultural land.

Tamado and Milberg (2000) during his research in Eastern Ethiopia found that 90% of the interviewed farmers rank parthenium as the first and most serious problem both in the rangeland and croplands. Similarly, parthenium has a great impact on palatable plant species in this study area as mentioned previously in the result from

the respondents of this present study. Moreover, this is evidenced by Oudhia (2000) who reported as because of its efficient biological activity and adaptability to varying soils and microenvironments, *Parthenium* weed has a tendency to replace the dominant flora in wide range of habitats cutting across state boundaries and agro-climatic regions. Very little or sometimes no other vegetation can be seen in *P. hysterophorus* dominated areas. Wherever it invades, it forms a territory of its own by replacing the indigenous natural flora including medicinal herbs utilized by man as a source of medicine. Its allelopathic properties, which cause inhibition of germination and suppression of the natural vegetation including many medicinal herbs, pose a strong threat to biodiversity.

The study conducted by Shabbir and Bajwa (2007) showed that impact of *P. hysterophorus* on livestock production is significant, both directly and indirectly, and affects grazing lands, animal health, and milk and meat quality, and the marketing of pasture seed and grain. The initial symptoms of allergy caused through Parthenium contact are described as itching, redness, swelling and blisters on eyelids, face and neck, which then spread to the elbows and knees. In the later stages, the skin thickens and darkens. The allergic reactions include hay fever, asthma or dermatitis and can be caused by the dust, debris or volatile fumes from the plant as well as its pollen (Kohli et al., 2004). However, from this present study, the interviewed people explained that they did not recognize the impact in the study area but heard about the impact of the weed on milk quality, cropland, human health and biodiversity.

The study done by Mohammed (2010) showed that some of the preventive methods that restrict the entrance of weed seed into a non-infested area are uprooting of the weed before flowering and seed setting. They indicated it as the most effective and less costly strategy to manage parthenium (and it is to be practiced only in small areas like in gardens, flower beds, intensively cultivated fields or high value crops). This present study reveals that there was a sharp decline of diversity index as the density of parthenium increased. This finding is similar with Kohli et al. (2004) findings where the Shannon index showed great plant diversity in uninfested area whereas the index was reduced by 36 to 51% in the weed infested areas. Therefore, the higher value of the diversity index indicates the variation in the type of species and the heterogeneity in the community, whereas the lesser value points the homogeneity in the community. In areas of high parthenium weed infestation there was also a high mean parthenium density per m^2 as depicted in Figure 2. This is probably due to high viability of the parthenium weed seed banks in soil. Thus parthenium weed density per m^2 is often increasing in every generation, unless intervention is taken to control its spread (Tamado, 2001).

The number of flower head/capitula for this present study is much higher than the number studied by Labrada (1988) where a single plant producing an average of 810 flower heads. This might be because of the fact that the season for data collection was more suitable for parthenium growth and development. In the Caribbean area parthenium flowers 30 to 45 days after germination and the whole plant cycle is completed within about 5 months. A photoperiod of 13 h and warm conditions are conducive to flowering (Tamado et al., 2002). Moreover, the above data showed certain similarity with the study done in North-Western Indian Himalia by Dogra et al. (2011). They described that the inflorescence bearing flowers started appearing in late April or early May. The flowering stage in some plants lasted till June and then seed setting started. In the rainy season (July, August and September), some new seedlings emerged and they also flowered in August. In September, the seed setting was observed in most of the plants. The seeds fully ripened in the month of October and November or it completed its life cycle by that time.

The plant species that were identified from this present study which have parthenium competitive ability include *Cyanodon dactylon, Ageratum conyzoides, Xanthium strumarium, Lantana camara and others mentioned in the result section.* They grow fast, have short life cycle and except *Lantana camara,* they have greater reproductive potential, competitive ability, and allelopathy that make them successful invaders of non native habitat (Grice, 2006). Due to its high growth rate, parthenium becomes competitive and develops the ability to exclude the growth of other species. The inhibitory allelopathic effect of parthenium on the germination and seedling development of *Ageratum conyzoides* (Singh et al., 2002), *Eragrostis tef* (Tadele, 2002), *Brassica* spp., *Cicer arietinum* and *Raphanus sativus* (Batish et al., 2005) were studied.

ACKNOWLEDGEMENTS

We thank Dilla University Research and Dissemination Office for financial support and Agricultural and Development Offices of *Gedeo Zone,* Dilla Zuria Woreda, Wonago Woreda, Yirga Cheffe Woreda and Urban Agriculture offices of Dilla Administrative town for providing invaluable information.

REFERENCES

Adkins SW (1996). The allelopathic potential of parthenium weed (*Parthenium hysterophorus* L.). Australia Plant Prot. Q. 11:20-23.

Amaha K (2003). Pastoralism and the need for the future intervention in pastoral areas of Ethiopia. Annual Review on National Dry land Agriculture Research System, Addis Ababa, Ethiopia.

Batish DR, Sing HP, Pandher JK, Kohli RK (2005). Phytotoxic effect of *Parthenium hysterophorus* L. residues on three *Brassica* species. Weed Biol. Manage. 5: 105-109.

Chellamuthu A, Nadanassababady P, Rammohan J (2005). Present status of *Parthenium hysterophorus* L. in the coastal regions of Pondicherry and Karaikal. In: Proceeding of the Second International Conference on Parthenium Management 5-7 December2005. University of Agricultural Science Bangalore, India, pp. 44-47.

Chippendale JF, Panetta J (1994). The costs of parthenium weed in Queensland cattle industry. Plant Prot. Q. 9:73-76.

Dogra KS, Sood SK, Sharma R (2011). Distribution, Biology and Ecology of *Parthenium hysterophorus* L. (Congress Grass) an invasive species in the North-Western Indian Himalaya (Himachal Pradesh). Afr. J. Plant Sci. 5(11):682-687.

EARO (2003). Ethiopian Agricultural Research Organization. National dry land research strategic planning pastoral and agro pastoral research program. Addis Abeba, Ethiopia. p. 60.

Evans HC (1997). *Parthenium hysterophorus*: A review of its weed status and the possibilities for biological control. Biocon. N. Info. 18:89-98.

Gemedo D, Maass BL, Isselstein J (2006). Rangeland condition and trend in the semiarid Borena low lands, Southern Oromia Ethiopia. Afr. J. Range. Forage Sci. 23:49-58.

Grice AC (2006). The impacts of invasive plant species on the biodiversity of Australian rangelands. Rangeland J. *28:1-27.*

Haseler WH (1976). *Parthenium hysterophorus* L. in Australia. PANS 22:515-517.

Hedberg I, Edwards S (1995). Flora of Ethiopia and Eritrea. Vol. 7. The National Herbarium, Addis Ababa, Ethiopia. 660 pp.

Hill MO (1973). Diversity and its evenness, a unifying notation and its consequences. J. Ecol. 54:427–432.

Kohli RK, Batish DR, Singh HP, Dogra K (2006).Status, invasiveness and environmental threats of three tropical American invasive Weeds (Parthenium hysterophorus L., Ageratum conyzoides L., Lantana camara L.). Biological Invasions 8:1501-1510.

Kohli RK, Dogra K, Daizy J, Singh, RB (2004). Impact of invasive plants on the structure and composition of natural vegetation of Northwestern India, Himalayas. Weed Technol. 18:1296-1300.

Labrada R (1988). Complemento al estudio biológico de Parthenium hysterophorus L. Resumenes IX Congreso ALAM, julio 26-30, Maracaibo, Venezuela.

Ludwing JA, Reyonlds JF (1988). Statistical Ecology. Wiley inter Science, New York. p. 111.

Mohammed W (2010). Prevalence and distribution survey of an invasive alien weed (*Parthenium hysterophorus* L.) in Sheka zone, Southwestern Ethiopia. Afr. J. Agric. Res. 5(9): 922-927.

Mulatu W, Gezahegn B, Solomon T (2009). Allelopathic effects of an invasive alien weed *Parthenium hysterophorus* L. compost on lettuce germination and growth. Afr. J. Agric. Res. 4(11):1325-1330.

Oudhia P (2000). Medicinal Herbs and Insects of Chhattisgarh India. Insect Environ. 6:138.

Pandey DK, Kauraw LP, Bhan, WM (1993). Inhibitory effect of *Parthenium* (*Parthenium hysterophorus*) residue on growth of water hyacanth (*Eizhhornia crassipes* Mart Solms.II): Effect of leaf residue. J. Chem. Ecol. 19:2651-2662.

Rezene F, Meckasha C, Mengistu H (2005). Spread and Ecological consequences of *Parthenium hysterophorus,* in Ethiopia. Arem 6:11-23.

Sakai AK, Allendorf FW, Holt JS, Lodge DM, Molofsky J, Baughman KA, Cabin S, Cohen, RJ, Allstrand JE, McCauley NC, O'Neil DE, Parker IM, Thompson JN, Waller SG (2001). "The population biology of invasive species" Ann. Rev. Ecol. Syst. 32:305-332.

Seifu W (1990). *Parthenium hysterophorus* L. A recently introduced weed to Ethiopia.Preliminary reconnaissance survey report on Eastern Ethiopia, East Harargae, Ministry of Agriculture, Ethiopia.

SERP. (1990). South Eastern Rangelands Development Project. Assessment of drought impact in Eastern Ethiopia, Jijiga. SERP Technical Report (Draft). p. 112.

Shabbir A, Bajwa R (2007). Parthenium Invasion in Pakistan – A Threat Still unrecognized. Pak. J. Bot. 39(7):2519-2526.

Shabbir A, Swhsana RB (2005). *Parthenium hysterophorus* spread and status on its management in Pakistan. pp. 28-35. In: Proceeding of the Second International Conference on Parthenium Management. 5-7 December. 2005. University of Agricultural Science, Bangalore, India.

Shannon CE, Wiener W (1949).The Mathematical Theory of Communication. University of Illinois Press, Urbana, Illinois.

Singh HP, Batish DR, Kohli RK, Saxena DB, Arora V (2002). Effect of pathenin - A sesquiterpene lactone from *Parthenium hysterophorus* L. - on early growth and physiology of *Ageratum conyzoides*. J.

Chem. Ecol. 28(11):2169-2179.

Tadele T (2002). Allelopathic effects of *Parthenium hysterophorus*. L. extracts on seed germination and seedling growth of *Eragrostis tef*. J. Agron. Crop Sci. 188:306-310.

Tadesse K (2002). Five thousand years sustainability. A case study in Gedeo Land Use (Southern Ethiopia). Treemail publishers, Heelsum, The Netherlands. p. 296.

Tamado T, Milberg P (2000). Weed flora in arable fields of Eastern Ethiopia with emphasis on the occurrence of *Parthenium hysterophorus L*. J. Weed Res. 40:507-521.

Tamado T, Schutz W, Milberg P (2002). Germination ecology of the weed *Parthenium hysterophorus* L. in eastern Ethopia. Ann. Appl. Biol. 140:263-270.

Taye T (2007). The prospects of biological control of weeds in Ethiopia. Ethiopian J. Weed Manage. 1(1):63-78.

Wittenberg R, Simons SA, Mauremootoo JR (2004). Instrument and tools for assessing the impact of invasive alien species in Africa. Report Procedures under the PDF-B phase of UNEP GEF Project-Removing Barriers to invasive plant Management in Africa. CAB. International. Nairobi, Kenya.

17

Economics of using biofertilisers and their influence on certain quantitative traits of mulberry

M. F. Baqual

Sher-e-Kashmir University of Agricultural, Sciences and Technology of Kashmir (K), Division of Sericulture, Mirgund, P. Box 674, G.P.O., Srinagar – 190001 (J & K), India.

Significant variation was recorded in various quantitative parameters of the mulberry with the application of various bio-inoculants like phosphate solubilising microorganisms, nitrogen fixing bacteria and arbuscular mycorrhizal fungi. The longest shoot length was recorded in T_4 (175.43 cm) receiving 350: 140 kg/ha/year N and P along with co-inoculation followed by T_6 (172.78 cm) receiving 175: 70 kg/ha/year N and P along with co-inoculation respectively. The treatment T_8 receiving 175 kg nitrogen and 70 kg phosphorus/ha/year (with rock phosphate as a source of P) has indicated significantly higher number of shoots per plant (6.43) as compared to T_0 control recording only (5.64) number of shoots per plant. Significantly higher number of leaves per shoot was recorded in treatment T_8 (27.53) as compared to the rest of the treatments including T_0 (26.37) as control. The study also clearly indicated that it is possible to curtail the application of nitrogenous and phosphatic fertilizers in mulberry cultivation to an extent of 25 to 50% without any adverse effect on leaf yield and quality by supplementing nitrogen and phosphorus through the use of nitrogen fixing bacteria *Azotobacter*, phosphate solubilizing bacteria, fungi and VA-mycorrhiza. The studies on economics of using bio-inoculants in mulberry also indicated that approximately Rs 2000 to 4500 /ha/year can be saved only on the input cost of nitrogen and phosphorus in mulberry cultivation.

Key words: Mulberry, biofertilisers, economics.

INTRODUCTION

With the increase in scientific knowledge regarding the negative effects of using various inorganic fertilizers, the application of alternative source viz., biofertilisers, in various agricultural cropping systems has gained more and more popularity. Frequent use of chemical fertilizers for a prolonged period deteriorates the surface soil characteristics and affects the availability and uptake of nutrients by the plants (Subbaswamy et al., 1994). The biofertilisers have been proved to enhance crop yield both, if applied singly or in combination with other inoculants. The role of VAM in the nutrition of agricultural and horticultural crops has received much attention (Tinker, 1978; Menge et al., 1977). Mycorrhizal plants of mulberry are found to be superior in survivability, plant growth, biomass production and leaf quality in comparison to non-mycorrhizal plants (Fatima et al., 1996). Umakant and Bagyaraj (1998) working on the response of mulberry saplings to inoculation with VAM and *Azotobacter* reported that dual inoculation of nursery bed with *Glomus fasciculatum* and *Azotobacter chroococum* considerably increased the plant growth and mulberry sapling development. Further, inoculation of soil with these organisms enriches rhizosphere microflora which can have a vital influence on plant growth through mineral uptake (Sukhada, 1988). The available information on the combined use (co-inoculation) of vesicular arbuscular mycorrhiza, nitrogen fixing bacteria, phosphate solubilizing bacteria and fungi have indicated their use as enhancement in soil fertility and thereby improved plant growth through their increased biological activity in the rhizosphere Subba Rao (1998). Mulberry plants, whose leaf is exclusively used for rearing silkworms (*Bombyx mori* L.) for subsequent cocoon production, have indicated their positive response towards the application of biofertilisers. The beneficial effect of inoculation of mulberry plants with *Azotobacter* bio-fertilizer and Vesicular Arbuscular mycorrhiza have been well documented by Das et al. (1994a, b), Katiyar et al. (1995) and Baqual et al. (2005).

In India, the mulberry is grown in abundance, mostly in the tropical belt where most of the regions are dominated by soils that are low in available nutrients and moisture Osonubi et al. (1991). Therefore, it becomes imperative to use low cost biofertilisers with added advantages in terms of crop production and sustainability of sericulture. With this background, the present, study involving combined use of Vesicular Arbuscular mycorrhiza fungi viz., *Glomus fasciculatum* and *Glomus mosseae*, nitrogen fixing bacteria viz., *Azotobacter* and phosphate solubilizing micro-organisms viz., *Bacillus megaterium* and *Aspergillus awamori* was taken in order to generate some information about their influence on quantitative traits of mulberry and to work out the economics of using biofertilisers viz-a-viz inorganic fertilizers in mulberry cultivation.

MATERIALS AND METHODS

The experiment which was laid in randomized block design was conducted under field conditions in one year old established irrigated mulberry garden of V1 variety, situated at Central Sericultural Research and Training Institute (CSR and TI farm) Mysore, India. The experiment comprised of twelve treatments and three replications for each treatment. The treatments comprised of two levels of inoculation (I_0: No inoculation and I_1: inoculation), two sources of phosphorus (S_1: Single super phosphate and S_2: Rock phosphate) and three fertilizer doses (F_1, F_2, F_3 as recommended, $3/4^{th}$ of recommended and ½ the recommended doses of nitrogen and phosphorus respectively). The recommended doses of nitrogen and phosphorus are 350 and 140 kg /ha/year respectively. The plots were inoculated with a mixed culture of mycorrhiza containing spores of G. fasciculatum and G. mosseae by intercropping technique with maize as mycorrhizal host (Katiyar et al., 1998).

For this purpose small furrows were dug in between the rows of mulberry and maize grains were evenly placed in these furrows enabling them to germinate and aid in colonization after establishing root system. All the plots which were inoculated with arbuscular mycorrhiza were then subsequently inoculated with *Azotobacter* at 20 kg/ha/year and phosphate solubilizing bacteria (*Bacillus megaterium*) at 5 kg /ha/year, phosphate solubilizing fungi (*Aspergillus awamori*) at 5 kg /ha/year in five equal splits corresponding to five crop harvests. The bacterial biofertilizer and

fungi were used by mixing with 200 kg of powdered farm yard manure and applied near the rhizosphere of mulberry by making furrows. The chemical fertilizers nitrogen and phosphorus were applied in five split doses as per the doses mentioned in the treatments above. However, potassium was applied at 140 kg /ha/year in the form of muriate of potash as common dose irrespective of treatments. The fertilizers were applied after a gap of 10 to 12 days of the application of bio-fertilizers. The various parameters were estimated and the economics of using biofertilisers was worked out.

RESULTS AND DISCUSSION

Effect on longest shoot length

The data revealed a significant variation in longest shoot length due to different treatments. Among different treatments, the treatments T_4 (175.43 cm), T_6 (172.78 cm) and T_9 (174.65) have indicated prominent effect of co-inoculation with different micro-organisms. All these treatments were found to be statistically at par with T_0 (173.27 cm) as control (Table 1). However, treatment T_6 received half the recommended dose of nitrogen and phosphorus where phosphorus was applied as single super phosphate while the treatment T_9 received full dose of nitrogen and phosphorus where phosphorus was applied in the form of rock phosphate. These indicate that the application of fertilizer nitrogen and phosphorus in mulberry cultivation can be reduced by 50% of the recommended dose through the use of beneficial microbes like phosphate solubilizing micro-organisms, VA-mycorrhiza and nitrogen fixing bacteria without any adverse effect on the growth of the plants. Further, the use of rock phosphate can also bring down the cost of mulberry cultivation, as it is much cheaper than the single super phosphate. Almost similar trend was recorded in the second year also. Among different treatments, the treatments T_6 (182.15 cm) and T_9 (182.38 cm) have indicated most prominent effect of co-inoculation. However, unlike first year all these treatments were found to be highly significant over T_0 (174.74 cm) as control (Table 1). Thus the data of the second year in respect of shoot growth confirms the beneficial effect and higher efficacy of co-inoculation of mulberry as compared to the first year.

Effect on number of shoots per plant

Regarding number of shoots per plant, in general the co-inoculation of mulberry has resulted significant increase over un-inoculated treatments. In the first year, the treatment T_8 has indicated significantly higher number of shoots per plant (6.43) as compared to all other treatments including T_0 (5.64) as control except the treatment T_4. Almost similar result was also recorded in the second year in respect of number of shoot per plant. Maximum number of shoots per plant was recorded in the

Table 1. Effect of microbial co-inoculation on various growth parameters of V_1 variety under different levels and sources of phosphorus and nitrogen.

Treatments	No. of leaves/shoot		Longest shoot length (cm)		No. of shoots/plant	
	I Year	II Year	I Year	II Year	I Year	II Year
T_0	26.37	26.42	173.27	174.74	5.64	5.54
T_1	25.29	24.23	172.78	175.88	5.72	5.76
T_2	25.17	26.59	164.23	167.92	5.83	5.58
T_3	24.73	25.31	167.72	173.35	5.57	5.56
T_4	27.31	28.07	175.43	175.25	6.14	6.33
T_5	25.50	26.13	168.36	173.01	5.36	5.78
T_6	25.47	26.62	172.78	182.15	5.59	6.39
T_7	24.61	26.14	167.62	172.60	5.62	6.19
T_8	27.53	28.70	163.76	166.07	6.43	6.58
T_9	25.66	27.75	174.65	182.38	5.53	6.48
T_{10}	23.40	24.35	161.18	172.41	5.56	5.63
T_{11}	23.78	27.42	173.35	182.67	5.60	5.63
Cd at 5%	0.657	0.906	2.56	5.25	0.304	0.409

Treatment details, T_0, $F_1S_1I_0$; T_1, $F_1S_2I_0$; T_2, $F_2S_1I_0$; T_3, $F_2S_2I_0$; T_4, $F_1S_1I_1$, T_5, $F_2S_1I_1$; T_6, $F_3S_1I_1$; T_7, $F_3S_2I_1$; T_8, $F_2S_2I_1$; T_9, $F_1S_2I_1$; T_{10}, $F_3S_2I_0$; T_{11}, $F_3S_1I_0$; F_1, Recommended dose of nitrogen and phosphorus; F_2, 3/4th of recommended dose of nitrogen and phosphorus; F_3, ½ of recommended dose of nitrogen and phosphorus; I_0, no inoculation; I_1, inoculation; S_1, single super phosphate (SSP); S_2, rock phosphate (RP).

treatment T_8 (6.58) followed by T_9 (6.40), T_6 (6.39) and T_7 (6.19). All these treatments have received co-inoculation and different levels and sources of fertilizer. All these treatments were also significantly higher over treatment T_0 (5.54) as control. Thus from the analyzed data especially after perusing the treatments T_7 and T_8 it was clearly observed that use of different micro-organisms in association with reduced doses of nitrogen and phosphorus is highly useful in mulberry cultivation. The use of rock phosphate can bring down the cost of cultivation (Table 1). This confirms the observations of the first year.

Effect on number of leaves per shoot

First year data revealed that maximum number of leaves per shoot was recorded in treatment T_8 (27.53) which was significantly higher as compared to the rest of all the treatments including T_0 (26.37) as control (Table 1). The second year data also revealed a similar trend of the effect of co-inoculation of mulberry on leaf production. Higher number of leaves per shoot was recorded in almost all the treatments which received co-inoculation. However, maximum number of leaves per shoot was recorded in the treatment T_8 (28.70) which was significantly higher as compared to all the treatments including T_0 (26.42) as control except for T_4 (28.07) with which it was at par. Thus the second year's data also confirms the observation of the first year.

The overall observations have indicated that there has been a significant improvement in quantitative parameters of mulberry at a reduced cost which will in turn improve the productivity of cocoons thereby attracting more and more farmers towards the venture. More so, in view of tremendous pressure on agricultural land due to urbanization and heavy industrialization increasing vertical productivity is or high relevance and need of the hour.

Economics of co-inoculation of mulberry using beneficial micro-organisms

Although chemical fertilizers have played a better role in increasing the yield of agricultural crops, yet enormous increases in their prices has resulted in creating a distress for the farmers who are stuck with multi-pronged problems like marginal land holdings, erratic availability of these fertilisers and finally soil fertility degradation due to their excessive and constant use. Further sole dependence upon chemical fertilisers will only aggravate the situation in the coming future as such the need of hour is to integrate the use of chemical as well as biofertilisers for their effective and economic utilization for mulberry crop growth.

The economic potentiality of using beneficial micro-organisms as co-inoculants is tremendous. Since, the use of chemical fertilizers in mulberry cultivation is highly expensive the use of beneficial micro-organisms like *Azotobacter*, phosphate solubilizing bacteria, VA-mycorrhiza can reduce the application of nitrogen and phosphorus by 25 to 50% of the recommended dose. The economics of mulberry cultivation with or without using phosphate solubilizing micro-organisms, nitrogen fixing bacteria and VA- mycorrhiza was also calculated keeping

Table 2. Economics of using phosphate solubilizing microorganisms (PSM) Azotobacter, va-Mycorrhiza and rock phosphate in mulberry cultivation using 3/4th dose of nitrogen and phosphorus.

Treatments			Control		
Inputs	Quantity	Cost/ha/year (Rs)	Inputs	Quantity	Cost/ha/year (Rs)
FYM	20 MT	9600.00	FYM	20 MT	9600.00
AMM. SUL.	1312.5 kg	6536.00	AMM. SUL.	1750 kg	8715.00
RP	525 kg	1365.00	SSP	875	2800.00
MOP	233 kg	962.00	MOP	233 kg	962.00
PSB	5 kg	275.00	-	-	-
VAM	1000 kg	200.00*	-	-	-
Azotobacter	23 kg	1265.00	-	-	-
Total		20,203.00			22,077.00

The economics are worked out for V -1 mulberry variety under 350: 140:140 fertilizer dosage. Difference (Control -Treatment) = 22,077 - 20,203 = 1874.00 net saving over control; FYM: Farm Yard Manure, Amm. Sul: Ammonium Sulphate,RP:Rock Phosphate, MOP: Muriate of Potash, PSB: Phosphate solubilising bacteria, VAM: Vesicular Arbuscular Mycorrhiza, *The depreciation cost of VAM for 15 years.

in view the prevailing cost of various fertilizers, manures and other inputs. The values arrived at indicated that with the use of these bio-fertilizers, an amount of Rs. 2000 to 4500 can be saved /ha/year on mulberry cultivation without adversely affecting the yield. The input cost on account of these fertilizers and manures in control is Rs. 22,077; it is only Rs. 20,203 (If 3/4th dose of recommended nitrogen and phosphorus is used) and Rs.17,570 only, using half of the recommended dose of N and P fertilizers (Tables 2).

Further, the fertilizer industry posses a major threat upon our fossil remains which demands major cut in their use for dual purpose of decreased health hazards associated with synthetic fertilizer use and for overall sustenance of natural resources.

REFRENCES

Baqual MF, Das PK, Katiyar RS (2005). Effect of arbuscular mycorrhizal fungi and other microbial inoculants on chlorophyll content of mulberry (Morus spp.). Mycorrhiza News. 17(3):12-14.

Das PK, Katiyar RS, Gowda MH, Choudhury PC, Datta RK (1994a). Effect of Vesicular arbuscular mycorrhizal inoculation on growth and development of mulberry (Morus spp.) saplings. Ind. J. Seric. 1:15-17.

Das PK, Choudhury PC, Gosh A, Katiyar RS, Mathur VB, Madhava Rao AR, Mazumder MK (1994b). Studies on the effect of bacterial biofertilizer in irrigated mulberry (Morus alba). Ind. J. Seric, 33(2):170-173.

Fatima PS, Das, PK, Katiyar, RS, Himantraj MT, Pallavi SN (1996). Effect of VAM inoculation in mulberry under different levels and sources of phosphorus on silkworm growth, cocoon yield and quality. Indian J. Seric. 35(2):99-103.

Katiyar RS, Das PK, Choudhury PC, Ghosh A, Singh GB, Datta, R.K (1995). Response of irrigated mulberry (Morus alba L.) to VA-mycorrhiza inoculation under graded doses of phosphorus. Plant. Soil 170:331-337.

Katiyar RS, Das PK, Choudhury PC (1998). VA-mycorrhizal inoculation of established mulberry garden through maize (Zea maize) intercropping. An effective technique. Abs: National Conference on Moriculture: Physiological, Biochemical and Molecular aspects of stress tolerance in mulberry (Trichy), Feb. 22-23 P. 54.

Menge JS, Lembright H, Johnson Elv (1977). Utilisation of mycorrhizal fungi in Citrus nurseries. Proc. Int. Soc. Citric. 1:129.

Osonubi O, Mulongoy K, Awotoye OO, Atyee MO, Okali DU (1991). Effect of ectomycorrhizal and vesicular arbuscular mycorrhizal fungi on drought tolerance of four leguminous woody seedlings. Plant Soil 136:131-143.

Subba Rao NS (1998). Biofertilisers in Agriulture. Oxford and IBH Publishing Co., New Delhi.

Subbaswamy MR, Reddy MM, Sinha AK (1994). Tank silt. A cheap name to manure mulberry. Indian Silk 32(10):10.

Sukhada K (1988). Response of papaya to inoculation with VA-mycorrhizal fungi. First Asian Conference on mycorrhizae, Madras, P. 34.

Umakant GC, Bagyaraj DJ (1998) Response of mulberry saplings to inoculation with VA mycorrhizal fungi and Azotobscter. Sericologia 38(4):669-675.

Tinker PB (1978). Effects of VAM in plant nutrition and Plant growth. Physiol. Veg. 16:743.

18

Insecticide activity of extract of *Piper nigrum* L. and *Annona squamosa* L. on *Sitophilus zeamais* in grains of maize

Pedro José da Silva Júnior[1], Francisco de Assis Cardoso Almeida[2] and Francisco Braga da Paz Júnior[3]

[1]Instituto Federal de Educação, Ciência e Tecnologia de Pernambuco (IFPE), *Campus* Belo Jardim, PE, Brasil.
[2]Departamento de Engenharia Agrícola, Universidade Federal de Campina Grande (UFCG), *Campus* I, Campina Grande, PB, Brasil.
[3]Instituto Federal de Educação, Ciência e Tecnologia de Pernambuco (IFPE), *Campus* Recife, PE, Brasil.

The aim of this paper was to evaluate repellency, attraction and mortality of *Sitophilus zeamais* in grains of maize treated with hydroalcoholic extract of *Piper nigrum* L. and *Annona squamosa* L. In order to experiment with repellency and attraction of adult insects an arena composed of five diagonal boxes was formed. Samples of maize treated with powder extract were put in two flasks, in two different flasks the untreated maize and, in the central flask of the arena, 30 insects non-distinguished by gender. Hydroalcoholic extracts were obtained from the dust of the seed of these species, in percolator with the solvent ethyl alcohol (70% v/v). The experiment fit a completely randomized design with treatments disposed in factorial scheme, in which quantitative factors were revealed by regression in variance analysis. Repellency percentage was 86.75% with powder extract of *Annona squamosa* and 79.25% with *Piper nigrum*. Mortality of 100% was obtained with extract of *A. squamosa* at a 5 ml dose and 98% to extract of *P. nigrum*, at a 14 ml dose, thus indicating the influence of dose and extract on mortality of adult *S. zeamais*.

Key words: Insecticide effect, mortality, maize weevil, repellency.

INTRODUCTION

Maize is a culture great importance to Brazil, occupying about 12 million acres throughout the national territory, with an annual production of approximately 57.7 million tons (IBGE, 2011). Concern for maize plantation is as important as caution regarding production storage as losses for inadequate storage reach 15% of grains (EMBRAPA, 2008). Several studies have been conducted to discover techniques to protect grains and seeds against insect attacks.

In Brazil, maize weevil, *Sitophilus Zeamais* Motschulsky (Coleoptera: Curculionidae), is considered the most important pest of stored grains as they present elevated biotic potential, crossed infestation, capacity of penetrating pasta derived from grains, the elevated number of hosts as well as the fact that both the larvae and adult can damage grains (Capps et al., 2010). These insects cause weight loss and nutritional value of the grains, commercial devaluation and reducing germination of seeds (Antunes et al, 2011; Canepelle et al., 2003).

Using chemical insecticides of high toxicity is a

common practice for controlling maize weevil presence in stored grains, but the indiscriminate and incorrect use has caused its efficiency to decrease, which increased the number of necessary applications and leaded to the generation of resistant populations and the elevation of costs per application (Bogorni and Vendramin, 2003). Such products are greatly disadvantageous for mankind and the environment, therefore research is necessary to identify new products with insecticide effect by studying plants natural defenses, particularly those rich in bioactive organic compounds of insecticide effect, fungicide, growth inhibitor and repellent, amongst others (Almeida et al., 2005). The need for new and less toxic molecules with lesser environmental impact is fundamental, thus stimulating and increasing the interest in research on insecticides plants (Pungitore at al., 2005).

Using plants with insecticide properties is an old practice (Machado et al., 2007). Until the discovery of organo-synthetic insecticides in the first half of the 20th century, substances extracted from vegetables were wildly used in pest control. The rebirth of research on insecticide plants has occurred due to the necessity of new biorational substances to control pests without endangering human and the environment. Insecticide plants have become particularly important for organic food segment which has rapidly grown in cultivation and consumption in the past few years all around the globe (Almeida et al., 2005). According to the same author research on insecticide plants are done basically with the purpose of finding active molecules against insects that allow synthesis of new insecticide products and the obtainment of natural insecticides for direct use in pest control. Based on these considerations, the damage caused by Sitophilus zeamais during maize storage, elevated costs of defensive chemicals and the problems derived from its misuse, it was a goal of this study to investigate the efficiency of two hydroalcoholic vegetal extracts on repellency, attraction and mortality of Sitophilus zeamais.

MATERIALS AND METHODS

Experiments were conducted at Laboratório de Armazenamento e Processamento de Produtos Agrícolas (LAPPA) [Laboratory of Storage and Processinf of Agricultural Products] in Unidade Acadêmica de Engenharia Agrícola (UAEAg) [Academic Unity of Agricultural Engineering] at Universidade Federal de Campina Grande (UFCG) [Federal University of Campina Grande], Campina Grande, PB.

Experiments

The collecting of maize weevil was previously conducted using grains from non-controlled environments in storage facilities located at the central public market in Campina Grande, PB. In order to multiply the insect, collected samples were put together with whole grains of maize, previously purged, into a glass recipient with capacity inferior to 300 ml, and opening sealed with voile fabric to allow ventilation, after which the flasks were taken to an incubator

at a temperature of 26°C and 95% relative air humidity. After inoculation 35 days were given for copula and egg-laying. Afterwards, adult weevils were taken from the grains with help from a 4 mesh sieve, leaving grains and eggs in the flask until adult insects appeared, being thus used in the experiments.

Maize seeds, São José type, used in the experiment derived from a production field, 2010 crop, cultivated by the Instituto Agronômico de Pernambuco [Agronomic Institute of Pernambuco] located in Vitória de Santo Antão, PE.

Obtainment of hydroalcoholic extracts of Piper nigrum L. and Annona squamosa L.

Natural extracts were obtained from seeds of black pepper (Piper nigrum L) and sugar apple (Annona squamosa L) both obtained in a public market in Campina Grande, PB. During production of extracts it was used 500 g of raw material (powder) of the product and 1000 ml of ethylic alcohol for both extracts and the methodology followed was described by Almeida (2003), who after drying seeds from both sugar apple and black pepper seeds in a greenhouse at 40± 2°C, grinded them in a Tecnal knife milling machine. Material (powder) was weighted in a precision balance; moistened with small quantities of ethylic alcohol at 70% (v.v^{-1}). Moistened vegetal material was conveniently settled in stainless steel percolator onto 3 cm of hydrophilic cotton covered with filter paper; after which a second paper filter with a perforated metal disc was placed on the material, in order to avoid the extractor liquid from leaking. The solvent was then added through the superior part of the percolator and a 24 h waiting window was given, by the end of which filtered and concentrated solutions were stored in proper recipients until total the solvent totally evaporated. Extracts obtained were conditioned in weak ember glass flasks until they were used in the experiments.

Tests of repellency and/or attraction

Experiment 1

Experiment 1 consisted in evaluating the effect of powder extract from the plants already mentioned on attraction and/or repellency of adult insects. To achieve such goal arenas were set up with five interconnected and diagonally disposed flasks, which measured 6.0 cm diameter and 2.0 cm tall with the central box interconnected to all the others. In two recipients 10 g grinded maize samples were put and treated with 3 g of powder extract and in two others grinded and non-treated maize was put, in the central recipient of the arena 30 insects non-distinguished by gender with eight repetitions.

Experiment 2

In this experiment black pepper (Piper nigrum L.) and sugar apple (Annona squamosa L.) hydroalcoholic extracts were taken to the insects in form of vapor using an equipment developed for this purpose, similar to Potter tower, in which insects were placed in plastic flasks (104 × 141 mm). Small holes were made on the lids for entrance and exit. Treatment consisted of 4 repetitions with 30 insects each, in addition to a witness one that did not receive any application of extracts. Quantity of extracts per application was 2; 5; 8; 11 and 14 ml and evaluation was done after 24 and 48 h after the application.

Statistical analysis

Data were evaluated with software ASSISTAT version 7.5 (Silva

Table 1. Variance analysis for the repellency and attraction test for *Sitophilus zeamais* to black pepper and sugar apple extracts.

F.V.	G.L	S.Q.	Q.M.	F
Extract (E)	1	0.00	0.00	0.00[ns]
Proced. (P)	1	34848.00	34848.00	4928.00**
E x P	1	450.00	450.00	63.63 **
Treatment	3	35298.00	11766.00	1663.87**
Residue	28	198.00	7.07	
Total	31	35496.00		

**Significant at 1% probability ($p < 0.01$); *significant at 5% probability ($0.01 =< p < 0.05$); [ns]not significant ($p >= 0.05$).

Table 2. Means for repellency and attraction (%) for interaction extracts x procedures for Sitophilus zeamais attracted in samples of maize treated with powder extract of black pepper grains and sugar apple.

Extracts	Procedures	
	Repellency	Attraction
Black pepper	79.25[bA]	20.75[aB]
Sugar apple	86.75[aA]	13.25[bB]

DMS for columns: 2.7252; DMS for lines: 2.7252; CV: 5.31; Means followed by the same letter, (lower case) in the row and (upper case) in the column, do not differ by Tukey's test.

and Azevedo, 2009), in a completely randomized design (CRD), in which experiments were disposed in factorials with four repetitions:

1. First experiment: 2×2 (two extracts and repellency or attraction);
2. Second experiment: 2×5 (two extracts, five doses);

Means were compared by Tukey test at 1 and 5% probability, using as quantitative factor regression in variance analysis.

RESULTS AND DISCUSSION

Repellency and attraction tests

Results for variance analysis corresponds to repellency and attraction for adults of S. *zeamais* attracted in maize samples treated with powder extract of grains of *P. nigrum* (black pepper) and *Annona squamosa* (sugar apple) are exposed in Table 1 in which is verified effects very significant for procedures and double interaction.

According to data showed in Table 1, it is verified that S. *zeamais* presents clear preference for untreated maize seeds when compared to seed that were treated with powder extract of grains of black pepper and sugar apple; in quantitative terms the extract of the latter (86.75%) was statistically superior to black pepper (79.25%), tested for untreated powder. That is, the insect presented repellency over 7.5% when seeds (powder) were treated with sugar apple extract.

It is observed by the results disposed in Table 2 that regarding the extracts attraction of S. *zeamais* sugar apple presented results statistically inverse to repellency. Regarding the procedure (line) sugar apple and black pepper demonstrated equal results, statistically.

Studies done by Miyakado and Ohno (1989) found repellent effect for grinded black pepper, proving it constitutes a promising source of natural insecticide. Fruits have alkaloids of the amylum group unsaturated with toxic action on pests of stored grains.

Leão (2007) in a study of bioactivity of vegetal extract for controlling S. *oryzae* in rice obtained 70.84% repellency with powder of *P. nigru m L.* confirming results found in the present experiment. Kani et al. (2011), observed in their studies about the efficiency of hydroalcoholic extract of Black pepper in controlling adults of S. *zeamis*, when applied in vapor form (nebulization).

Lima (2007) affirms that most photochemical studies of Annonaceae do not concentrate on alkaloids but in a newer class of extremely bioactive substances referred to as acetogenins of Annonaceae. They are recognized for being source of compounds with various biological actions and relevant citotoxic, antitumor, anti-parasite, immunosuppressive, and pesticide activities.

According to Souza (2003), *Annona squamosa* is characterized mainly for presenting a class of substances denominated acetogenins, derived from long chain fatty acid combined with a unit 2-propanol, apparently originated from poliketids (C35-C37), with or without the presence of tetrahydrofuran and a valerolactone terminal. Such substance acts as potent inhibitor of breathing leading to a programmed cellular dead. The chemical structure of acetogenins seems crucial for a potent inhibitor activity.

Results obtained with ethanol extract of leaves from *A. squamosa* for Brito et al. (2008), confirmed the presence of secondary metabolism classes such as resins, tannins, condensed, saponins and alkaloids, and have presented moderately positive intensity, while flavonoids, steroids, triterpenoids and flavonones showed strongly positive. Such results suggested the existence of different classes of secondary metabolites in the species *A. squamosa*

Table 3. Variance analysis of mortality for *S. zeamais* to hydroalcoholic extracts of black pepper and sugar apple, applied through vaporization after 48 h.

F.V.	G.L	S.Q.	Q.M.	F
Extract (E)	1	115.60	115.60	130.86**
Dose (D)	4	1415.75	353.93	400.68**
E x D	4	189.65	47.41	53.67**
Treatment	9	1721.00	191.22	216.47**
Residue	30	26.50	0.88	
Total	39	1747.50		

**Significant at 1% probability (p < 0.01).

Table 4. Means for mortality (%) for *S. zeamais* for interaction extract x doses due to application of extracts of black pepper and sugar apple through vaporization after 48 h.

Extracts	Dose (ml)				
	2	5	8	11	14
Black pepper	45[a]	60[b]	92[b]	97.50[a]	98[a]
Sugar apple	49[a]	100[a]	100[a]	100[a]	100[a]

DMS for columns: 1.3581; DMS for lines: 1.9314; CV%: 3.72221. Means followed by the same letter in the column do not differ statistically through Tukey at 5% of probability.

Potenza et al. (2005) obtained 91.48% and 86.15% efficiency for controlling *Tetranychus urticae* with watery extracts of *A. squamosa* and *R. graveolens*.

Furthermore, in agreement to the theme, Guimarães et al. (2010), in their considerations of repellency and attraction of maize grains treated with vegetal extracts using the arena, proved that 0.01% of *Memora nodosa* (Silva Manso), a plant from Brazilian Cerrado, used popularly as healing for ulcers and external wounds and 0.1% of *Vernonia áurea* presented good repellency for adults of *S. zeamais*. Equally, according to Procópio et al. (2003), powder of eucalypt leaves (*Eucalyptus citriodora*) amongst seven vegetal species evaluated, was the only one what caused repellency on adults of *S. zeamais*.

Mortality of *S. zeamais*

Effects of factors considered for the present study regarding mortality of maize weevil, *S. zeamais*, according to variance analysis were significant at 1% probability (Table 3). In Table 4 and Figure 1 are exposed data on the interaction of factors in regard to mortality of *S. zeamais* after receiving hydroalcoholic extracts of *P. nigrum* and *A. squamosa*.

Regarding extracts within each dose all doses showed the same behavior except for 5 and 8 ml doses in which sugar apple extract controlled adults of *S. zeamais* at 100% being statistically different from black pepper, in which mortality reached 60% for 5 ml dose and 92% for 8 ml. This means to say they presented the same statistical mean (Table 4).

Regarding each extract within different doses it is observed by results of the same Table and Figure 1, higher mortality of insects as the number of doses taken to the same increase and that for black pepper higher mortalities took place with doses 11 ml (97.50%) and 14 ml (98%) which was statistically equal to control. Sugar apple extract showed mortality of 100% starting with 5 ml doses. These results partially confirm what was observed for this extract in repellency tests, previously discussed: The superiority of sugar apple extract over black pepper.

Data have been submitted to regression analysis and when significant, the best equation to represent them was studied, in Figure 1, which in the present study the equation of second order represents them with R^2 superior to 85% for the studied extracts.

Superiority of sugar apple is probably due to the amount of oil and chemical compounds in *Annona* extract, especially acetogenins (Lima, 2007).

The powder extract of *P. nigrum* L. studied by several authors (Leão, 2007; Procópio et al., 2003; Khani et al 2011) and of proven efficiency against *S. zeamais* presented in this study less repellent than *A. scamosa* L. which does not diminish its importance as product to be used in this insect pest control.

Pessoa (2004), analyzing mortality percentage of *S. zeamais* when exposed to extract of *Ocimum basilicum* in doses of 4, 8, 12 and 16 ml via vaporization it was noted that for 4 ml doses mortality reached 52.5% and as the applied extract increased percentages were 97% for 8 ml, 98% for 12 ml and 96.5% for 16 ml. Such results converge to the ones found in the present research (except for 16 ml doses) as higher were the doses, higher

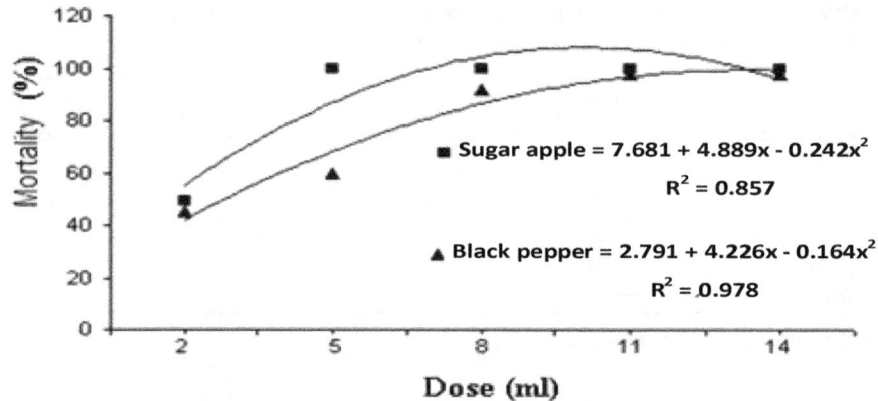

Figure 1. Mortality (%) of *S. zeamais* adults due to applications of extracts of black pepper and sugar apple in different doses through vapor after 48 h.

were mortality rates.

According to to Khani et al. (2011), mortality of adults *S. oryzae* increases with the time of exposition to extracts. According to Khani et al. (2011), mortality of adults S. oryzae increased with increasing concentration and the time of exposition to extracts. The petroleum ether and chloroform extracts of P. nigrum were efficient in control of rice weevil, with 99.65 and 93.56% mortality.

Shortly, results show that only the repellent effect is not sufficient to promote effective control of S. *zeamais*, due to the possibility of bioactive volatile compounds rapidly dissipated, according to the properties of each compound and physical conditions of storage structures, direct effects on biology and physiology of the insect are also necessary, in order to justify its use in the alternative control of this pest (Oliveira and Vendramim, 1999). Black pepper extracts at doses of 11 and 14 ml and sugar apple at 5 and 14 ml for 500 g of seed were effective in controlling adults S. *zeamais*, generating mortality of 97.85 and 100%, respectively.

Due to insecticide properties these natural extracts can be of great utility in handling S. *zeamais* in stores seeds of maize, particularly in small rural properties, which highlights the need for standardization of collection processes, vegetal draining as well as quantifying of bioactive compounds, to the point in which results obtained might be reproduced and/or compared. However, recommendation for using the treatment of maize grains destined to human and animal consumption requires additional studies, to provide the user a safe and efficient product from a toxicological point of view.

Conclusions

1. *S. zeamais* adult insects are repelled with extract of *P. nigrum* (79.25%), and *A. squamosa* (86.75%).
2. Mortality rates are even higher with increased number of doses applied to the same.

3. Black pepper presented higher mortality rated at 11 ml doses (97.50%) and 14 ml (98.00%); sugar apple presented mortality of 100% observed from 5 ml doses.

ACKNOWLEDGEMENT

We thank FACEPE (Fundação de Amparo à Ciência e Tecnologia do Estado de Pernambuco – Foundation of Support to Sciences and Technology for the State of Pernambuco) for granting the scholarship (BCT modality) that allowed the present research to be conducted.

REFERENCES

Almeida AS (2003). Extratos vegetais no controle do Callosobruchus maculatus e seus efeitos na conservação o feijão Vigna unguiculata. 2003. 80f. Dissertação (Mestrado em Engenharia Agrícola – Área de Concentração em Processamento e Armazenamento, Universidade Federal de Campina Grande (UFCG), Campina Grande, 2003.

Almeida FAC, Almeida AS, Santos NR, Gomes JP, Araújo MER (2005). Efeitos de extratos alcoólicos de plantas sobre o caruncho do feijão Vigna (Callosobruchus maculatus). Rev. Bras. Eng. Agríc. Ambient 9(4):585-590.

Antunes LEG, Viebrantz PC, Gottardi R, Dionello RG (2011). Physicochemical characteristics of corn damaged by Sitophilus zeamais during storage. Rev. Bras. Eng. Agríc. Ambient. [online]. 15(6):615-620.

Bogorni PC, Vendramim JD (2003). Bioactivity of aqueous extracts of Trichilia spp. on Spodoptera frugiperda (J.E. Smith) (Lepidoptera: Noctuidae) development on maize. Neotrop. Entomol. 32(4):665-669.

Brito HO, Noronha EP, França LM, Brito LMO, Prado MSA (2008). Phytochemical analysis composition from Annona squamosa (ATA) ethanolic extract leaves. Braz. J. Pharm. 89(3):180-184.

Capps ALAP, Novo JPS, Novo MCSS (2010). Repelência e toxicidade de Cyperus iria L., em início de florescimento, ao gorgulho Sitophilus oryzae. Rev. Bras. Eng. Agríc. Ambient 14(2):203-209.

Canepelle MAB, Caneppele C, Lázzari FA, Lázzari SMN (2003). Correlation between the infestation level of Sitophilus zeamais Motschulsky, 1855 (Coleoptera, Curculionidae) and the quality factors of stored corn, Zea mays L. (Poaceae). Rev. Bras. Entomol. 47:625-630.

EMBRAPA Jornal Eletrônico Milho e Sorgo (2008). Available at: http://www.cnpms.embrapa.br/grao/5_edicao/index.htm. Accessed: Sept. 28, 2010.

Insecticide activity of extract of Piper nigrum L. and Annona squamosa L. on Sitophilus zeamais...

143

Guimarães CG, Tavares WS, Moreira CO, Teixeira MFF, Hany, Mahmoud HAFH, Ribeiro RC, Petacci F (2010). *Sitophilus zeamais* (Coleoptera: Curculionidae) em grãos de milho com extratos botânicos do cerrado. In: XXVIII Congresso Nacional de Milho e Sorgo, 2010, Goiânia: Associação Brasileira de Milho e Sorgo. CD-Rom.

IBGE – Instituto Brasileiro de Geografia e Estatística – Levantamento sistemático da produção agrícola (2011). Available: *http://www.ibge.gov.br/home/estatistica/indicadores/* agropecuaria/lspa/default.shtm. Accessed: May. 22, 2011.

Khani M, Awang RM, Omar R, Rahmani R, Rezazadeh S. (2011)Tropical medicinal plant extracts against rice weevil, *Sitophilus oryzae* L. J. Med. Plant Res. 5 (2): 259-265.

Machado LA, Silva VB, OI MM (2007). Uso de extratos vegetais no controle de pragas em horticultura. Biológico São Paulo v.69, n.2):103-106,

Miyakado M, Nakayama I, Ohno N (1989). Insecticidal unsaturated isobutylamides: from natural products to agrochemical leads, p.183-187. In: Amason JT, Philogène BJR, Morand, P. (Ed.). Insecticides of plant origin. ACS Symposium Series 387, Am. Chem. Soc. New York, P. 320.

Oliveira AM, Vendramim JD (1999). Repellency of essencial oils and powders from plants on adults of *Zabrotes subfasciatus* (Boh.) (Coleoptera: Bruchidae) on bean seeds. An. Soc. Entomol. Bras. 28(3):549-555.

Pessoa EB (2004). Controle do *Sitoplilus zeamais* em milho pipoca nas fases adulta imatura com extratos vegetais. 2004. 57f. Dissertação (Mestrado em Engenharia Agrícola – Área de Concentração em Processamento e Armazenamento, Universidade Federal de Campina Grande (UFCG), Campina Grande.

Potenza MR, Takematsu AP, Jocys T, Felicio JDF, Rossi MH, Nakaoka-Sakita M (2005). Evaluation of plant extracts for the control of coffee red mite Oligonychus ilicis (McGregor) (Acari: Tetranychidae). Arq. Inst. Biol. 72(4):499-503.

Pungitore CR, García M, Gianello JC, Sosa ME, Tonn CE (2005). Insecticidal and antifeedant effects of *Junellia aspera* (Verbenaceae) triterpenes and derivatives on *Sitophilus oryzae* (Coleoptera: Curculionidae). J. Stored Prod. Res. 41:434-443.

Procópio SO, Vendramin JD, Ribeiro Júnior JI, Santos JB (2003). Bioactivity of powders from some plants on *Sitophilus zeamais* Mots. (Coleoptera: Curculionidae). Ciênc. Agrotec. 27(6):1231-1236.

Silva FAS, Azevedo CAV (2009). Principal components analysis in the software assistat-statistical attendance. In: World Congress on computers in Agriculture, 7, Reno-NV-USA: American Society of Agricultural and Biological Engineers.

Leão JDJ (2007). Bioatividade de extratos vegetais no controle de *Sitophilus oryzae* (LINNÉ, 1763) em arroz. 2007. 91f. Tese (Doutorado em agronomia) – Universidade Federal de Santa Maria, Santa Maria.

Lima MD (2007). Perfil cromatográfico dos extratos brutos das sementes de *Annona Muricata* L. *Annona scamosa* L. através da cromatografia líquida de alta eficiência. 2007. 102f. Dissertação (Mestre em Química e Biotecnologia) – Universidade Federal de Alagoas, Alagoas.

Souza MMC (2003). Avaliação da atividade ovicida de *Annona squamosa* Linnaeus sobre o nematóide *Haemonchus contortus* Rudolphi e toxicidade em camundongos. 2003. 95f. Dissertação (Mestrado em Ciências Veterinárias, Área de concentração Medicina Veterinária Preventiva) – Universidade Estadual do Ceará, Fortaleza.

19

In vitro studies of biowastes on growth and sporulation of fungal bioagents

Mucksood Ahmad Ganaie and Tabreiz Ahmad Khan

Section of Plant Pathology and Nematology, Department of Botany, Aligarh Muslim University, Aligarh, India.

Biowastes of mango, carrot, papaya, banana, chukandar, pomegranate, orange, mosambi, chickpea and wheat were used to study the colony formation and sporulation of *Trichoderma harzianum* and *Paecilomyces lilacinus*. In the present study, it was observed that the initiation of colony formation of the fungal bioagents was recorded after 12 h only in biowaste of carrot, mango, chukandar, banana and papaya. In general, the growth of colony formation increased with the increase in duration. The whole diameter of Petri dish was occupied by fungal mycelium after 120 h in the biowaste of carrot, mango, chukandar, banana, and papaya. Further, it was observed that the maximum number of spores/ml of *T. harzianum* were obtained on biowaste of carrot followed by mango, chukandar, banana, papaya, orange, mosambi, gram, pomegranate and wheat. Similarly, in case of *P. lilacinus,* the highest numbers of spores were recorded on biowaste of mango followed by carrot, papaya, banana, chukandar, orange, mosambi, gram, pomegranate and wheat. Evaluation of sporulation of *T. harzianum* and *P. lilacinus* on different biowastes revealed that the tested fungi showed significant increase in sporulation on biowastes of mango, carrot, banana, papaya and chukandar as compared to control (Potato Dextrose Agar [PDA]). However, on the other hand, significant reduction in fungal sporulation was observed on biowastes of orange, mosambi, pomegranate, gram and wheat as compared to control.

Key words: Biowaste, *Trichoderma harzianum, Paecilomyces lilacinus,* fungal sporulation, mycellial growth.

INTRODUCTION

During the last few decades fruits have attracted the attention of biologists not only because of their nutritive value but also their use in mass multiplication of bioagents (Zaki and Bhatti, 1991; Tewari and Bhanu, 2003). Biowastes of different fruits are used as bioagents either directly or indirectly. Greater emphasis has been given to biological control of plant parasitic nematodes through the use of nematode parasitic fungi *Paecilomyces lilacinus* (Morgan-Jones et al., 1981; Jatala, 1985; Khan and Saxena, 1996; Anver and Alam, 1998). *P. lilacinus* is a common soil hyphomycete with a worldwide distribution, parasitizes eggs of *Meloidogyne incognita*. The fungus has been used as bioagent for the management of plant parasitic nematodes. Similarly, research has been done on the biocontrol nature of *Trichoderma* species (Papavizas, 1985; Chet, 1987) of these *T. harzianum* showed highest potential against many soil born fungal pathogens. *T. harzianum* have been known to suppress many soil born fungi. It has been found to antagonize fungal plant pathogens as well as plant parasitic nematodes (Siddiqui et al., 1999).

Since, *P. lilacinus* and *Trichoderma harzianum* are important bioagents against plant parasitic nematodes and fungal pathogens. Mass production of these fungi is an important prerequisite for any large scale field application. Our emphasis in the present study was to

Table 1. Effect of different Bio-wastes on mycellial growth and sporulation of *Paecilomyces lilacinus* at different intervals.

Bio- waste used	Diameter of mycelia growth (cm) after (h)						*Spores/ml	Percentage increase or decrease over control
	12	24	48	72	96	120		
Mango	0.5	2.0	4.2	7.4	9.5	10	3.30×10^{7a}	+68.36
carrot	0.4	1.8	4.0	6.9	9.2	10	3.16×10^{7a}	+62.22
Wheat	0.3	1.7	3.8	6.7	8.9	10	2.97×10^{7b}	+32
Banana	0.2	1.7	3.5	6.5	8.6	10	2.54×10^{7c}	+14
Chukander	0.2	1.6	3.3	6.5	8.4	10	2.50×107^{c}	+12
Papaya	0.0	1.3	2.5	5.4	7.4	8.1	2.33×10^{7e}	-16
Mosambi	0.0	1.2	2.4	5.2	7.4	8.o	2.15×10^{7e}	-26
Gram	0.0	1.2	2.3	3.7	4.5	7.6	0.97×10^{7f}	-50
Pomegranate	0.0	0.8	1.4	2.9	4.3	6.7	0.92×10^{7f}	-53
Orange	0.0	0.6	1.3	2.7	4.2	5.8	0.68×10^{7g}	-65
PDA	0.2	1.5	3.1	6.5	7.5	9.6	1.96×10^{7d}	
LSD at 5%						2.08		
LSD at 1%						2.82		

*Values with same alphabetical letters are non significant, while values with different letters are significant according to data analysis by SPSS version 16.

find out the highly productive, cost effective and locally available biowastes for mass multiplication of bioagents viz., *P. lilacinus* and *T. harzianum*.

MATERIALS AND METHODS

Locally available fresh biowastes of mango, carrot, papaya, banana, chukandar, orange, mosambi, and pomegranate were obtained from fruit juice shop. In addition to the above, gram husk and wheat bran were also taken to evaluate the mass multiplication of the tested fungus viz. *T. harzianum* and *P. lilacinus*. The pericarp of selected biowastes viz., mango, carrot, papaya, banana, chukandar, orange, mosambi, and pomegranate were cut into small pieces with the help of electric grinder. Ten grams of each biowaste was taken and spread on Petri plate. Biowaste of gram husk and wheat bran were moistened with distilled water before spreading on Petri plates. These Petri plates with different biowastes were autoclaved at 20 lbs for about 15 min. After sterilization, each Petri plate was inoculated with 1 ml of spore suspension prepared from 15 days old culture of *P. lilacinus* and *T. harzianum* grown on potato dextrose agar (PDA). Each set of biowaste was replicated three times. The inoculated Petri plates were kept in biochemical oxygen demand (BOD) incubator at 28 ± 1 °C. The Petri plates were observed at different intervals to observe the colony growth of fungal bioagents viz., *T. harzianum* and *P. lilacinus* on different biowastes. The spore suspension of each sample was prepared by dispersing 1 g of each inoculum in 10 ml distilled water and spores were counted by using a haemocytometer.

RESULTS AND DISCUSSION

The results of the present study revealed that both the fungi grew profusely on some of the tested biowastes. The highest spore count in case of *P. lilacinus* was recorded on biowaste of mango (3.30×10^7) followed by carrot (3.16×10^7), papaya (2.60×10^7), banana (2.1×10^7) and chukandar (2.1×10^7). The fungus *P. lilacinus* has

also been previously grown on different fruit grains (Zaki and Bhatti, 1991) and other agro industrial wastes (Leena et al., 2003). *P. lilacinus* is an important fungal biocontrol agent and has been reported to be effective in controlling plant parasitic nematodes in different crops (Jatala, 1986; Nagesh and Parvatha Reddy, 2003; Ashraf et al., 2005). Similarly in case of *T. harzianum*, the maximum spore count was observed on biowaste of carrot (3.14×10^7) followed by mango (3.07×10^7), chukandar (2.97×10^7), banana (2.94×10^7) and papaya (2.86×10^7). The high spore count of the tested fungi on these biowastes may be possibly due to high carbohydrate content in them. However, in case of both tested fungi, the lower spore count was observed on biowastes of orange, mosambi, pomegranate, gram and wheat in comparison with control (PDA). The lower spore count on these biowastes might be due to the inhibitory effect of some chemicals and/or lower carbohydrate content present in these biowastes.

Furthermore, significant increase in spore count of *P. lilacinus* was observed on biowaste of mango (+68.36%) followed by carrot (+62.22%), papaya (+32%), banana (+14%) and chukandar (+12%). A significant decrease in spore count of *P. lilacinus* was observed on biowaste of wheat (-65%) followed by pomegranate (-53%), gram (-50%), mosambi (-26%) and orange (-16%). In case of *T. harzianum*, significant increase in spore count was observed on biowaste of carrot (+67%) followed by mango (+64%), chukandar (+52%), banana (+44%) and papaya (+41%) and significant decrease in spore count of *T. harzianum* was observed on biowaste of wheat (-77%) followed by pomegranate (-68%), gram (-58%), mosambi (-40%) and orange (-35%).

The data presented in Tables 1 and 2 reveal that, in general mycellial growth of both tested fungi on different biowastes increased with increase in incubation period.

Table 2. Effect of different Bio-wastes on mycellial growth and sporulation of *Trichoderma harzanium* at different intervals.

Bio-waste used	Diameter of mycelia growth (cm) after (h)						*Spores/ml	Percentage increase or decrease over control
	12	24	48	72	96	120		
Wheat	0.4	2.0	4.3	6.2	8.2	10	$3.64×10^{7a}$	+67
Mango	0.4	1.7	3.6	6.7	8.3	10	$3.62×10^{7a}$	+64
Chukandar	0.3	1.7	3.4	6.2	7.8	10	$3.50×10^{7a,b}$	+52
Banana	0.2	1.6	3.5	6.4	7.7	10	$2.71×10^{7b}$	+44
Papaya	0.2	1.5	2.7	5.8	6.9	10	$2.65×10^{7b}$	+41
Carrot	0.0	1.3	2.5	3.2	5.6	8.6	$2.61×10^{7d}$	-35
Mosambi	0.0	1.0	1.6	2.5	4.7	7.7	$1.95×10^{7d}$	-40
Gram	0.0	1.0	1.3	2.4	4.6	7.2	$1.38×10^{7e}$	-58
Pomegranate	0.0	0.6	1.2	2.5	3.2	5.8	$1.20×107^{e,f}$	-68
Orange	0.0	0.5	1.1	2.3	3.1	5.2	$0.92×10^{7f}$	-77
PDA	0.3	1.6	3.4	6.5	7.6	9.4	$1.87×10^{7c}$	
LSD at 5%						2.02		
LSD at 1%						2.74		

*Values with same alphabetical letters are non significant, while values with different letters are significant according to data analysis by SPSS version 16.

The colony morphology of all the isolates was more or less similar showing spare to thin cottony mycellial mass. These results also support the earlier findings made by Rifai (1969), Domsch et al. (1980), Martha (1992) and Mazumdar (1993). Initiation of sporulation was observed after 12 h on biowastes of mango, carrot, papaya, banana and chukandar for both tested fungi whereas for the rest of biowastes initiation of sporulation was observed after 48 h.

The present study concludes that biowastes of mango, carrot, papaya, banana and chukandar are efficient for the mass multiplication of bioagents *viz., T. harzianum* and *P. lilacinus* as compared to control (PDA). Therefore, these biowastes may be used as supplements for the mass multiplication of these fungi. Elad et al. (1980) have reported the mixture of bran, saw dust and water as suitable substrates for mass multiplication of *T. harzianum*. Biowastes of pomegranate, orange, mosambi, gram and wheat although resulted in low sporulation may serve as substrate in combination with other nutrient rich supplements for mass multiplication of *T. harzianum* and *P. lilacinus*.

REFRENCES

Anver S, Alam MM (1998). Control of *Meloidogyne incognita* and *Rotylenchus reniformis* singly and concomitantly on pigeonpea with *Paecilomyces lilacinus*. Indian J. Nematol. 27(2):209-213.

Ashraf MS, Khan TA, Nisar S (2005). Integrated management of reniform nematode *Rotylenchulus reniformis* infecting okra by oil cakes and biocontrol agent *Paecilomyces lilacinus*. Pak. J. Nematol. 23:3005-309.

Domsch KH, Gams W, Anderson TH (1980). *Trichoderma* pers, ex. Fr. 1821. In compendium of soil Fungi, .1:368-77. Academic Press, N.Y.

Chet I (1987).Trichoderma: application, mode of action and potential as a biocontrol agent of soilborne plant pathogenic fungi, pp. 137–160.In

I. Chet (ed.), Innovative approaches to plant disease control. John Wiley & Sons, New York.

Jatala P (1985). Biological control of nematode *Isn.* :Sasser, J. N. & Carter C. C. (Eds). *An advanced treatise on Meloidogyne. Biology and Control*. Raleigh, Department of Plant Pathology, North Carolina State University & USAID. pp. 302-308.

Jatala P (1986). Biological control of plant parasitic nematodes. Ann. Rev. Phytopathol. 24:453-489.

Khan TA, Saxena SK (1996). Comparative efficacy of *Paecilomyces lilacinus* in the control of *Meloidogyne* Spp. And *Rotylenchus reniformis* on tomato. Pak. J. Nematol. 14(2):111-116.

Leena MD, Easwaramoorthy S, Nirmala R (2003). In vitro production of entomopathogenic fungi *Paecilomyces farinosus* (Hotmskiold) and *Paecilomyces lilacinus* (Thom) Samson using byproducts of sugar industry and other agro industrial byproducts and wastes. Sugar Tech. 5(4):231-236.

Martha PK (1992). Influence of some physic chemical factors on the germination and growth of biotype of *Trichoderma harzianum* and *Gliocladium virens*. M.Sc. dissertation. B.C.K.V., Mohanpur.

Mazumdar D (1993). Hyperparasitic potential of a few biotypes of *T. harzianum* and *G. virens* against two major pathogens of betelvine (Piper betle L.) M. Sc. Dissertation, B.C.K.V., Mohanpur.

Morgan-Jones G, Godoy, Rodriguez-Kabana R (1981). *Verticillium chlamydosporium*, fungal parasite of *Meloidogyne arenaria* females. Nematropica 11:115-120.

Nagesh M, Parvatha Reddy P (2003). Current trends in biological control of nematode pests in india. In: Nematode management of plants. Trivedi, P.C. (Ed.). Scientific Publishers, India, pp. 203-239.

Papavizas GC (1985). *Trichoderma* and *Gliocladium* their biology, ecology and potential of biocontrol. Ann. Rev. Phytopathol. 18:389-413.

Rifai MA (1969). A revision of the genus *Trichoderma* common W. Mycol. Inst. Mycol. 116:5-6.

Siddiqui IA, Ehteshamul-Haque S, Ghaffar A (1999). Root dip treatment with *Pseudomonas aeruginosa* and *Trichoderma* spp., in the control of root rot and root knot disease complex in chili (*Capsicum annum* L.). Pak. J. Nematol. 17:67-75.

Tewari L, Bhanu C (2003). Screening of various substrates for sporulation and mass multiplication of biocontrol agent Trichoderma harzianum through solid state fermentation. Indian Phytopathol. 56(4):476-478

Zaki FA, Bhatti DS (1991). Effect of culture media on sporulation of Paecilomyces lilacinus and its Efficacy against Meloidogyne javanica in tomato. Nematol. Medit. 19:211-212.

Elad Y, Chet I, Katan J (1980). *Trichoderma harzianum* a biocontrol agent of *Sclerotium rolfsii* and *Rhizoctonia solani*. Phytopathology 70:119-112.

20

Weed interference periods on potato crop in Botucatu region, Brazil

D. Martins[1], S. R. Marchi[2] and N. V. Costa[3]

[1]São Paulo State University, Botucatu/SP, Brazil.
[2]Federal University of Mato Grosso, Barra do Garça/MT, Brazil.
[3]University of Paraná Weast, Marechal Cândido Rondon/PR, Brazil.

This study was carried out in Botucatu, State of São Paulo, Brazil in order to evaluate the interference periods of weeds on potato crop. The experimental design was a randomized complete blocks, with four replications, and treatments were arranged in two groups: 1) the crop was kept free from weeds through the periods of 7, 14, 21, 28 and 35 days after emergence, after each period weeds were allowed to grow; 2) the crop was kept weedy for the same periods of the first group, and afterwards the crop was kept weed-free besides a control maintained weed free and another maintained in coexistence with the weeds at 98 days (harvest). *Urochloa plantaginea, Cyperus esculentus, Raphanus raphanistrum, Sida rhombifolia,* **and** *Galinsoga parviflora* **were the main weeds in the experimental area, being** *U. plantaginea* **the weed with the highest dry matter accumulation. Tuber size and yield were affected by interference of weed community. The total period of weed interference was 35, while the period previous of interference was 7; consequently, the critical period of weed interference was from 7 to 35 after crop emergence.**

Key words: Competition, *Solanum tuberosum,* weed periods, weed management.

INTRODUCTION

The potato crop is exposed to a range of biotic and abiotic factors that affect its growth, development, and economic productivity. Considering these factors, weeds are very important and, according to Lutman (1992), compete by water, nutrients, and light, being the competition degree dependent on the type of weed community, species density and ability to compete for these environmental factors.

From all interference components, competition and allelopathy are the most significant processes, happening with high frequency (Velini, 1997). Furthermore, weeds can affect the quality of tubers (Vangessel and Renner, 1990; Monteiro et al., 2011), reducing their size, modifying their density, causing deformation, and thus hindering their commercialization. Interference degree usually is measured in relation to the crop yield and can be defined as the reduction in the percentage on economical crop yield caused by the interference of weed community. Interference degree among cultivated plants and weeds is dependent on factors related to the weed community (specific composition, density, and distribution), to the crop (genus, species or cultivars, spacing between plants and density of sowing), to the environment (weather, soil, crop management) and to the periods they are kept together (Pitelli, 1985). Losses between 12 and 86% due to competition with different weeds have been found in potato yield (Nelson and Thoreson, 1981; Tripathi et al., 1989; Muhammad, 1993; Liebman et al., 1996; Ciuberkis et al., 2007; Costa et al., 2008; Monteiro et al., 2011; Ahmadvand et al., 2009).

Pitelli and Durigan (1984) established three periods in relation to time and duration of coexistence period with weeds: period previous of interference (PPI), total period of weed interference (TPWI) and critical period of weed interference (CPWI). During PPI, the crop can grow with weed community before interference; during TPWI the crop can control and avoid the growth of weeds; during CPWI, the most important period, weeds and crop compete more intensively for limiting resources, weed control is critical, and weed community development should not be allowed (Pitelli, 1985).

The present study aimed to determine the weed interference periods of potato crop (cultivar Atlantic), by means of determining total period of weed interference (TPWI), the period previous of interference (PPI), and the critical period of weed interference (CPWI).

MATERIALS AND METHODS

An experiment was carried out in the Lageado Experimental Field, at the Department of Crop Science, College of Agricultural Science, UNESP – São Paulo State University, Botucatu/SP, Brazil (22° 51' 09" S and 48' 25' 89" WGr.) with 740 m of altitude).

The tuber was planted in clay soil area, presenting the following properties: pH ($CaC1_2$) = 4.4; organic matter (g dm^{-3}) = 24; P (g dm^{-3}) = 14; H+Al, K, Ca, Mg, SB, CTC = 58, 5.0, 18, 6, 29, and 87 mmol$_c$.dm^{-3}, and V% = 33. The area was prepared by a moldboard plow, a heavy harrow, two leveling harrow, a rotative harrow, and a furrowing (20 cm depth) and the soil was fertilized and corrected according to Miranda-Filho (1996): 3.2 t ha^{-1} dolomitic limestone, 1.0 t ha^{-1} phosphorus, and 1.5 t ha^{-1} manure 8-28-16.

For planting, seed tubers of cultivars Atlantic were used and the plots were arranged in 4 rows of 5 m, spaced by 0.7 m, where seed tubers were placed at 0.25 m from each other. For evaluation, only two central rows in each plot were considered as useful area. Sprouting occurred in 19 after planting.

The experimental design was a randomized complete blocks, with four replications. The treatments were arranged in two groups: 1) the crop was kept free from weeds through the periods of 7, 14, 21, 28, and 35 days after emergence, after each period weeds were allowed to grow; 2) the crop was kept weedy for the same periods of the first group, and afterwards the crop was kept weed-free besides a control maintained weed free and another maintained in coexistence with the weeds at 98 days. All treatments were harvested at 98 days after planting. Weed control was performed through manual weeding.

Weed community was evaluated at the end of each coexistence period, when all weeds present in 0.5 m^2 of useful area of each plot which corresponded to two sub-samples of 0.25 m^2 were collected. Species were identified, quantified, and taken to the laboratory to be washed and oven dried at 70°C, until reaching constant weight. After that, dry matter mass from aerial parts of collected weeds was determined using 0.01 g precision scales.

Weed community phytosociological indices were determined by the following variables: absolute density, relative frequency, and index of importance value according to Mueller-Dombois and Ellenberg (1974).

Also, tubers were classified according to their sizes as follow: Type 1 > 54 mm; Type 2 > 48 to 54 mm, Type 3 > 41 to 47 mm, Type 4 > 34 to 40 mm, and Type 5 < 33 mm of diameter. After classification, tubers were weighed, and percentage of each tuber type was calculated.

Results of dry matter mass, absolute density, and tuber size were submitted to analysis of variance (ANOVA) and F test; the treatment means were compared by the t test at 5% of probability.

For determination of the critical period of interference prevention, yield data were obtained through the different coexistence and weed control periods, which were adjusted using the following model of non-linear regression:

$$y = a+b/[1+(x/c)^d]$$

Where: Y = yield tubers; x = days after crop emergence; a = minimum yield in the initial periods without weeds and the end of the trial to initial weed competition periods; b = differences between maximum and minimum yields; c = number of days which occurred 50% of reduction on maximum yield; d = slope of curve.

The limits of interference period were determined allowing maximum yield losses of 5% in relation to the yield obtained in plots that were kept free of weed competition during crop cycle.

RESULTS AND DISCUSSION

In Table 1, it can be noted that the weed community was composed by 15 weed species, with predominance of broad leaved (10 species). Among them, the family Asteraceae had the highest number of species (three species), while other families had only one species. Among the monocots, it was observed one species belonging to the Family Cyperaceae and four species to the family Poaceae, being the latter family, which presented the highest plant density, representing 27% of weed community. In other study of weed interference periods on potato, however in a different planting time, Costa et al. (2008) observed a different weed community. The weed community varied because it is influenced by seed germination and weather condition.

Urochloa plantaginea (Link) Hithc. was the specie that presented the highest dry biomass accumulation during the crop cycle (Table 2). Controlling weeds for 7 days reduced markedly the number of plants and the dry matter mass of *U. plantaginea*. Once the periods with no weed competition were increasing, the number and dry matter were decreasing, although without statistical difference among periods. Already for *Cyperus rotundus* L. and *Raphanus raphanistrum* L., a control for 14 days reduced both the density and dry matter accumulation in plant.

During the period of crop coexistence with weeds, from 14 to 28 days after crop emergence, it was observed that, the occurrence of a high frequency of *U. plantaginea* is therefore, the highest accumulation of dry matter mass observed at 35 days. Reduction of plant density and consequent increase in dry matter mass accumulation of weed community, during crop cycle were also verified by Martins (1994). Environment factors become restrictive, resulting in intraspecific competition with death of less capable individuals, followed by vigorous development of survivors.

In general, all species in the area have presented similar behavior, in which the control through 7 days caused a very significant reduction on weed dry matter mass accumulation, although density of *C. rotundus*, *Sida*

Table 1. Weed community of potato crop during the experimental period. Botucatu/SP, Brazil.

Family	Species	Common name	Code
Dicotiledoneae			
Brassicaceae	*Raphanus raphanistrum* L.	Wild Radish	RAPRA
Malvaceae	*Sida rhombifolia* L.	Country mallow	SIDRH
Asteraceae	*Galinsoga parviflora* Cav.	Gallant-soldier	GASPA
Asteraceae	*Bidens pilosa* L.	Hairy beggarticks	BIDPI
Asteraceae	*Emilia sonchifolia* (L.) DC.	Cupid's shaving	EMISO
Amaranthaceae	*Amaranthus deflexus* L.	Largefruit	AMADE
Rubiaceae	*Richardia brasiliensis* Gómez	Brazil puzley	RCHBR
Portulacaceae	*Portulaca oleracea* L.	Little hogweed	POROL
Convolvulaceae	*Ipomoea purpurea* (L.) Roth	Morning glory	PHBPU
Oxalidaceae	*Oxalis latifolia* Kunth	Woodsorrel	OXALA
Monocotiledoneae			
Poaceae	*Urochloa plantaginea* (Link) Hithc.	Alessandergrass	URPL
Poaceae	*Panicum maximum* Jacq	Guinea grass	PANMA
Poaceae	*Eleusine indica* (L.) Gaert.	Indian goosegrass	ELEIN
Poaceae	*Digitaria horizontalis* Willd	Hay grass	DIGHO
Cyperaceae	*Cyperus esculentus* L.	Purple nutsedge	CYPES

Table 2. Effect of different periods of control or coexistence with weeds on density and dry matter accumulation of *Urochola plantaginea* (URPL), *Cyerus esculentus* (CYPES), *Raphanus raphanistrum* (RAPRA), and other weed species present in experimental area, Botucatu/SP, Brazil.

Treatment	URPL Density plants m^{-2}	URPL Dry matter (g)	CYPES Density plants m^{-2}	CYPES Dry matter (g)	RAPRA Density plants m^{-2}	RAPRA Dry matter (g)
Weed free						
0-7[1]	15.5^c	8.1^c	20.5^{cde}	2.0^{cd}	11.5^c	2.3^c
0-14	9.0^c	1.6^c	8.5^{de}	1.0^d	8.0^c	0.4^c
0-21	3.5^c	0.35^c	10.0^{cde}	0.2^d	3.0^c	0.1^c
0-28	2.5^c	0.1^c	6.5^e	0.1^d	5.0^c	0.0^c
0-35	1.5^c	0.0^c	4.0^e	0.0^d	3.0^c	0.0^c
0-harvest	0.0^c	0.0^c	0.0^e	0.0^d	0.0^c	0.0^c
Weedy						
0-7	59.0^a	10.6^c	48.0^{ab}	8.1^{bc}	25.5^{abc}	5.8^c
0-14	34.5^b	19.6^c	36.5^{abc}	8.1^{bc}	42.0^a	16.6^{bc}
0-21	58.0^a	50.1^b	55.5^a	16.5^a	34.5^{ab}	32.0^{ab}
0-28	52.5^a	67.3^b	55.0^a	18.84^a	37.0^{ab}	47.9^a
0-35	44.0^{ab}	112.4^a	38.0^{abc}	12.94^a	15.0^{bc}	26.8^b
0-harvest	50.0^{ab}	99.3^a	31.5^{bcd}	14.98^a	13.5^{bc}	34.7^{ab}
F. treatment	17.32**	22.04**	6.31**	10.64**	2.59*	5.72**
F.block	1.01^{ns}	1.77^{ns}	4.28*	4.31*	1.55^{ns}	2.66*
C.V. (%)	42.4	57.0	62.3	64.5	110.2	103.4
L.S.D.	16.8	25.28	23.46	6.42	26.16	20.67

Averages followed by the same letter in column do not differ statistically by t test ($p \leq 0.05$); [1] days.

rhonbifolia L., *Galinsoga parviflora* Cav., and of other species group increased until 21 (Table 3).

Table 3. Effect of different periods of control or coexistence with weeds on density and dry matter accumulation of *Sida rhombifolia* (SIDRH), *Galinsoga parviflora* (GASPA), and other weed species present in experimental area, Botucatu/SP, Brazil.

Treatment	SIDRO		GASPA		Other species	
	Density plants m^{-2}	Dry matter (g)	Density plants m^{-2}	Dry matter (g)	Density plants m^{-2}	Dry matter (g)
Weed free						
0-7[1]	6.0^{de}	0.6^b	33.5^{abc}	3.4^{bcd}	13.0^{de}	1.0^c
0-14	9.0^{de}	0.3^b	27.0^{abc}	1.4^d	3.5^e	0.1^c
0-21	7.5^{de}	0.1^b	24.0^{abc}	0.2^b	16.5^{cde}	0.5^c
0-28	4.0^e	0.0^b	5.5^{bc}	0.0^d	5.0^e	0.0^c
0-35	3.0^e	0.0^b	4.5^{bc}	0.0^d	0.5^e	0.0^c
0-harvest	0.0^e	0.0^b	0.0^c	0.0^d	0.0^e	0.0^c
Weedy						
0-7	65.5^{bcd}	1.2^b	30.0^{abc}	1.6^d	38.0^{bcd}	2.9^{bc}
0-14	148.5^a	5.8^b	42.0^a	2.0^{cd}	83.5^a	8.2^{bc}
0-21	111.5^{ab}	5.5^b	52.0^a	8.3^{abcd}	42.0^{bcd}	6.3^{bc}
0-28	41.5^{cde}	2.0^b	35.0^{abc}	11.6^{abc}	45.5^{bc}	6.5^{bc}
0-35	73.5^{bc}	5.5^b	31.0^{abc}	17.1^a	49.5^b	20.0^a
0-harvest	114.5^{ab}	15.8^a	37.5^{ab}	13.0^{ab}	49.5^b	10.5^b
F. treatment	6.42**	3.22**	1.72^{ns}	3.07**	5.55**	3.49**
F. block	1.44**	0.82^{ns}	7.19**	5.33**	2.87*	1.29^{ns}
C.V. (%)	85.6	167.4	90.7	140.7	77.4	139.2
L.S.D.	60.03	7.41	35.03	9.88	32.17	9.37

Averages followed by the same letter in a column do not differ statistically by t test ($p \leq 0.05$); [1]days.

In the coexistence periods between crop and weed, it was observed that *R. raphanistrum* L., *S. rhombifolia,* and the group of less frequent species were in the highest plant density through 14 days; while the highest biomass accumulation was observed at 21 days for *R. raphanistrum* and at 35 days for the other species, except in the case of *S. rhombifolia*, which presented higher accumulation of dry biomass than the control treatment with coexistence with weeds during all crop cycle (Tables 2 and 3).

The highest density of *C. rotundus* plants was verified at 21 and 28 days, and the highest biomass accumulation at 21 days (Table 2). For *G. parviflora*, the highest density occurred at 14 and 21 days, and the highest dry matter accumulation was also observed at 21 days (Table 3). It was observed that biomass accumulation increased in the longest periods, while density was higher in the initial periods. Density of plants in the weed community is also dependent on the bank of seeds present in the soil and, according to environment factors, it can performe differently in relation to seed dormancy (Nogushi, 1983; Martins and Silva, 1994).

Regarding the relative density of weeds (Figure 1) and the density of each species in relation to the rest of them present in that area, the *G. parviflora* distinguished from the other species in all evaluated periods, except at 28

days. In the periods of coexistence between weed community and crop, *S. rhombifolia* plants presented the highest relative density. These results indicate that *G. parviflora* presented low competitiveness when compared with weed community in the area, once its infestation was low through the coexistence periods. Thus, initial periods of weed community control, even in the short ones, as 7 days, influenced *G. parviflora* plants survival rate.

For *S. rhombifolia*, it was observed that species should present aggressiveness that can help on survival, even by interspecific competition condition occurring in the experimental area. *S. rhombifolia* is common in areas where there is small perturbance of soil, as in no-till system, orchards, and pastures. Therefore, minimum soil movement is more suitable for its development.

Relative frequency data showed that all species presented similar results, except for *R. raphanistrum* and *C. rotundus* at 35 days, which reached the relative frequency mean of 27% (Figure 2). Also, the index of importance (Figure 3) showed that *G. parviflora* was the species with the highest value in the initial periods of control, while in the coexistence periods *S. rhombifolia* was the most important species. Costa et al. (2008) observed that *Commelina benghalensis* L. was the most important species in the initial periods of control and, *U. plantaginea* was in coexistence periods.

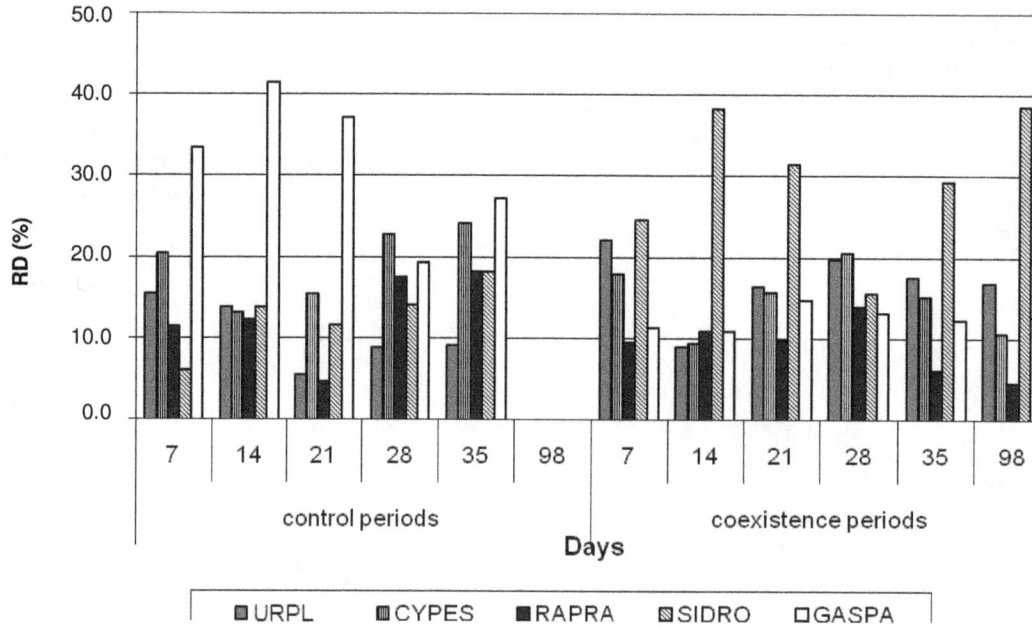

Figure 1. Relative density (RD) of *Uroclhoa plantaginea* (URPL), *Cyerus esculentus* (CYPES), *Raphanus raphanistrum* (RAPRA), *Sida rhombifolia* (SIDRH), *Galinsoga parviflora* (GASPA) present in experimental area, as a function of the number of days after potato emergence with control or coexistence with weeds, Botucatu/SP, Brazil.

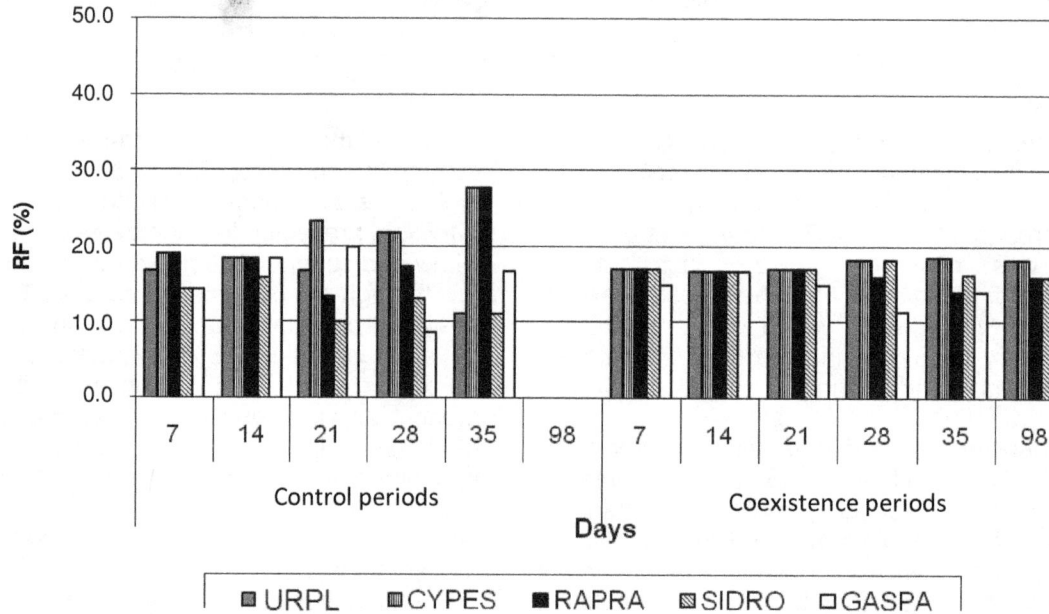

Figure 2. Relative frequency (RF) of *Urochloa plantaginea* (URPL), *Cyerus esculentus* (CYPES), *Raphanus raphanistrum* (RAPRA), *Sida rhombifolia* (SIDRH), *Galinsoga parviflora* (GASPA) present in experimental area, as a function of the number of days after potato emergence with control or coexistence with weeds, Botucatu/SP, Brazil.

In Table 4, it was verified that for tuber weight classification, to obtain potato of Type 1, or rather, with the highest diameter and weight, control for 35 was enough to assure the same percentage of potato of Type

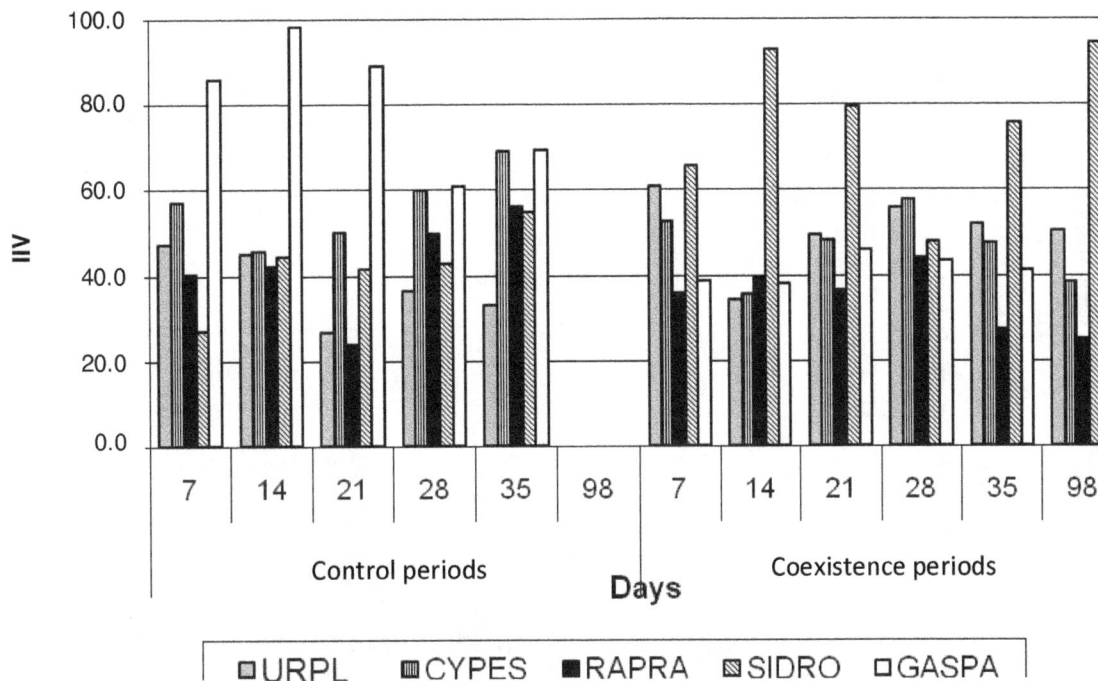

Figure 3. Index of importance value (IIV) of *Urochloa plantaginea* (URPL), *Cyerus esculentus* (CYPES), *Raphanus raphanistrum* (RAPRA), *Sida rhombifolia* (SIDRH), *Galinsoga parviflora* (GASPA) present in experimental area, as a function of the number of days after potato emergence with control or coexistence with weeds, Botucatu/SP, Brazil.

Table 4. Effect of different periods of control or coexistence with weeds on percentage of tuber classified in five different size types (diameter). Botucatu/SP, Brazil.

Treatment	Tuberclassification (%)				
	Type 1 >70 mm	Type 2 >42≤70 mm	Type 3 >33≤42 mm	Type 4 >28≤33 mm	Type 5 ≤28 mm
Weed free					
0-7[1]	18.3^{dc}	15.5^{dc}	29.5^{abc}	25.8^{bc}	10.9^{abc}
0-14	17.1^{cde}	16.5^{bcd}	31.6^{abc}	25.5^{bc}	8.1^{bc}
0-21	14.8^{def}	27.8^{ab}	28.9^{abc}	21.6^{bc}	6.9^{c}
0-28	21.2^{bcd}	21.3^{bc}	28.1^{abc}	21.1^{bc}	8.2^{bc}
0-35	34.3^{ab}	16.7^{bcd}	18.8^{c}	20.8^{bc}	9.5^{abc}
0-harvest	38.8^{a}	22.1^{bc}	18.8^{c}	14.3^{c}	6.1^{c}
Weedy					
0-7	29.2^{abc}	20.6^{bc}	20.6^{bc}	20.2^{bc}	9.5^{abc}
0-14	13.8^{def}	23.8^{bc}	32.3^{ab}	20.6^{bc}	9.5^{abc}
0-21	11.4^{def}	27.3^{abc}	27.5^{abc}	27.8^{abc}	5.9^{c}
0-28	4.1^{ef}	16.6^{bcd}	25..3^{abc}	40.2^{a}	13.8^{a}
0-35	2.9^{f}	37.3^{a}	19.1^{c}	30.8^{ab}	10.2^{abc}
0-harvest	9.9d^{ef}	7.7^{d}	36.6^{a}	32.6^{ab}	13.2^{ab}
F. treatment	5.20**	3.18**	1.81*	2.07*	1.86^{ns}
F.block	0.42**	2.48^{ns}	0.85^{ns}	0.84^{ns}	0.79^{ns}
C.V. (%)	54.9	39.9	33.3	38.6	39.5
L.S.D.	14.21	12.13	12.60	13.91	5.34

Averages followed by the same letter in a column do not differ statistically by t test ($p ≤ 0.05$); [1]days.

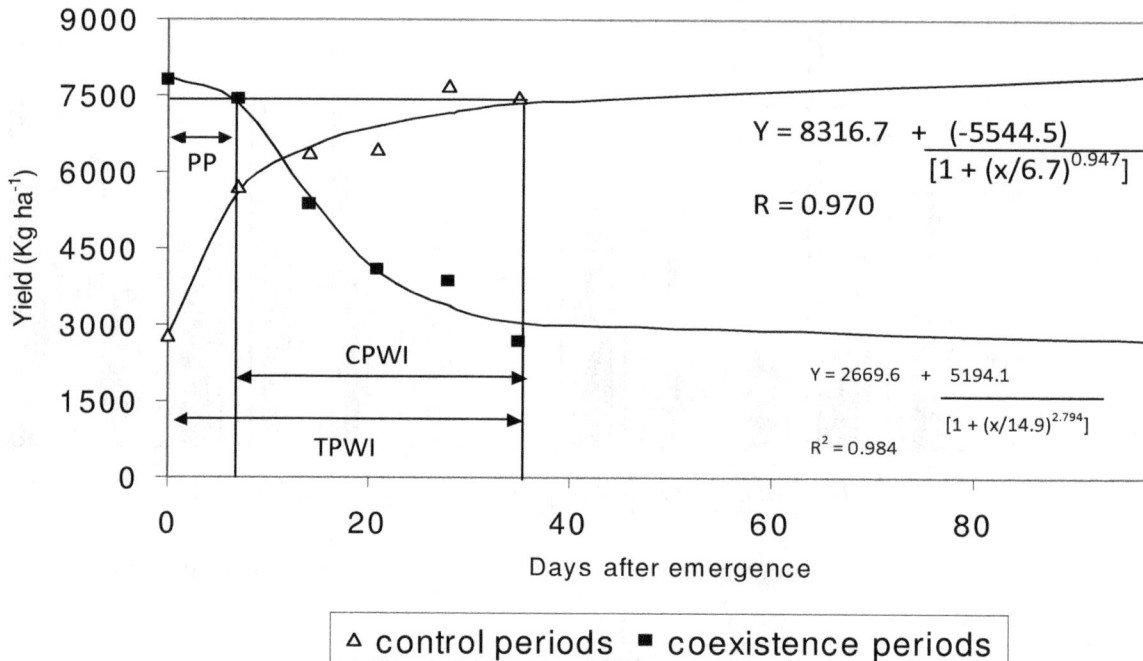

Figure 4. Yield of potato tubers as a function of coexistence and weed control periods in experimental area, Botucatu/SP, Brazil.

1 as the control treatment without weed competition. Percentage of Type 1 tubers was inferior to the control treatment when crop was under weed competition for a period of 14 days or more, however, about 25% of losses was observed in the 7-days weedy treatment. Although, not statistically differing from the control treatment, this loss can be considered important. In general, it was observed that the longer is the coexistence period, the higher is the percentage of tubers with inferior quality, with low weight and diameter. Vangessel and Renner (1990) and Monteiro et al. (2011) also found that the quantity and quality of marketable tubers was reduced when there was competition in potato crops against weeds.

Regarding tuber yield (Figure 4), it was observed that the coexistence with weeds during all crop cycle caused a reduction of 65% in relation to the control treatment without weed competition. It was verified, according to the terminology proposed by Pitelli and Durigan (1984) that (PPI) was of 7 days after sprouting, with 5% of yield reduction of potato crop, and (TPWI) was 35 days and, thus (CPWI) was from 7 to 35 (Figure 4). In fact, this shows that the control period between 7 and 35 resulted in yield as the control treatment. However, according to Jaiswal (1992) for the potato crop would be the critical period between 25 and 35 days after planting and Singh et al. (2005) found a period between 15 and 45 days. It should be noted that these differences are related to variety, planting date, differences in weeds, soil, and climatic conditions.

Treatments (control or coexistence periods) were started from sprouting, 19 days after seed tuber planting. That fact determined that an interval of only 7 had formed already the (PPI), because the interval of time between planting and sprouting was long enough for weed community to interfere on crop yield already by the first period studied.

In future studies, the beginning of treatment should be from planting of seed tubers, because the period until sprouting can be too long as observed, what could give advantage to the weed community in relation to the crop. It is also noteworthy that the time to sprouting depends on the cultivar, the temperature, the soil type, and others factors.

In summary, the present work determined that the total period of weed interference (TPWI) was 35, while the period previous of interference (PPI) was 7 and, consequently, the critical period of weed interference (CPWI) was from 7 to 35 after crop emergence.

REFERENCES

Ahmadvand G, Mondani F, Golzardi F (2009). Effect of crop plant density on critical of weed competition in potato. Sci. Hort. 121:249-254.

Ciuberkis S, Bernotas S, Raudonius S, Felix J (2007). Effect of weed emergence time and intervals of weed and crop competition on potato yield. Weed Tecnol. 21:612-617.

Costa NV, Cardoso LA, Rodrigues ACP, Martins D (2008). Períodos de interferência de uma comunidade de plantas daninhas na cultura da batata. Planta Daninha 26:83-91.

Jaiswal VP (1992). Crop weed competition studies in potato. J. Indian Potato Assoc. 18:131-134.

Liebman M, Drummond FA, Corson S, Zhang JX (1996). Tillage and rotation crop effects on weed dynamics in potato productions systems. Agron. J. 88:18-26.

Lutman PJW (1992). Weeds in Potatoes. In: Harris, PM. The Potatoes Crop: The Scientific Basis for Improvement. Chapman and Hall. London. England. pp. 373-379.

Martins CC, Silva WR (1994). Estudos de bancos de sementes do solo. Informativo Abrates. 4:49-56.

Martins D (1994). Interferência de capim-marmelada na cultura da soja. Planta daninha. 12:93-99.

Miranda-Filho HS (1996). Batata. In: Raij, BV, Cantarella H, Quaggio JA, Furlani, AMC (eds). Recomendações de adubação e calagem para o estado de São Paulo. Instituto Agronômico and Fundação IAC (Boletim técnico 100). Campinas. Br. pp. 123-145.

Monteiro A, Henriques I, Moreira I (2011). Critical período for weed control in potatoes in the Huambo Province (Angola). Planta Daninha. 29:351-362.

Mueller-Dombois D, Ellenberg HA (1974). Aims and Methods of Vegetation Ecology. John Wiley. New York. pp. 168-197.

Muhammad B (1993). Impact of weed competition on potato production. Pak. J. Agric. Res. 14:64-71.

Nelson DC, Thorenson HC (1981). Competition between potatoes (*Solanum tuberosum*) and weeds. Weed Sci. 29:672-677.

Nogushi K (1983). Ecological tudy on ligth competition between upland crops and weeds. B. Natl. Agric. Res. Center. 1:37-103.

Pitelli RA (1985). Interferência de plantas daninhas em culturas agrícolas. Inf. Agropec. 11:16-27.

Pitelli RA, Durigan JC (1984). Terminologia para períodos de controle e de convivência de plantas daninhas em culturas anuais e bianuais. In: XV Congresso Brasileiro de Herbicidas e Plantas Daninhas. Belo Horizonte. Brasil P. 37.

Singh VP, Mishra JS, Gogoi AK (2005). Effect of weed interference and fertilizer levels on weeds and productivity of potato. Indian J. Weed Sci. 37:225-227.

Tripathi B, Singh CM, Bhargava A (1989). Comparative efficacy of herbicides in potato under the conditions of North-Western Himalayas. Pesticides 23:37-38.

Vangessel MJ, Renner KA (1990). Effect of soil type, hilling time, and weed interference on potato (*Solanum tuberosum*) development and yield. Weed Technol. 4:299-305.

Velini ED (1997). Interferência entre plantas daninhas e cultivadas. In: I Simpósio sobre Herbicidas e Plantas Daninhas, Dourados/MS, Brasil. pp. 29-49.

21

Effect of glyphosate on weed management and grain yield in *Kharif* maize of transgenic stacked and conventional maize hybrids for higher productivity

Sivagamy Kannan and Chinnusamy Chinnagounder

Department of Agronomy, Tamil Nadu Agricultural University, Coimbatore, India.

A field experiment was carried out at Tamil Nadu Agricultural University, Coimbatore during *kharif* seasons of 2010 and 2011 to study the effect of Glyphosate on weed management and grain yield in *kharif* maize of transgenic stacked and conventional maize hybrid. This investigation was conducted with the following objectives to evaluate the weed control efficiency and crop productivity with K salt of glyphosate formulations under field conditions. Treatments consisted of two transgenic stacked hybrids named 30V92 and 30B11 applied with glyphosate as early post emergence at 900 and 1800 g a.e ha^{-1} during *kharif* season of 2010 and conventional maize hybrids named 30V92 and 30B11 applied with glyphosate by controlled droplet application method at 900, 1350 and 1800 g a.e ha^{-1} during *kharif* season of 2011 compared with non-transgenic counterpart maize hybrids applied with pre emergence atrazine at 0.5 kg ha^{-1} followed by one hand weeding on 40 days after sowing (DAS) with and without insect management. Among the treatments, early POE application of glyphosate at 1800 g a.e ha^{-1} registered lower weed density and higher weed control efficiency in transgenic and non-transgenic maize hybrids at all the intervals. Higher grain yield was registered with post emergence application of glyphosate at 1800 g a.e ha^{-1} in transgenic and non transgenic maize hybrid of 30V92 during both the *kharif* seasons

Key words: Glyphosate, transgenic maize, weed control efficiency, weed index, yield.

INTRODUCTION

Herbicide tolerance has been introduced through genetic modification into a number of crops including maize. The development of crop cultivars with resistance to selected herbicides has the positive impact on agricultural production systems and food safety. Roundup Ready® crop varieties that can be safely treated with glyphosate herbicide to control weeds were first commercialized for soybeans in 1996, for cotton in 1997, and for corn in 1998 (Green et al., 2008). Herbicide tolerance has been introduced through genetic modification into a number of crops including corn. Glyphosate, the active ingredient in the Roundup family (ROUNDUP, ROUNDUP ULTRA AND ROUNDUP READY) were registered trademarks of Monsanto Technology of agricultural herbicides is one of the most widely used herbicides in the world. Glyphosate is highly effective against the majority of annual, perennial grasses, and broad-leaf weeds and has superior environmental and toxicological characteristics such as rapid soil binding and biodegradation as well as extremely low toxicity to mammals, birds, and fish. Glyphosate is a foliar applied, broad spectrum, post emergence herbicide capable of controlling annual, perennial grasses and dicotyledonous weeds.

The introduction of glyphosate resistant crops has created new opportunities for the use of effective, non selective herbicides like glyphosate as selective weed control in crop production. Now, it can be used as post emergence herbicide in glyphosate resistant crops (Norsworthy et al., 2001). Roundup Ready corn event NK603 was produced by the stable insertion of two gene cassettes that express 5-enolpyruvylshikimate-3-phosphate synthases from *Agrobacterium* sp. strain CP4 (*CP4 EPSPS*). Corn event NK603 differs from Roundup Ready corn event GA21 that expresses a modified corn EPSPS. While TC1507 maize expresses a Bt insecticidal protein (Cry1F) for control of certain lepidopteron (stem borers) pests and NK603 corn expresses a modified maize 5-enolpyruvylshikimate-3-phosphate synthase enzyme (*CP4 EPSPS*) that confers tolerance to herbicide products containing glyphosate. Post emergence herbicides have been achieved in adequate weed control programmes, due to its broad spectrum of activity, excellent crop safety, convenience and flexibility was reported by Ferrel and Witt (2002).

Post emergence application of glyphosate at 1800 g a.e ha^{-1} in transgenic maize and post emergence control droplet application method of glyphosate at 1800 g a.e ha^{-1} in conventional maize hybrid (30V92) recorded high productivity and profitability. In view of the above facts, an experiment on "Effect of Glyphosate on weed management and grain yield in *kharif* maize of transgenic Stacked and conventional maize Hybrids" taken up during *kharif* season of 2010 and 2011 (Tables 1 and 2). Target pests viz., stem borer and cob borer were effectively controlled in transgenic maize hybrids during both the *kharif* seasons.

MATERIALS AND METHODS

Experimental site

Field experiments were laid out during *kharif* seasons of 2010 and 2011 in Eastern bloc farm of Tamil Nadu Agricultural University, located at Coimbatore, India. The geographical location of the experimental site is situated in western agro climatic zone of Tamil Nadu at 11°N longitude and 77°E latitude with an altitude of 426.7 m above MSL and the farm receives the total annual rainfall of 674 mm in 45.8 rainy days. The soil of the experimental site was sandy clay loam in texture (32.48% clay, 18.50% silt and 28.96% coarse sand) with low available nitrogen, medium in available phosphorous and high in available potassium. The soil analysed 260, 11.90 and 490 Kg ha^{-1} of KMnO4-N, Olsen–P and NH$_4$OAC-K, respectively with EC of 0.16 dSm^{-1}, pH of 8.11 and organic carbon of 0.31%.

Selection of cultivar and sowing

The experiment was laid out in randomized complete block design (RBD) with sixteen treatments and replicated thrice. The gross plot size adopted was (5 × 3.6 m) 18 m^2. The adopted spacing between the rows and plants were 60 and 25 cm respectively. Herbicide tolerant transgenic maize test hybrids namely 30V92, 30B11, and conventional hybrids of 30V92, 30B11, BIO 9681, and COHM5 during the *kharif* season of 2010. Conventional maize hybrids 30V92, 30B11, BIO 9681 and COHM5 were raised during *kharif* season of 2011. After sowing the seed, immediate light irrigation was given to the crop for uniform germination

The herbicides as per the treatments schedule were applied as pre emergence at third day after sowing, glyphosate application at 2 to 4 leaf stage of weeds [20 to 25 days after sowing (DAS) of maize]. Hand operated knapsack sprayer fitted with a flat fan type nozzle (WFN 40) was used for spraying the herbicides adopting a spray volume of 250 L ha^{-1}. The recommended dose of 150:75:75 Kg of NPK ha^{-1} are in the form of urea, single super phosphate and muriate of potash.

RESULTS AND DISCUSSION

Predominant weed flora of the experimental field

Effect on weeds

Weed flora of the experimental field in maize predominantly consist of 12 species of broad leaved weeds, 5 species of grasses and a sedge weed. The dominant among broadleaved weeds were *Trianthema portulacastrum, Datura stramonium, Cleome gynandra, Digera arvensis, Physallis minima*, and *Corchorus olitorius*. The dominant grass weeds were *Setaria verticillata* and *Cynodon dactylon*. *C. rotundus* was the only sedge present in the experimental field. With respect to individual weed species during both the years, density of *T. portulacastrum* recorded about 162.80% No.m^{-2} before spraying of glyphosate. Higher weed flora composition registered during both years might be due to adequate rainfall during cropping period which favoured a conducive field environment for weed growth: *T. portulacastrum, D. stramonium, C. gynandra, P. minima, D. arvensis, S. verticillata*, and *C. dactylon*. The results are in line with the findings of (Nadeem et al., 2008) who reported that *T. portulacastrum, D. arvensis* were the most common weeds which compete with maize and assimilate faster biomass than maize.

Weed control rating in maize

Weed control rating score was done at 7, 15, and 21 days after sowing (DAS) in transgenic maize hybrids with POE application of glyphosate at various rates of application. At 7 DAS, moderate control of broad leaved weeds and grass (score = 6) and poor to deficient control of sedges (score = 3) were observed with glyphosate at 900 g a.e ha^{-1}. Satisfactory control of broad leaved weeds and grass (score = 7), deficient control of sedges (score = 4) were observed under glyphosate at 1800 g a.e ha^{-1}. Glyphosate at 900 and 1800 g a.e ha^{-1} at 15 DAS resulted in good control of broad leaved and grass weeds (score = 8), moderate control of sedges (score = 6) (Table 2). Whereas at 21 DAS, complete control of broad leaved weeds and grass (score = 10), good control of sedges (score = 9) were noticed under glyphosate at 1800 g a.e ha^{-1}. Satisfactory control of sedges was observed

Table 1. Weed control rating in transgenic maize – *kharif* season of 2010.

Treatment	7 DAS			15 DAS			21 DAS		
	BLW	Grass	Sedge	BLW	Grass	Sedge	BLW	Grass	Sedge
T_1 - 30V92 HR Glyphosate at 900 g a.e ha^{-1}	6.0	6.0	3.0	8.0	7.0	6.0	10.0	8.0	8.0
T_2 - 30V92 HR Glyphosate at 1800 g a.e ha^{-1}	7.0	7.0	4.0	9.0	8.0	6.0	10.0	10.0	9.0
T_4 - 30B11HR Glyphosate at 900 g a.e ha^{-1}	6.0	6.0	3.0	7.0	7.0	6.0	9.0	9.0	8.0
T_5 - 30B11HR Glyphosate at 1800 g a.e ha^{-1}	7.0	7.0	4.0	9.0	8.0	6.0	10.0	10.0	9.0

Data not statistically analysed BLW: Broad leaved weeds.

Table 2. Weed control rating in non transgenic maize – *kharif* season of 2011.

Treatment	7 DAS			15 DAS			21 DAS		
	BLW	Grass	Sedge	BLW	Grass	Sedge	BLW	Grass	Sedge
T_1 - 30V92 POE Glyphosate at 900 g a.e ha^{-1}	6.0	6.0	3.0	7.0	7.0	5.0	9.0	9.0	8.0
T_2 - 30V92 POE Glyphosate at 1350 g a.e ha^{-1}	7.0	7.0	4.0	8.0	8.0	6.0	10.0	10.0	9.0
T_3 - 30B11 POE Glyphosate at 1800 g a.e ha	8.0	7.0	4.0	8.0	8.0	6.0	10.0	10.0	9.0
T_4 - 30B11 POE Glyphosate at 900 g a.e ha^{-1}	6.0	6.0	3.0	7.0	7.0	5.0	9.0	9.0	8.0
T_5 - 30B11 POE Glyphosate at 1350 g a.e ha^{-1}	7.0	7.0	4.0	8.0	8.0	6.0	10.0	10.0	9.0
T_6 - 30B11 POE Glyphosate at 1800 g a.eha^{-1}	8.0	7.0	4.0	8.0	8.0	6.0	10.0	10.0	9.0

Data not statistically analysed BLW: Broad leaved weeds.

with glyphosate at 900 g a.e ha^{-1}.In non-transgenic maize hybrids at 3 DAS, deficient to moderate control of broad leaved weeds, poor to deficit control of grass and poor control of sedge (scoring = 5, 3, and 2), respectively were observed with glyphosate application at 900 g a.e ha^{-1}. Whereas, POE application of glyphosate at 1350 and 1800 g a.e ha^{-1} was observed a deficient to moderate control of broad leaved weeds and grasses (score = 5), poor control of sedge (score = 2). Glyphosate at 1350 and 1800 g a.e ha^{-1} at 15 DAS resulted in good control of broad leaved and grass weeds (score = 8), moderate control of sedges (score = 6). Whereas, at 21 DAS, complete control of broad leaved weeds and grass (score = 10) good control of sedges (score = 9) were noticed under glyphosate at 1350 and 1800 g a.e ha^{-1} (Table 2).

Weed density

The weed control methods effectively controlled the density of all the weeds under both transgenic and non-transgenic maize hybrids at different stages of crop growth as compared to unweeded control. During *kharif* season of 2010, lower weed density was achieved under non transgenic maize hybrid BIO 9681 and 30B11 with pre emergence application of atrazine at 0.5 Kg ha^{-1} followed by hand weeding at 20 DAS. Relatively, a higher density was observed under unweeded checks and transgenic maize before imposing post emergence application of glyphosate. Atrazine effectively controlled majority of broad leaved and grassy weeds at earlier stages of maize growth. Mundra et al. (2003) reported

that, application of atrazine at 0.5 kg ha^{-1} as pre-emergence fb inter cultivation at 35 DAS in maize significantly reduced the total weed density (Table 3).

At 40 and 60 DAS, lower weed density (2.04 and 2.35) was observed under transgenic maize hybrid 30V92 with post emergence application of glyphosate at 1800 g a.e ha^{-1} resulted in effective control of broad leaved weeds, grasses and sedges due to its broad spectrum action (Wilcut et al., 1996). This may due to more impressive control of broadleaved weeds like *T. portulacastrum, D. stramonium, C. gynandra* and *P. minima*. Foliar application of glyphosate was readily and rapidly translocated throughout the actively growing aerial and underground portions at active growing stage of broadleaved weeds might have blocked the 5-Enulpyruvate shikimate-3-phosphate synthase enzyme and arrest the amino acid synthesis which led to complete control (Summons et al., 1995). During kharif season of 2011, post emergence controlled droplet application of glyphosate at conventional maize hybrid of 30V92 at 1800 g a.e ha^{-1} (1.84 Nos m^{-2}) observed lesser total weed density at 40 DAS. Thus, glyphosate effectively controlled a broad spectrum of annual and perennial grasses, sedges and broadleaved weeds could be due to increased translocation of glyphosate inside the plant tissues Suwunnamek and Parker (1975) (Table 4).

Effect on crop

High persistence nature of weeds was attributed to their ability of high seed production and seed viability. Post

Table 3. Effect of glyphosate application on total weed density in transgenic maize.

Treatment	Total weed density (No. m^{-2})		
	Kharif season of 2010		
	20 DAS	40 DAS	60 DAS
T_1 - T.30V92 HR Glyphosate at 900 g a.e ha^{-1}	15.43 (236.22)	2.78 (5.75)	3.41 (9.63)
T_2 - T.30V92HR Glyphosate at 1800 g a.e ha^{-1}	15.33 (233.08)	2.04 (2.15)	2.35 (3.52)
T_3 - T.30V92HR (Weedy check)	15.74 (245.60)	14.32 (202.93)	13.81 (188.75)
T_4 - T.30B11HR Glyphosate at 900 g a.e ha^{-1}	15.78 (246.89)	3.31 (8.98)	3.84 (12.74)
T_5 - T.30B11HR Glyphosate at 1800 g a.e ha^{-1}	16.06 (256.07)	2.55 (4.50)	3.06 (7.35)
T_6 - T.30B11HR (Weedy check)	15.81 (248.10)	14.54 (209.43)	14.42 (205.99)
T_7 - N.T.30V92 PE atrazine 0.5 kg ha^{-1} + HW+ IC	7.99 (61.85)	7.81 (59.00)	5.79 (31.48)
T_8 - N.T.30V92 No WC and only IC	15.45 (236.55)	13.64 (183.99	12.74 (160.36)
T_9 - N.T.30V92 No WC and no IC	16.05 (255.75)	14.37 (204.37)	14.38 (204.69)
T_{10} - N.T.30B11 PE atrazine 0.5 kg ha^{-1} + HW+ IC	7.55 (55.00)	8.14 (64.34)	5.87 (32.43)
T_{11} - N.T.30B11No WC and only IC	15.51 (238.44)	13.58 (182.38)	13.12 (170.11)
T_{12} - N.T.30B11 No WC and no IC	16.25 (262.00)	15.05 (224.47)	15.05 (224.57)
T_{13} - BIO9681 PE atrazine 0.5 kg ha^{-1}+ HW + IC	7.15 (49.14)	7.52 (54.58)	5.96 (33.49)
T_{14} - BIO9681No WC and no IC	14.69 (213.70)	13.85 (189.93)	14.52 (208.94)
T_{15} - CoHM5 PE atrazine 0.5 kg ha^{-1}+ HW + IC	7.83 (59.37)	8.32 (67.3)	6.20 (36.44)
T_{16} - CoHM5 No WC and no IC	16.38 (266.19)	15.24 (230.37)	15.79 (247.44)
SEd	1.34	1.11	1.06
CD(P = 0.05)	2.74	2.27	2.17

T.30V92-Transgenic stacked 30V92, N.T.30V92-Non transgenic 30V92, T.30B11– Transgenic30B11, N.T.30B11-non transgenic 30B11, HW-hand weeding, IC-insect control, WC-weed control.

Table 4. Effect of glyphosate application on total weed density in non transgenic maize.

Treatment	Total weed density (No. m^{-2})		
	Kharif season of 2011		
	20 DAS	40 DAS	60 DAS
T_1 - N.T.30V92 POE Glyphosate at 900 g a.e ha^{-1}	16.61 (273.97)	4.11 (14.89)	4.61 (19.29)
T_2 - N.T.30V92 POE Glyphosate at 1350 g a.e ha^{-1}	16.25 (262.05)	2.91 (6.45)	3.69 (11.62)
T_3 - N.T.30V92 POE Glyphosate at 1800 g a.e ha^{-1}	16.52 (271.05)	1.84 (1.4)	2.85 (6.10)
T_4 - N.T.30B11 POE Glyphosate at 900 g a.e ha^{-1}	16.41 (267.29)	4.32 (16.65)	4.84 (21.41)
T_5 - N.T.30B11 POE Glyphosate at 1350 g a.e ha^{-1}	16.60 (273.46)	3.16 (8.01)	4.16 (15.27)
T_6 - N.T.30B11 POE Glyphosate at 1800 g a.e ha^{-1}	16.93 (284.57)	2.23 (2.99)	3.36 (9.32)
T_7 - 30V92 PE atrazine 0.5 kg ha^{-1} + HW+ IC	7.37 (52.27)	8.78 (75.16)	6.81 (44.43)
T_8 - 30V92 No WC and only IC	16.35 (265.46)	14.83 (217.99)	14.58 (210.68)
T_9 - 30V92 No WC and no IC	17.03 (287.95	15.49 (238.01)	15.35 (233.48)
T_{10} - 30B11 PE atrazine 0.5 kg ha^{-1} + HW+ IC	8.10 (63.62)	9.36 (85.67)	7.47 (53.85)
T_{11} - 30B11No WC and only IC	15.74 (245.85)	15.13 (226.78)	14.97 (222.00)
T_{12} - 30B11 No WC and no IC	17.12 (291.03)	15.91 (251.15)	16.06 (255.96)
T_{13} - BIO9681 PE atrazine 0.5 kg ha^{-1} + HW+ IC	7.95 (61.21)	8.84 (76.16)	6.86 (45.02)
T_{14} - BIO9681No WC and no IC	16.56 (272.3)	15.53 (239.32)	15.32 (232.73)
T_{15} - CoHM5 PE atrazine 0.5 kg ha^{-1} + HW+ IC	8.49 (70.03)	9.82 (94.53)	7.20 (49.79)
T_{16} - CoHM5 No WC and no IC	17.21 (294.18)	17.10 (290.48)	16.98 (286.30)
SEd	1.41	1.10	1.05
CD(P = 0.05)	2.89	2.26	2.14

T_1-T_{16}- Non Transgenic maize hybrids ; HW-Hand weeding; IC-Insect control; WC-Weed control.

emergence herbicides have been achieved adequate weed control programmes. During both years of study, among

Table 5. Effect of glyphosate application on weed control efficiency, weed index and grain yield of transgenic maize.

Treatment	Kharif season of 2010			
	WCE (%), weed index (%), yield (kg ha^{-1})			
	20 DAS	40 DAS	90 DAS	90 DAS
T_1 - T.30V92 HR Glyphosate at 900 g a.e ha^{-1}	0.00	98.56	9.09	11.10
T_2 - T.30V92HR Glyphosate at 1800 g a.e ha^{-1}	0.00	99.53	0.00	12.21
T_3 - T.30V92HR (Weedy check)	0.00	0.00	27.60	8.84
T_4 - T.30B11HR Glyphosate at 900 g a.e ha^{-1}	0.00	97.72	10.15	10.97
T_5 - T.30B11HR Glyphosate at 1800 g a.e ha^{-1}	0.00	98.97	1.88	11.98
T_6 - T.30B11HR (Weedy check)	0.00	0.00	25.30	9.12
T_7 - N.T.30V92 PE atrazine 0.5 kg ha^{-1} + HW+ IC	80.28	72.57	16.21	10.23
T_8 - N.T.30V92 No WC and only IC	0.00	14.66	31.77	8.33
T_9 - N.T.30V92 No WC and no IC	0.00	0.00	38.41	7.52
T_{10} - N.T.30B11 PE atrazine 0.5 kg ha^{-1} + HW+ IC	79.66	70.33	20.06	9.76
T_{11} - N.T.30B11No WC and only IC	0.00	11.92	32.84	8.20
T_{12} - N.T.30B11 No WC and no IC	0.00	0.00	39.80	7.35
T_{13} - BIO9681 PE atrazine 0.5 kg ha^{-1}+HW+ IC	77.27	68.73	34.47	8.00
T_{14} - BIO9681No WC and no IC	0.00	0.00	49.87	6.12
T_{15} - CoHM5 PE atrazine 0.5 kg ha^{-1}+HW+ IC	79.28	68.56	39.96	7.33
T_{16} - CoHM5 No WC and no IC	0.00	0.00	58.39	5.08
SEd	-	-	-	0.41
CD(P = 0.05)	-	-	-	0.84

T.30V92-Transgenic stacked 30V92 , N.T.30V92- non transgenic 30V92,T.30B11 – transgenic30B11, N.T.30B11-non transgenic 30B11, HW-hand weeding, IC-insect control; WC-weed control.

the weed control treatments, post emergence application of glyphosate at 1800 g a.e ha^{-1} in transgenic corn hybrid recorded higher grain yield of 12.21 t ha^{-1} this was 36.64 % higher than the unweeded check plot of transgenic 30V92 during kharif season of 2010 (Table 5). Whereas during kharif season of 2011, post emergence controlled droplet application of glyphosate at 1800 g a.e ha^{-1} in conventional maize hybrid of 30V92 resulted in higher grain yield of 11.23 t ha^{-1} (Table 6). This was 44.79% higher than the unweeded check plot of conventional maize hybrid. This could be the achieved control of weeds with non selective, translocated herbicide, provided the favourable crop growth environment at the establishment stage of the crop itself by minimizing the perennial and annual weeds and increased the seed and stalk yields (Tharp et al., 1999). This might be due to the fact that, the perennial weeds like Cyperus rotundus, C. dactylon, troublesome broadleaved weeds like T. portulacastrum weeds were effectively controlled and might increase the maize yield may be due to better light utilization of narrow row zone and faster canopy closure (Murphy et al., 1996). This might be also improved yield components viz., higher number of grains per cob, grain weight per plant and test weight. This improvement in turn was due to improved growth attributes such as higher total dry matter production and distribution in different parts, higher leaf area index. Thus, the improvement in crop growth and yield components was the consequence of lower crop weed competition, which

shifted the balance in favour of crop in the utilization of nutrients, moisture, light and space. These results are in conformity with the findings of Kamble et al. (2005).

Maize grain yield of POE application of glyphosate at 1800 g a.e ha^{-1} in transgenic 30V92 (T_2) was taken as basis to work out the weed index (WI) during kharif season of 2010. In transgenic maize hybrids, among the different rates of glyphosate, 900 g a.e ha^{-1} recorded lesser weed index of (9.09 and 10.15 per cent) in transgenic 30V92 (T_1) and 30B11 (T_4) respectively. In non-transgenic maize hybrids, PE application of atrazine 0.5 kg ha^{-1} + HW in 30V92 recorded lesser weed index (16.21%) compared all other non-transgenic hybrids with same treatment. During kharif of the 2011 among the different rates of glyphosate by controlled droplet application method of glyphosate at 1350 g a.e ha^{-1} recorded lower weed index of 7.75 and 15.23% in non transgenic maize hybrids of 30V92 (T_2) and 30B11(T_5). It was followed by POE application of glyphosate at 900 g a.e ha^{-1} in both non transgenic maize hybrids viz., 30V92 and 30B11. However, in PE application of atrazine at 0.5 kg ha^{-1} fb HW in 30V92 (T_7) maize hybrid recorded least weed index compared all other non-transgenic hybrids with same treatment. Unweeded check plots resulted in higher weed index and performed poorly during both the years.

Weed control efficiency which indicates the comparative magnitude of reduction in weed dry matter, was highly influenced by different weed control treatments. Pre emergence application of atrazine at 0.5 Kg ha^{-1} followed

Table 6. Effect of glyphosate application on weed control efficiency, weed index and grain yield of transgenic maize.

| Treatment | *Kharif* season of 2011 WCE (%), weed index (%), yield (Kg ha^{-1}) | | | |
	20 DAS	40 DAS	90 DAS	90 DAS
T_1 - N.T.30V92 POE Glyphosate at 900 g a.e ha^{-1}	5.14	96.15	9.09	9.12
T_2 - N.T.30V92 POE Glyphosate at 1350 g a.e ha^{-1}	14.29	97.66	0.00	10.36
T_3 - N.T.30V92 POE Glyphosate at 1800 g a.e ha^{-1}	8.73	99.14	27.60	11.23
T_4 - N.T.30B11 POE Glyphosate at 900 g a.e ha^{-1}	21.41	95.86	10.15	8.25
T_5 - N.T.30B11 POE Glyphosate at 1350 g a.e ha^{-1}	14.16	97.17	1.88	9.52
T_6 - N.T.30B11 POE Glyphosate at 1800 g a.e ha^{-1}	11.15	98.87	25.30	10.39
T_7 - 30V92 PE atrazine 0.5 kg ha^{-1} + HW+ IC	82.26	68.96	16.21	8.72
T_8 - 30V92 No WC and only IC	13.97	10.25	31.77	7.40
T_9 - 30V92 No WC and no IC	0.00	0.00	38.41	6.20
T_{10} - 30B11 PE atrazine 0.5 kg ha^{-1} + HW+ IC	80.03	65.71	20.06	8.01
T_{11} - 30B11 No WC and only IC	13.57	8.31	32.84	6.80
T_{12} - 30B11 No WC and no IC	0.00	0.00	39.80	6.22
T_{13} - BIO9681 PE atrazine 0.5 kg ha^{-1} + HW+ IC	78.97	63.82	34.47	7.10
T_{14} - BIO9681 No WC and no IC	0.00	0.00	49.87	5.60
T_{15} - CoHM5 PE atrazine 0.5 kg ha^{-1} + HW+ IC	73.19	61.68	39.96	6.10
T_{16} - CoHM5 No WC and no IC	0.00	0.00	58.39	4.80
SEd	-	-	-	0.80
CD(P = 0.05)	-	-	-	1.64

T_1-T_{16}- Non Transgenic maize hybrids, HW-Hand weeding; IC-Insect control; WC-Weed control.

by hand weeding recorded higher weed control efficiency of 80.28% in non transgenic maize hybrid 30V92 at 20 DAS. Whereas at 40 DAS, after spraying of herbicide, higher weed control efficiency of 99.53% was recorded in glyphosate at 1800 g a.e ha^{-1} followed by 30B11 was observed 98.97% during *kharif* season of 2010 (Table 5). Whereas, during *kharif* season of 2011, higher weed control efficiency was observed with glyphosate at 1800 g a.e ha^{-1} in conventional maize hybrid of 30V92 registered maximum weed control efficiency of 99.14% owing to the fact that registered the lesser weed density (Table 6).

Different rates of glyphosate under transgenic maize hybrids recorded more than 90% control efficiency at 40 DAS. Whereas, at the same time PE application of atrazine in non transgenic hybrids recorded only 70 to 80. This might be due to the application of glyphosate which did not allow weeds to accumulate sufficient biomass and ultimately resulted in higher weed control efficiency. Properly timed sequential application of glyphosate was effective in season-long control of common waterhemp (*Amaranthus rudis*), giant foxtail (*Setaria faberi*), velvetleaf (*Abutilion theophrasti*), common cocklebur (*Xanthum strumarium*) and common lambsquarters (*Chenopodium album*) at levels more than 90 per cent through the season was reported by (Hellwig et al., 2002).

Conclusion

The results of this experiment revealed that, lesser weed

dry weight and higher weed control efficiency were achieved with post emergence application of glyphosate at 1800 g a.e ha^{-1} in transgenic and post emergence controlled application of glyphosate at 1800 g a.e ha^{-1} in non transgenic hybrid of 30V92 during *kharif* season of 2010 and 2011 seasons, respectively. These enhanced the productivity of kharif maize resulting in higher economic returns.

REFERENCES

Ferrel JA, Witt WW (2002). Comparison of glyphosate with other herbicides for weed control in corn (*Zea mays*): Efficacy and economics. Weed Technol. 16:701-706.
Green JM, Hazel CB, Forney DR, Pugh LM (2008). New multiple-herbicide crop resistance and formulation technology to augment the utility of glyphosate. Pest Manage. Sci. 64(4):332-339.
Kamble TC, Kakade SU, Nemade SU, Pawar RV, Apotikar VA (2005). An integrated weed management in hybrid maize. Crop Res. 29(3):396-400.
Mundra SL, Vyas AK, Maliwal PL (2003). Effect of weed and nutrient management on weed growth and productivity of maize (*Zea mays* L.). Indian J. Weed Sci. 35(1-2):57-61.
Murphy SD, Yakuba Y, Weise SF, Swanton CJ (1996). Effect of planting patterns on inter row and competition between corn and late emerging weeds. Weed Sci. 44:865-887.
Hellwig KB, Johnson WG, Scharf PC (2002). Grass weeds interference and nitrogen accumulation in no tillage corn. Weed Sci. 50:757-762.
Nadeem MA, Ahmad R, Khalid M, Naveed M, Tanveer A, Ahmad JN (2008). Growth and yield response of autumn planted maize (*Zea mays*) and its weeds to reduced doses of herbicide application in combination with urea. Pak. J. Bot. 40(2):667-676.
Norsworthy JK, Burgos NR, Oliver LR (2001). Differences in weed

tolerance to glyphosate involve different mechanisms. Weed Technol. 15:725-731.

Summons RD, Grugs KJ, Andersen KS, Johnson KA, Sikors JA (1995). Re-evaluating glyphosate as a transition-state inhibitor of EPSP synthase: Identification of an EPSP synthase-EPSP glyphosate ternary complex. Biochemistry 34:6433-6440.

Suwunnamek U, Parker C (1975). Control of Cyperus rotundus with glyphosate,The influence of ammonium sulfate and other additives. Weed Res. 15:13-19.

Tharp BE, Schabenberger O, Kells JJ (1999). Response of annual weed species to glufosinate and glyphosate. Weed Technol. 13:542-547.

Wilcut JW, Coble HD, York AC, Monks DW (1996). The niche for herbicide-resistant crops in U.S. agriculture. In: Herbicide-Resistant Crops: Agricultural, Environmental, Economic, Regulatory, and Technical Aspects, S.O. Duke (ed.) CRC Press, Boca Raton. pp. 213-230.

22

Relationship between potato yield and the degree of weed infestation

Marek Gugała and Krystyna Zarzecka

Department of Plant Cultivation, Faculty of Life Sciences, University of Natural Sciences and Humanities in Siedlce, Siedlce, Poland.

Field experiment was carried out in the years 2002 to 2004 to study the effect of soil tillage methods and weed control techniques with herbicides. Potato yield and degree of weed infestation were investigated using correlation coefficients and linear regression analysis. High negative associations were found between the number of weeds, their fresh and air dry matter, and total potato tuber yield as well as marketable fraction tuber yield. The relationships were negative linear. A higher negative influence of weed infestation on potato yield was found before harvest, compared with early stage of crop growth before closing of potato rows. The presence of weeds was increasing by 1 ton of air weed dry matter per 1 ha furthermore it reduced the marketable fraction tubers yield by, respectively, 0.94 and 1.34 t per 1 ha when recorded before closing of potato rows, and by 2.62 and 3.76 t ha^{-1} when recorded before potato harvest. To conclude, weeds can affect yield and harvesting of potatoes and may encourage certain pests and diseases. Consequently weed reduction by using two techniques, either mechanical or combination of mechanical and chemical weed control is surely required.

Key words: Potato, weed infestation, yield, correlation coefficient, analysis of regression, herbicides.

INTRODUCTION

Yield losses worldwide caused by pests are estimated at 43 to 85%, and worldwide use of herbicides reaches 47% of all pesticides (Aydin and Uzunören, 2006; Dowley et al., 2008; Tolman et al., 1986; Wilson et al., 1997). Potato crop losses due to weed infestation are estimated at 20 to 80% (Jaiswal and Lal, 1996; Jan et al., 2004; Knezevic et al., 1995; Souza and Eberlein, 1997). Ekeberg and Riley (1996) and Lisińska and Leszczyński (1989) found that potato uptake of nutrients supplied with fertilizers was as low as 20 to 30%, the remaining 70 to 80% being taken up by weeds as a result of reciprocal competition. According to Lehoczky et al. (2003) under conditions of substantial weed infestation, nutrient uptake of nitrogen, phosphorus, potassium and calcium, by potato was by: 10, 17, 22 and 30% lower, respectively. Radecki (1977) calculated that weed infestation of 2 t ha^{-1} weed air-dry

matter use approximately 115 NPK, which could give 10tonnes of potato yield. Moreover, weeds worsen yield structure and tuber quality, make harvest more difficult and reduce profitability of potato cultivation (Hashim et al., 2003; Eberlein et al., 1997; Gugała and Zarzecka, 2009). Thus the purpose of the study was to find out what losses of tuber total and commercial fraction of tubers yield result from the number and amount of weeds found in potato stands.

MATERIALS AND METHODS

A field experiment was conducted at the University of Natural Sciences and Humanities Experimental farm in Zawady during the years 2002 to 2004. It was set up on a soil belonging to the rye very good complex, slightly acidic (pH=4, 8-5.2). The experiment was

Table 1. Decrease weed infestation under the effect of herbicides (values negative – decrease, average for 3 years and tillage systems).

Index of weed infestation	Herbicides							LSD$_{0.05}$
	Control	Plateen	Plateen + Fusilade Forte	Plateen + Fusilade Forte + Atpolan	Barox	Barox + Fusilade Forte	Barox + Fusilade Forte + Atpolan	
Number of weeds per 1 m^2 before closing of potato rows	21.7	-11.7	-12.4	-13.6	-9.2	-10.7	-12.3	3.8
Number of weeds per 1 m^2 before harvest of tubers	18.0	-7.6	-8.2	-10.0	-6.1	-6.6	-7.8	2.9
Fresh matter of weeds before closing of potato rows, t ha^{-1}	0.91	-0.46	-0.74	-0.76	-0.38	-0.52	-0.56	0.17
Fresh matter of weeds before harvest of tubers, t ha^{-1}	3.63	-1.01	-1.76	-2.10	-0.95	-1.34	-1.67	0.72
Air-dry matter of weeds before closing of potato rows, t ha^{-1}	0.24	-0.11	-0.18	-0.20	-0.08	-0.13	-0.14	0.05
Air-dry matter of weeds before harvest, t ha^{-1}	0.99	-0.31	-0.46	-0.59	-0.26	-034	-0.43	0.28

established as a randomised complete block in a split-plot arrangement with two soil tillage methods as a main plots and seven weed control technologies, design in three replications and it included the following factors:

(I) Two soil tillage methods – conventional farming system (skimming in September + fall ploughing in October + harrowing + cultivating + harrowing in April) and simplified farming system (either skimming and cultivating in April),

(II) Seven weed control methods including an application of herbicides:

(1) Control – mechanical weed control prior to and following potato emergence,
(2) Plateen 41.5 WG (metribuzin + flufenacet) 2.0 kg ha^{-1}
(3) Plateen 41.5 WG (metribuzin + flufenacet) 2.0 kg ha^{-1} + Fusilade Forte 150 EC (fluazifop-P-butyl) 2.5 dm^3 ha^{-1} (mixture),
(4) Plateen 41.5 WG (metribuzin + flufenacet) 1.6 kg ha^{-1} + Fusilade Forte 150 EC (fluazifop-P-butyl) 2.0 dm^3 ha^{-1} + adjuvant Atpolan 80 EC 1.5 dm^3 ha^{-1} (mixture),
(5) Barox 460 SL (bentazone + MCPA) 3.0 dm^3 ha^{-1},
(6) Barox 460 SL (bentazone + MCPA) 3.0 dm^3 ha^{-1} + Fusilade Forte 150 EC (fluazifop-P-butyl) 2.5 dm^3 ha^{-1} (mixture),
(7) Barox 460 SL (bentazone + MCPA) 2.4 dm^3 ha^{-1} + Fusilade Forte 150 EC (fluazifop-P-butyl) 2.0 dm^3 ha^{-1} + adjuvant Atpolan 80 EC 1.5 dm^3 ha^{-1} (mixture).

Treatments 2 to 7 was used to performed pre-emergence mechanical weed control. Herbicides were applied just

prior to potato emergence (Treatments 2, 3 and 4) and post-emergence to 10 to 15 cm of potato plant height (Treatments 5, 6 and 7). In the experiment, farmyard manure and mineral fertilizers were applied at the respective rates of 25 t/ha and 90 kg N, 90 kg P$_2$O$_5$, and 135 kg K$_2$O per hectare. Potato tubers of Wiking cultivar were planted in the third decade of April at the spacing of 67.5 × 37 cm.

Weed infestation was determined by the square frame and gravimetrical method before closing of potato rows and before tubers harvest. There were determined number of weeds and their fresh and air dry matter per m^2. Large subsamples of 10 plants were taken from the produce of each plot, then graded to set sizes as follows: less than 30 mm, 30 to 40 mm, 40 to 50 mm, 50 to 60 mm and tubers greater than 60 mm, and subsequently weighted. The weight of tubers greater than 40 mm was counted to estimate the marketable fraction tubers yield. The yield of potato tubers data have been presented in the paper by Gugała and Zarzecka (2009).

A significant effect of weed control methods on weed infestation characteristics as well as potato yields stimulated us to calculate relationships of the number and amount of weeds with potato yields. In order to determine the relationships, correlation coefficients were calculated and linear regression equations were developed at the significance level p=0.05, the significance being checked by t-Student test. Statistical calculations were performed based on average values from three years of studies and means for soil tillage methods.

RESULTS AND DISCUSSION

Application of herbicides and their mixtures in the experiment caused significantly lower number of weeds, fresh matter and air-dry matter of weeds determined at the beginning and towards the end of potato growth compared with the control object (Table 1). Correlation coefficients reflected a significant negative impact of weed infestation indicators (number, fresh and air dry matter of weeds) determined at the beginning and end of potato vegetation on tuber total and marketable fraction yield of tubers (Table 2). Fresh and air dry matters of weeds were more strongly correlated with tuber total and commercial fraction yields than the number of weeds. It indicates that the amount of weed mass threatens potato fields much more than number of weeds. Results observed by Pomykalska (1991) and Zarzecka (2004) support this inference. Also Chistaz and Nelson (1983) and Hashim et al. (2003) showed that correlation between tuber yield and air dry matter of weeds was significant and negative, and the correlation coefficient amounted to -0.970 and -0.498, respectively. Moreover, Hashim et al. (2003) found that tuber yield was positively

Table 2. Significant values of linear correlation coefficients between the weediness index and the total and marketable fraction of potato tubers yield (average for 3 years and tillage systems).

Index of weed infestation	Total yield of tubers t ha^{-1}	Marketable fraction of tubers yield t ha^{-1}
Number of weeds per 1 m^2 before closing of potato rows	- 0.913[*]	- 0.935[*]
Number of weeds per 1 m^2 before harvest of tubers	- 0.959[*]	- 0.968[*]
Fresh matter of weeds before closing of potato rows, t ha^{-1}	- 0.969*	- 0.976*
Fresh matter of weeds before harvest of tubers, t ha^{-1}	- 0.981[*]	- 0.997[*]
Air-dry matter of weeds before closing of potato rows, t ha^{-1}	- 0.981[*]	- 0.986[*]
Air-dry matter of weeds before harvest, t ha^{-1}	- 0.995[*]	- 0.997[*]

*Significant at P<0.05.

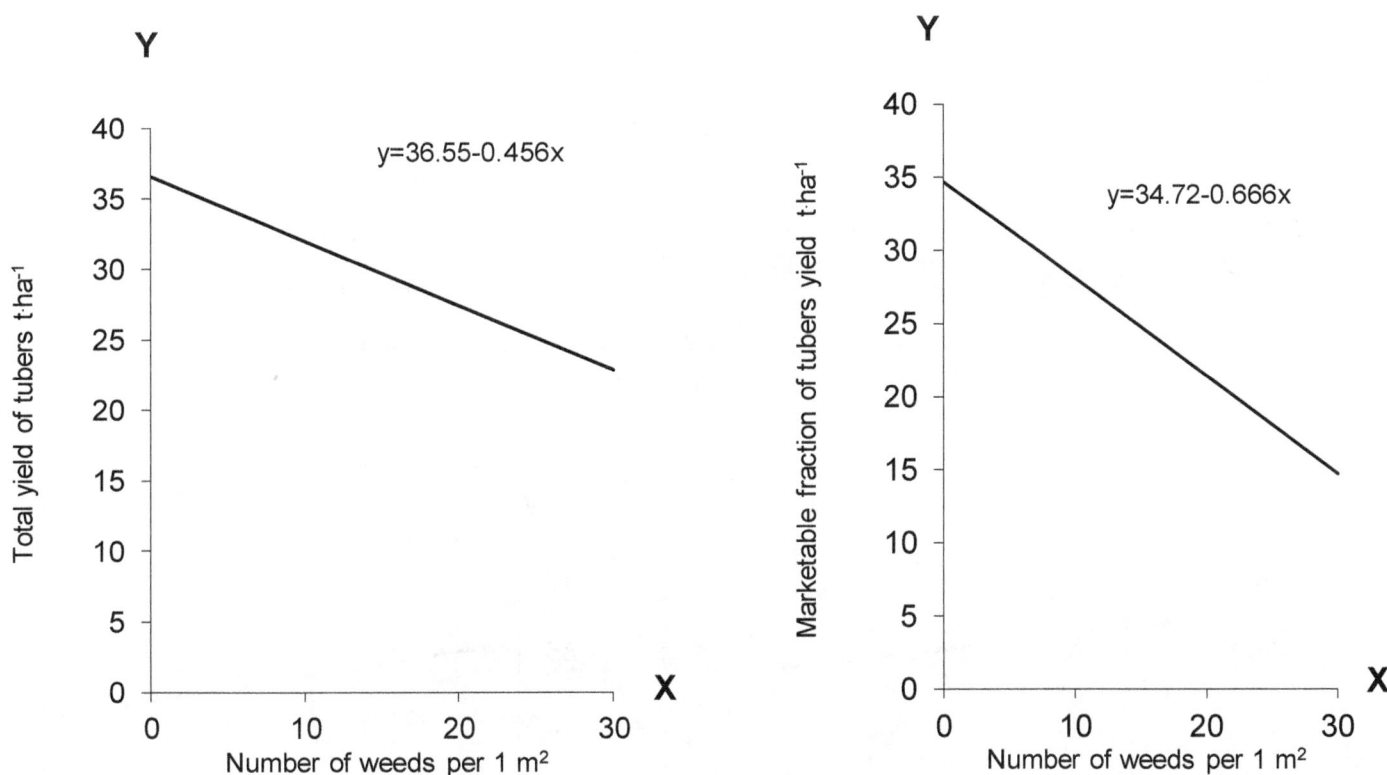

Figure 1. Relationship between total and marketable fraction yield of tubers and the number of weeds before closing of potato rows.

associated with potato plant height and number of tubers harvested from an area of 1 ha, but the differences were statistically insignificant. In turn, Mišovic et al. (1997) proved, on the basis of values of correlation coefficients, that of weeds biomass (r = 0.868) had less effect on yield of potato tubers than number of weed plants (r = 0.902).

In the present study it was also shown that, at the second date of weed infestation determination (before tubers harvest) potato tuber yields were more correlated with weed infestation indices (number of weeds, fresh and air dry matter of weeds) than at the beginning of vegetation (before closing of potato rows). Similar regression relationships between tuber yield and weed

infestation were obtained by Pomykalska (1991) for potatoes grown on soil originating from sand and from loess.

An application of regression analysis revealed a linear negative relationship of total and marketable fraction yield of tubers with number of weeds, fresh and air dry matter of weeds determined both at the beginning at towards the end of vegetation (Figures 1 to 6). Increasing weed infestation by 1 plant per 1 m^2 reduced total and marketable fraction yields by, respectively, 0.46 and 0.67 t per 1 ha before closing of potato rows, and 0.54 and0.77 t per 1 ha before tuber harvest (Figures 1 and 2). In the studies by Pomykalska (1991) the reduction

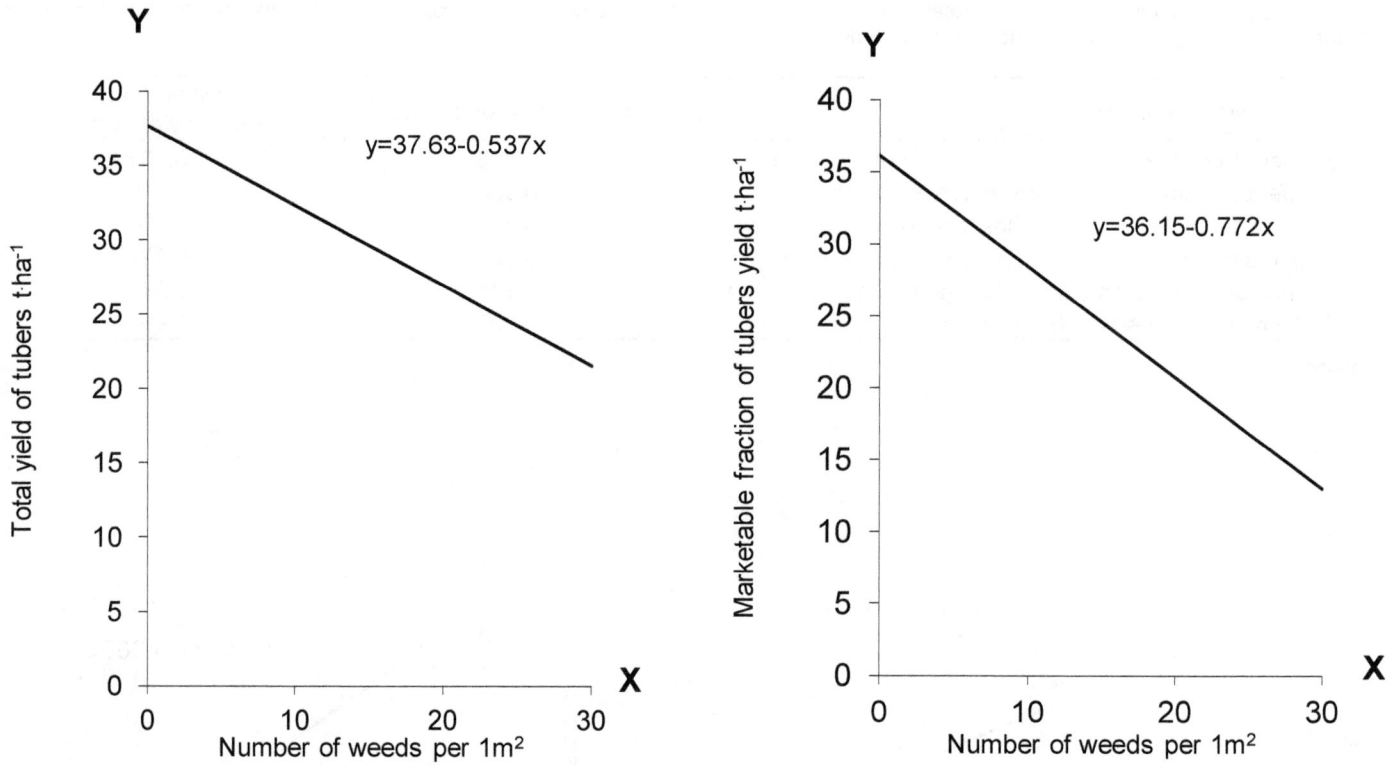

Figure 2. Relationship between total and marketable fraction yield of tubers and the number of weeds before harvest of tubers.

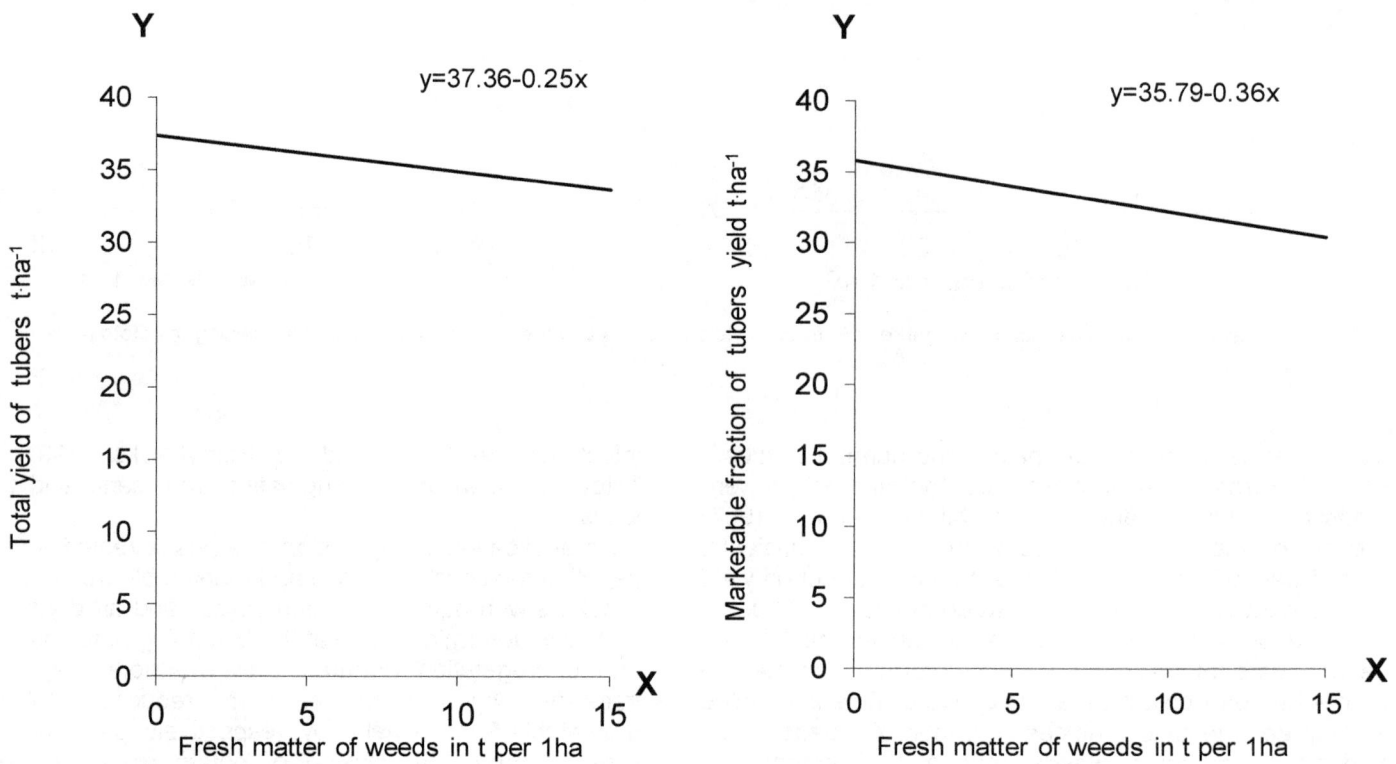

Figure 3. Relationship between total and marketable fraction yield of tubers and the fresh matter of weeds before closing of potato rows.

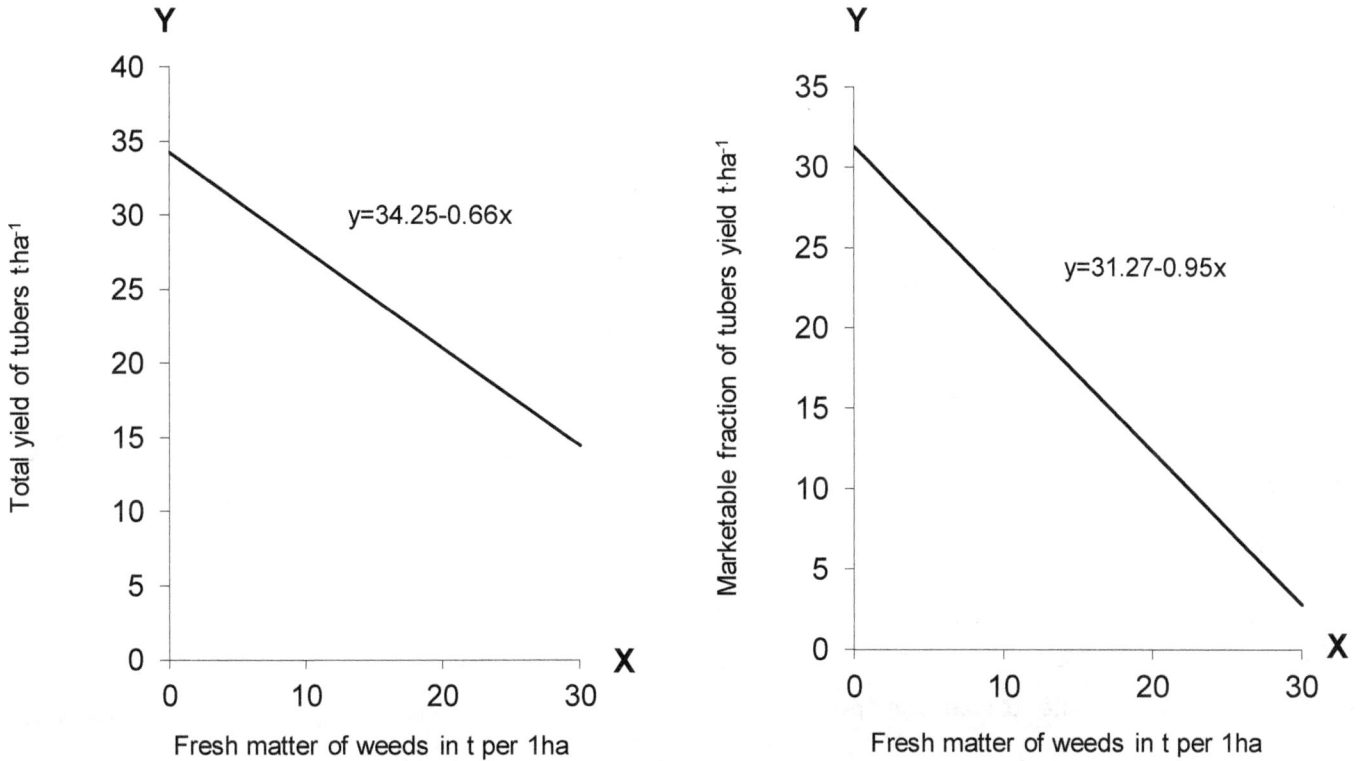

Figure 4. Relationship between total and marketable fraction yield of tubers and the fresh matter of weeds before harvest of tubers.

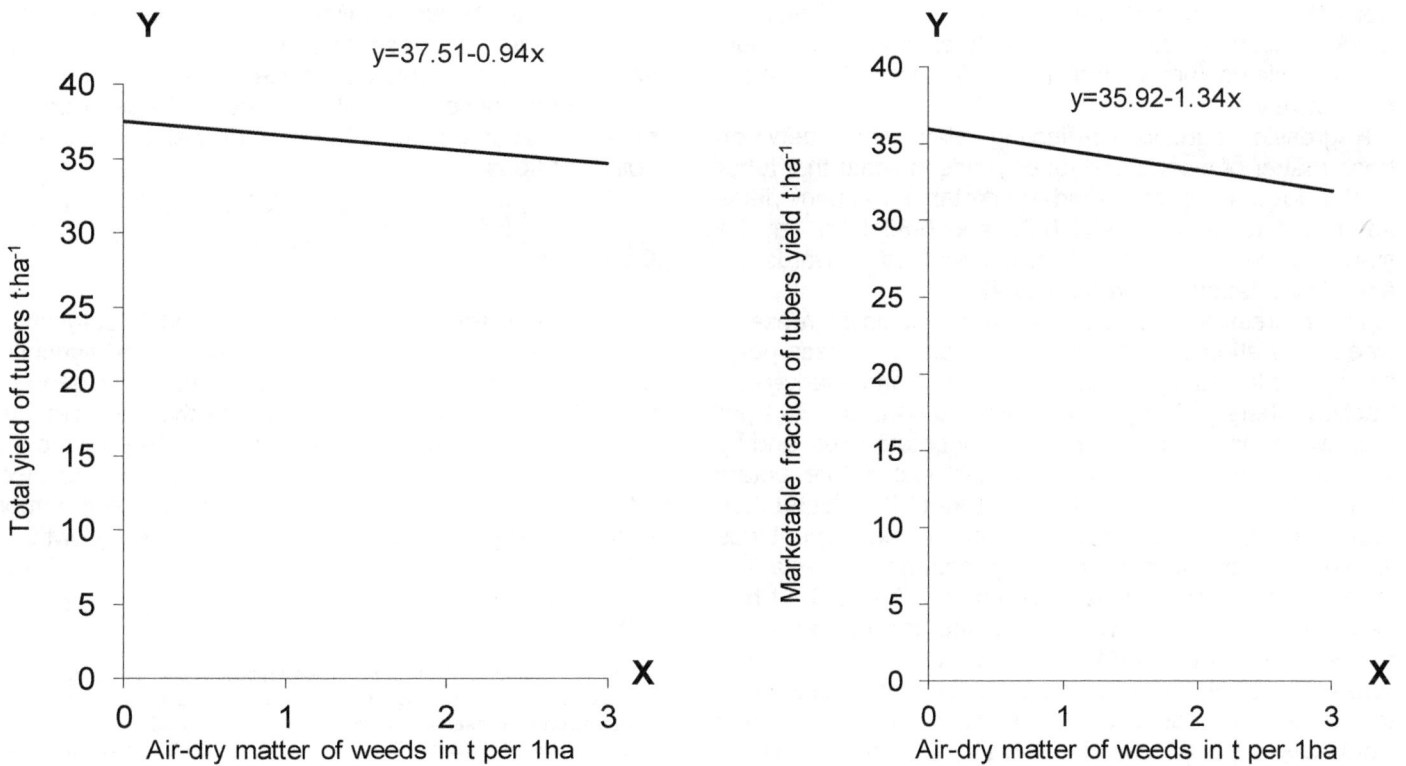

Figure 5. Relationship between total and marketable fraction yield of tubers and the air-dry matter of weeds before closing of potato rows.

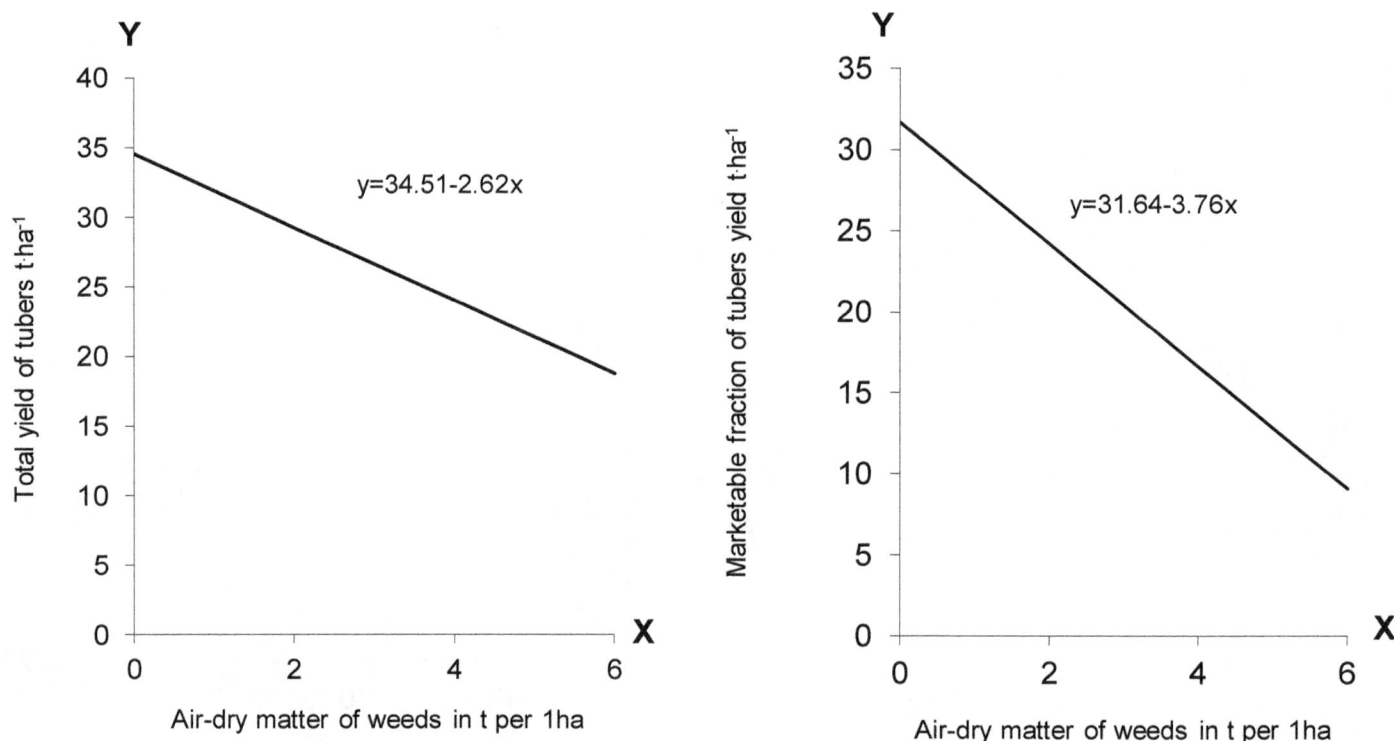

Figure 6. Relationship between total and marketable fraction yield of tubers and the air-dry matter of weeds before harvest of tubers.

amounted to 0.10 to 0.12 t ha^{-1} for total yield, and Maykuhus (1988) says the respective total yield drops were 0.10-0.46 and 0.02-0.06 t ha^{-1}. In turn, Zarzecka (2004) recorded reductions in total and marketable fraction yields amounting to 0.23 and 0.28 t ha^{-1}, respectively.

Regression equations reflecting relationships between fresh matter of weeds and tuber yields indicate that tuber matter losses for total and marketable fraction yields amounted to, respectively, 0.25 and 0.36 t ha^{-1} at the start of vegetation, and 0.66 and 0.95 t ha^{-1} towards the end of vegetation (Figures 3 and 4).

The regression analysis indicated that an increase in weed infestation by 1 ton of air dry matter of weeds per 1 ha was followed by reductions in total and marketable fraction tubers yield by, respectively, 0.94 and 1.34 t per 1 ha when recorded before closing of potato rows, and by 2.62 and 3.76 t ha^{-1} when recorded just before potato harvest (Figures 5 and 6). Pomykalska (1991) found that, depending on a type of soil on which the experiment was located, a 1-ton increase in air dry matter of weeds at the start of potato growth was followed by a 5.5 to 6.7 t ha^{-1} decrease in total yield whereas at the end of vegetation the reduction amounted to 5.2 to 6.0 t ha^{-1}. According to Zarzecka (2004), an average total yield decrease per 1 ton of air-dry matter of weeds was 2.6 t ha^{-1}. Also Pomykalska (1991) and Radecki (1977) showed a linear negative association between yield and air dry matter of weeds, and found that 1 ton of weeds decreased

tuber total yield by 5.0 to 7.0 t ha^{-1}. According to Pomykalska (1991), Radecki (1977), Boydston (2007), Buddoi et al. (1998) and Renner (1998) weed weight is a more precise and accurate indicator, than number of weeds of determining potato yield decrease. Moreover, the workers observed that the level of losses depended on species composition of weeds, as well as weather and soil conditions.

Conclusions

The issue of relationship between weed infestation and potato yielding is of importance as far as agricultural practice is concerned because it reflects the degree to which the yield is at risk, and explains the purposefulness of applying mechanical-chemical weed control operations. A higher negative influence of weed infestation on potato yielding was found before tubers harvest, compared with the beginning of crop growth.

REFERENCES

Aydin H, Uzunören N (2006). Investigation of herbicide group drug residues used in the agricultural field in soil and *Oligochaeta* class earthworms. Fresenius Environ. Bull. 15(5):335-337.
Buddoi GH, Penescu HC, Maillet J (1998). Integrated weed management in potato crop: combination between mechanical and chemical weeding. Comptes – rendus 6 eme symposium Mediterraneen EWRS, Montpellier, France. pp. 325-326.

Chistaz M, Nelson DC (1983). Comparison of various weed control programs for potatoes. Potato Res. 60(4):271-280.

Dowley LJ, Grant J, Griffin D (2008). Yield losses caused by late blight (*Phytophthora infestans* (Mont.) de Bary) in potato crops in Ireland. Irish J. Agric. Food Res. 47:69-78.

Eberlein CV, Petersom PE, Guttieri MJ, Stark JC (1997). Eonomics of cultivation weed control in potato. Weed Technol. 11(2):257-264.

Ekeberg E, Riley HCF (1996). Effect of mouldboard ploughing and direct planting on yield and nutrient uptake of potatoes in Norway. Soil Till. Res. 39:131-142.

Gugała M, Zarzecka K (2009). Weed infestation and potato yielding under the conditions of diversifield tillage operations. Pestycydy/Pesticides 1-4:41-50.

Hashim S, Marwat KB, Hassan G (2003). Chemical weed control efficiency in potato (*Solanum tuberosum* L.) under agro-climatic conditions of Peshawar, Pakistan. Pak. J. Weed Sci. Res. 9(1-2):105-110.

Jaiswal VP, Lal SS (1996). Efficacy of cultural and chemical weed control methods in potato (*Solanum tuberosum*). Indian J. Agron. 41(3):454-456.

Jan H, Muhammad A, Ali A (2004). Studies on weed control in potato in Pakhal plains of Mansehra. Pak. J. Weed Sci. Res. 10(3-4):157-160.

Knezevic M, Durkic M, Samota D (1995). Chemical and mechanical weed control in potatoes. Fragmenta Phytomedica et Herbologica 23(2):61-67.

Lehoczky É, Dobozi M, Gyüre K (2003). Competition between weeds and potato with special regard to competition for nutrients. Magyar Gyomkutatás és Technológia 4(1):19-30.

Lisińska G, Leszczyński W (1989). Potato Science and technology. Elsevier Applied Science. London-New York.

Maykuhus F (1988). Erfahrungen mit Racer bei der Unkraut bekämpfung in Kartoffeln. Kartoffelbau 39(2):46-48.

Mišovic MM, Brocic ZA, Momirovic NM, Šinzar BC (1997). Herbicide combination efficacy and potato yield in agro-ecological conditions of Dragacevo. Acta Horticulturae 462:363-368.

Pomykalska A (1991). Investigations on determination of the harmfulness threshold of weeds in the potato stand. Roczniki Nauk Rolniczych, Seria A 109:21-35.

Radecki A (1977). The studies possibility limitations cultivation of treatments in potato cultivation. Part III. The studies relationship between yielding of potato weed infestation degree. Roczniki Nauk Rolniczych. Seria A 102:21-32.

Renner KA (1998). Weed control in potato with rimsulfuron and metribuzin. Weed Technol. 12(2):406-409.

Souza EJ, Eberlein CE (1997). Cover crop system. United States Patent 5606823. Issued on March 4.

Tolman JH, McLeod DGR, Harris CR (1986). Yield losses in potatoes, onions and rutabagas in Southwestern Ontario, Canada- the case for pest control. Crop Protect. 5(4):227-237.

Wilson RD, Geronima J, Armbruster JA (1997). 2,4-D dissipation in field soils after application of 2,4-D dimethylamine salt and 2,4-D 2-ethylhexyl ester. Environ. Toxicol. Chem. 16(6):1239-1246.

Zarzecka K (2004). Evaluation of different methods of potato weed control. Part II. Relationship between weed infestation and yielding. Acta Sci. Polonorum, Agric. 3(2):195-202.

23

South African seaweed aquaculture: A sustainable development example for other African coastal countries

Albert O. Amosu[1], Deborah V. Robertson-Andersson[1], Gavin W. Maneveldt[1], Robert J. Anderson[2] and John J. Bolton[3]

[1]Department of Biodiversity and Conservation Biology, Faculty of Natural Sciences, University of the Western Cape, Private Bag X17, Bellville 7535, South Africa.
[2]Fisheries Research, Department of Agriculture, Forestry and Fisheries, Private Bag X2, Roggebaai 8012, South Africa.
[3]Department of Biological Sciences, Faculty of Science, University of Cape Town, Cape Town. Private Bag X3, Rondebosch 7701, South Africa.

The green seaweed *Ulva* is one of South Africa's most important aquaculture products, constituting an important feed source particularly for abalone (*Haliotis midae* L.), and utilized as a bioremediation tool and other benefits such as biomass for biofuel production and for integrated aquaculture. Besides *Ulva* spp, *Gracilaria* spp. are also cultivated. Wild seaweed harvest in South Africa totals 7,602 mt, compared to 2,015 mt of cultivated *Ulva*. To mitigate for the reliance on wild harvesting, the South African seaweed aquaculture industry has grown rapidly over the past few decades. On-land integrated culture units, with paddle-wheel raceways, are now widely viewed as the preferred method of production for the industry. The success of seaweed aquaculture in South Africa is due to a number of natural and human (industrial) factors. The development of the seaweed aquaculture industry has paralleled the growth of the abalone industry, and has been successful largely because of bilateral technology transfer and innovation between commercial abalone farms and research institutions. In South Africa seaweeds have been used commercially as feedstock for phycocolloid production, for the production of abalone feed, and the production of Kelpak® and Afrikelp®, which are plant-growth stimulants used in the agricultural sector. Additionally, *Ulva* is being investigated for large-scale biogas production. The South African seaweed industry provides a template that could be used by other coastal African nations to further their undeveloped aquaculture potential.

Key words: Aquaculture, resources, seaweed, *Ulva*, South Africa.

INTRODUCTION

Fisheries and aquaculture provide significant food and income for the world's coastal countries, constituting the livelihoods of over 3 billion people (FAO, 2009; Smith et al., 2010; Amosu et al., 2012). Fisheries rely upon renewable harvest from the aquatic environment, while aquaculture is the cultivation of desirable aquatic organisms in open, closed, or semi enclosed bodies of water (Lorentzen et al.,2001). Aquaculture is currently the fastest growing primary industry (FAO, 2009) and is likely to overtake wild capture fisheries, providing a source of animal protein and food fish (FAO, 2010a, 2012). Active coastal aquaculture development has been practiced in Asia, Europe and South America for several decades, whereas only minimal production has thus far been achieved in Africa (FAO, 2012). Currently the African continent accounts for less than 1% of the annual total global aquaculture production (FAO, 2010a, 2012) and the vast majority of Africa's aquaculture is in freshwater.

Seaweed is currently the most significant aquatic plant that has contributed to the development of fisheries and the aquaculture industry (FAO, 2010b). Since 1970, the production of aquatic plants (seaweed and angiosperm) worldwide has consistently increased at an annual rate of 7.7%; 93.8% of the total world seaweed production is now from aquaculture (McHugh, 2001; FAO, 2003, 2009, 2010b, 2011), a higher figure than for any other group of marine organism. Globally, the production of seaweed increased from 11.66, 16.83 and 19.9 million mt in 2002, 2008 and 2010 respectively, while seaweed biomass accounted for 23% of the world aquaculture output in 2007 (FAO, 2012; Paul and Tseng, 2012). In recent years the total global annual seaweed harvest, produced by over 30 countries, ranges from 3.1 to 3.8 million mt (FAO, 2010b). Aquaculture production of aquatic plants in 2008 was estimated at US$7.4 billion (99.6% quantity and 99.3% value) (FAO, 2009). Even though Africa is the second largest continent and has a shoreline of about 30,000 km, it has yet to contribute significantly to the development of the seaweed industry despite its rich seaweed diversity (FAO, 2002, 2010b). Abalone farming in South Africa has developed rapidly and the country is now the largest producer outside Asia, partly achieved due to seaweed production. From a seaweed perspective, Eucheuma farming is well established in Zanzibar where commercial interests have assisted the establishment and development of the industry (FAO, 2002), making Tanzania the largest seaweed producing country in Africa, and among the top ten producers and one of only a few countries around the world producing more than 8,000 mt of seaweed per annum (FAO, 2012).

Africa offers numerous aquaculture opportunities, including integrated seaweed aquaculture production, which has been on the increase since 2001 (Table 1). The continent produced 138,989 mt of farmed seaweeds in 2010, with Tanzania (mainly Zanzibar), Madagascar, South Africa, Mozambique, and Namibia as the leading producers (FAO, 2012). This paper examines the current status of seaweed aquaculture in South Africa, the philosophy behind the country's achievements, prospects for the future, and the lessons for other African coastal countries.

THE SOUTH AFRICAN SEAWEED INDUSTRY

South Africa has had a seaweed industry for over 60 years. The commercial exploitation of seaweeds in South Africa is based largely on beach-cast collecting and cutting of kelp. Harvesting of Ecklonia maxima (Osbeck) Papenfuss and Laminaria pallida Greville ex J. Agardh started in the 1940's as a result of the scarcity of kelp during the Second World War (McHugh, 1987). When supplies of agar from Japan became unavailable, various potential resources were identified. However, commercial exploitation only began in the early 1950s (McHugh, 1987), followed by hand-picking of Gelidium sp. in the Eastern Cape since 1957. Most of this harvest was shipped to Europe, North America, and Asia for alginate extraction (Anderson et al., 1989). South African kelps yield alginate concentrations of between 22 and 40% (Anderson et al., 1989). Some trade figures showed that powdered kelp was also exported to Japan for use in formulated fish-feed (Zhang et al., 2004). Since 1975, wet kelp has been harvested from Concession Area 9 along the west coast solely for the production of Kelpak®, which is a plant-growth stimulant[1] and soil conditioner (Khan et al., 2009). Similar harvesting of wet kelp in small quantities started in 1979 on the west coast and later on the south coast for the production of Afrikelp®, which is also a plant-growth stimulant. This harvesting continues today (Anderson et al., 1989, 2003; Robertson-Anderson et al., 2006; Troell et al., 2006).

The bulk of the wet harvested kelp forms the major fresh feed ingredient for cultured abalone in South Africa. The use of kelp for abalone feed has fluctuated since 2005 (Table 2), but in 2010a total of 5,542 mt of fresh kelp fronds were supplied to farmers (DAFF, 2011a). Despite the large quantities of kelp supplied for use as abalone feed, some farmers also use formulated feeds, such as Abfeed™ and Midae Meal™, and some do not use kelp at all.

In 2010 commercial quantities of Gelidium were collected only from Concession Area 1; G. pristoides (Turner) Kuetzing comprises more than 90% of the harvest. Abundant endemic species such as Gelidium pristiodes, Gelidium pteridifolium Norris, Hommersand & Fredericq and Gelidium abbottiorum R. E. Norris have been harvested from Eastern Cape intertidal areas since the mid 1950s and have been identified for possible exploitation from other Concession Areas[2] (viz. 1, 20, 21, 22 and 23). Yields vary with demand from a few to about 120 mt dry weight annually. The sheltered waters of Gracilaria gracilis (Stackhouse) Greville wash-ups from Saldanha Bay on the west coast were exported for extraction of agar. Although some Gracilaria cultivation was attempted in the 1990's in Saldanha Bay and St Helena Bay, these commercial ventures failed (Anderson

---

[1] Kelp contains active ingredients (cytokinins and auxins) that have been shown to improve the growth performance and efficacy of many food and agricultural crops.

[2] The coastline between the Orange and Mtamvuna Rivers is divided into 23 seaweed rights areas. In each area, the rights to each group of seaweeds (e.g. kelp, Gelidium, or Gracilarioids) can be held by only one company, to prevent competitive overexploitation of these resources. Different companies may hold the rights to different resources in the same area (Figure 2).

Table 1. World aquaculture production of aquatic plants by producers in 2010.

Country		2001	2002	2003	2004	2005	2006	2007	2008	2009	2010
World total	Q	9,715,105	10,597,528	11,401,733	12,664,638	13,518,933	14,091,266	14,993,619	15,877,626	17,355,766	19,007,053
	V	3,042,257	3,127,449	3,352,420	3,850,013	3,916,781	3,985,305	4,262,670	4,377,832	4,961,197	5,651,167
China	Q	7,167,171	7,720,529	8,580,314	9,374,297	9,494,591	9,502,403	9,752,745	9,933,885	10,495,905	11,092,270
	V	1,361,096	1,487,034	1,809,558	2,099,267	2,038,868	2,027,812	2,072,273	2,311,139	2,357,839	2,533,196
Indonesia	Q	212,473	223,080	231,900	410,570	910,636	1,170,000	1,728,475	2,145,061	2,963,556	3,915,017
	V	21,247	36,636	17,059	32,846	127,489	210,600	392,980	300,309	811,822	1,268,367
Philippines	Q	785,795	894,857	988,889	1,204,808	1,338,597	1,468,905	1,505,070	1,666,556	1,739,995	1,801,272
	V	58,017	62,664	67,674	93,863	109,801	173,963	136,850	291,039	201,154	256,715
Korea Rep	Q	373,538	497,557	452,054	547,108	621,154	765,595	792,953	921,024	858,659	901,672
	V	127,981	156,797	169,819	231,917	262,523	269,657	332,524	311,305	252,112	327,823
Korea D P Rp	Q	444,295	444,295	444,295	444,295	444,295	444,300	444,300	444,300	444,300	444,300
	V	244,362	244,362	244,362	244,362	244,362	244,365	244,365	66,645	66,645	66,645
Japan	Q	511,448	557,951	477,705	484,389	507,742	490,062	513,964	456,337	456,426	432,796
	V	1,169,140	1,073,392	987,696	1,105,796	1,097,189	969,620	1,005,664	1,020,354	1,121,388	1,138,184
Malaysia	Q	18,863	18,871	25,000	30,948	40,000	60,000	80,000	111,298	138,857	207,892
	V	1,986	1,986	2,632	4,072	1,584	2,454	3,492	6,686	7,884	17,444
Zanzibar	Q	81,860	111,830	94,640	71,860	73,620	76,760	84,850	107,925	102,682	125,157
	V	963	1,086	697	542	638	740	579	1,265	1,327	1,781
Viet Nam	Q	20,000	25,000	30,000	30,000	30,000	36,000	38,000	35,700	33,600	35,000
	V	10,000	12,500	15,000	15,000	15,000	18,000	19,000	17,850	16,800	17,500
Chile	Q	65,538	71,648	40,079	20,273	15,493	38,219	26,387	27,703	88,193	12,179
	V	29,492	46,571	32,098	14,200	11,621	61,660	43,307	46,731	114,678	15,841
Solomon Is	Q	...	6	400	2,140	3,260	1,690	1,080	1,440	5,100	8,000
	V	...	...	11	57	87	33	21	58	253	397
Tanzania	Q	1,000	2,000	2,000	3,000	3,000	3,200	4,000	5,000	5,520	6,885

Table 1. Contd.

		9	18	14	21	24	31	27	65	168	196
	V										
China, Taiwan	Q	15,628	16,799	12,250	9,164	2,438	5,949	9,390	6,879	4,383	4,888
	V	15,177	2,235	2,200	1,461	502	447	8,311	1,206	5,161	3,158
Kiribati	Q	9,264	4,248	3,904	3,904	5,000	8,837	1,112	1,083	1,788	4,745
	V	287	139	153	156	200	353	44	91	139	428
South Africa	Q	12	1,050	2,824	2,845	3,000	3,000	3,000	1,834	1,900	2,015
	V	49	525	1,065	1,252	1,340	1,265	1,208	756	807	744
Timor-Leste	Q	…	…	…	…	…	…	370	1,000	1,500	1,500
	V	…	…	…	…	…	…	28	75	113	113
Brazil	Q	…	…	…	…	…	…	…	320	520	730
	V	…	…	…	…	…	…	…	26	39	62
Mozambique	Q	…	1,570	5,230	920	560	150	690	700	700	700
	V	…	31	120	18	11	3	14	14	14	14
Russian Fed	Q	504	143	67	216	245	818	300	260	739	614
	V	1,260	286	101	259	294	982	360	312	887	737
Fiji	Q	3,200	800	250	450	450	1,190	650	660	440	560
	V	176	44	14	25	25	65	36	46	31	56
Myanmar	Q	…	…	…	…	…	…	…	36	150	262
	V	…	…	…	…	…	…	…	11	45	79
Namibia	Q	20	38	67	67	67	70	27	132	130	130
	V	10	35	62	62	62	65	25	97	93	107
France	Q	35	38	37	37	45	32	35	53	125	120
	V	2	2	2	16	16	17	16	30	66	61
Burkina Faso	Q	…	…	…	…	10	20	20	70	70	70
	V	…	…	…	…	5	10	10	37	37	37

Table 1. Contd.

Cent Africa Rep	Q	⋯	⋯	⋯	⋯	⋯	⋯	⋯	⋯	⋯	⋯	⋯	⋯	30
	V	⋯	⋯	⋯	⋯	⋯	⋯	⋯	⋯	⋯	⋯	⋯	⋯	15
Ireland	Q	⋯	⋯	⋯	⋯	3	⋯	⋯	3	⋯	⋯	⋯	⋯	3
	V	⋯	⋯	⋯	⋯	2	⋯	⋯	2	⋯	⋯	⋯	⋯	2
Senegal	Q	⋯	⋯	⋯	1	⋯	1	⋯	⋯	⋯	⋯	⋯	⋯	2
	V	⋯	⋯	⋯	⋯	⋯	⋯	⋯	⋯	⋯	⋯	⋯	⋯	6
St Lucia	Q	3	2	2	1	1	1	1	1	1	1	1	1	2
	V	12	8	8	13	16	10	5	7	9	9	7	9	13
Spain	Q	⋯	⋯	25	0	1	0	25	1	14	5	5	943	1
	V	⋯	⋯	951	0	678	⋯	951	⋯	1,009	943	943	1,009	746
Other countries	Q	3,590	3,860	7,850	16,870	18,080	⋯	⋯	6,810	⋯	⋯	⋯	⋯	1
	V	895	996	1,955	4,216	4,516	⋯	⋯	1,703	⋯	⋯	⋯	⋯	6

V = Value in USD 1 000; Q = Quantity.

Table 2. Annual yields of commercial seaweeds in South Africa, 2000-2010.

Year	Gelidium (kg dry wt)	Gracilaria (kg dry wt)	Kelp Beach cast (kg dry wt)	Kelp fronds harvest (kg fresh wt)	Kelp fresh beach cast (kg fresh wt)	Kelpak (kg fresh wt)
2001	144 997	247 900	845 233	5 924 489	0	641 375
2002	137 766	65 461	745 773	5 334 474	0	701 270
2003	113 869	92 215	1 102 384	4 050 654	1 866 344	957 063
2004	119 143	157 161	1 874 654	3 119 579	1 235 153	1 168 703
2005	84 885	19 382	590 691	3 508 269	126 894	1 089 565
2006	104 456	50 370	440 632	3 602 410	242 798	918 365
2007	95 606	600	580 806	4 795 381	510 326	1 224 310
2008	120 247	0	550 496	5 060 148	369 131	809 862
2009	115 502	0	606 709	4 762 626	346 685	1 232 760
2010	103 903	0	696 811	5 336 503	205 707	1 264 739
Total	1 140 374	633 089	8 034 189	45 494 533	4 903 038	10 008 012

Kelp beach cast' (column 4) refers to material that is collected in a semi-dry state, whereas 'kelp fresh beach cast' (column 6) refers to clean wet kelp fronds that, together with 'kelp fronds harvest' are supplied as abalone feed. 'Kelp fresh beach cast' was only recorded separately since 2003. Source: DAFF 2011a.

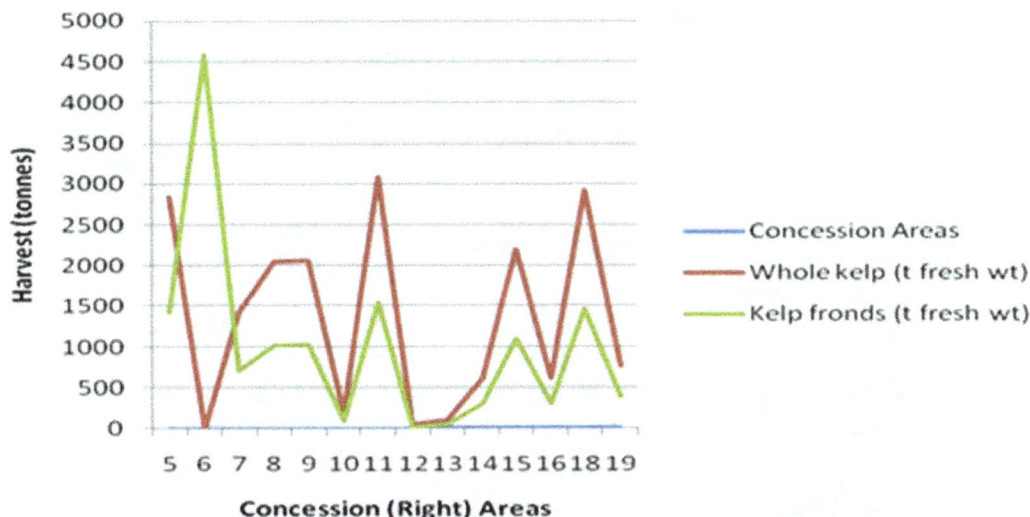

Figure 1. Maximum sustainable yield (MSY) of harvested kelp for all areas for 2010.

et al., 1989, 2003). Only beach-cast *Gracilaria* material may be collected commercially, because harvesting of the living beds is not sustainable. In Saldanha Bay, large yields (up to 2,000 mt dry weights in 1967) were obtained until the construction of the iron ore jetty and breakwater in 1974, after which yields fell dramatically. Occasional small wash-ups are obtained in St Helena Bay. Because total annual yields of Gracilarioids range from zero to a few hundred tonnes dry weight, this resource is regarded as unreliable. Accordingly, no gracilarioids have been collected commercially since 2007. From the start of commercial seaweed exploitation in South Africa in the 1950's, only six seaweed genera (*Ecklonia*, *Laminaria*, *Gracilaria*, *Gelidium*, *Gigartina* and *Porphyra*) have been harvested, with most of this material being exported for use in the phycocolloid industry. *Ulva* has also been harvested in small amounts, but mostly for seaweed salt.

Seaweed aquaculture in South Africa started as an offshoot of the abalone (*Haliotis midae* L.) farming industry in the 1990's and has increased accordingly. Within South Africa twelve seaweed species are currently being exploited: *Ulva* sp., *Porphyra* sp., *E. maxima*, *Laminaria pallida*, *Gracilaria gracilis*, *Gracilariopsis longissima* (S. G. Gmelin) M. Steentoft, L. M. Irvine & W. F. Farnham, *G. abbottiorum*, *G. pteridifolium*, *G. pristoides*, *G. capense* (S.G. Gmelin) P. C. Silva, and *Plocamium corallorhiza* (Turner) Harvey (ESS, 2005; Troell et al., 2006; Robertson-Andersson, 2007).

South Africa's seaweed resources are well protected under the Marine Living Resources Act of 1998 and are conserved from a concessional perspective[3] (Anderson et al., 1989, 2003; GPR, 2005; Anderson et al., 2006). In certain Concession Areas, limitations are placed on the quantity that can be harvested. These sustainable limits

are termed Maximum Sustainable Yields (MSY) and equate to 10% of the estimated kelp accessible (non-reserve) biomass, a value that was estimated to equal the annual mortality rate for the kelp *E. maxima* (Simons and Jarman, 1981). A large amount of this harvested seaweed is exported for the extraction of gums and 42.7% of the total harvest of fresh kelp fronds was supplied to abalone farmers as feed, this harvested kelp fetching a market value of R6 million (~ US$750 000) in 2010 (DAFF, 2012). Within the 23 Concession Areas, currently 14 areas are for kelp rights (Figure 1); no commercial activity was reported in five of these areas (DAFF, 2012).

IMPORTANCE, USES AND BENEFITS

Seaweed is produced in 25 countries globally, comprising 145 species used in food production and 101 species used in phycocolloid production (Dhargalkar and Verlecar, 2009). There are 20 commercial seaweed species being cultivated in each of the main genera, which includes *Caulerpa*, *Chondrus*, *Eucheuma* and *Kappaphycus*, *Gracilaria* and *Gracilariopsis*), *Palmaria*, *Pyropia* (formerly *Porphyra*), *Monostroma*, *Saccharina* (formerly *Laminaria*), *Ulva* and *Undaria* (Zemke-White et al., 1999, Fleurence, 2004; Bruton et al., 2009; Klaus et al., 2009; Mohammad and Chakrabarti, 2009; Pia et al., 2009; FAO, 2010b; Paul and Tseng, 2012).

Agar and carrageenan are commercially valuable substances. The best quality agar, and its associated derivative agarose, comes from red algae belonging to the family Gelidiaceae, while lower-quality agars are mainly found in other families, mainly the Gracilariaceae. Globally agarose is used extensively in gels for electrophoresis in molecular biology. Carrageenans are generally employed for their viscous properties in

---

[3]The seaweed resources is managed in terms of both a Total Applied Effort (TAE) and a Total Allowable Catch (TAC)

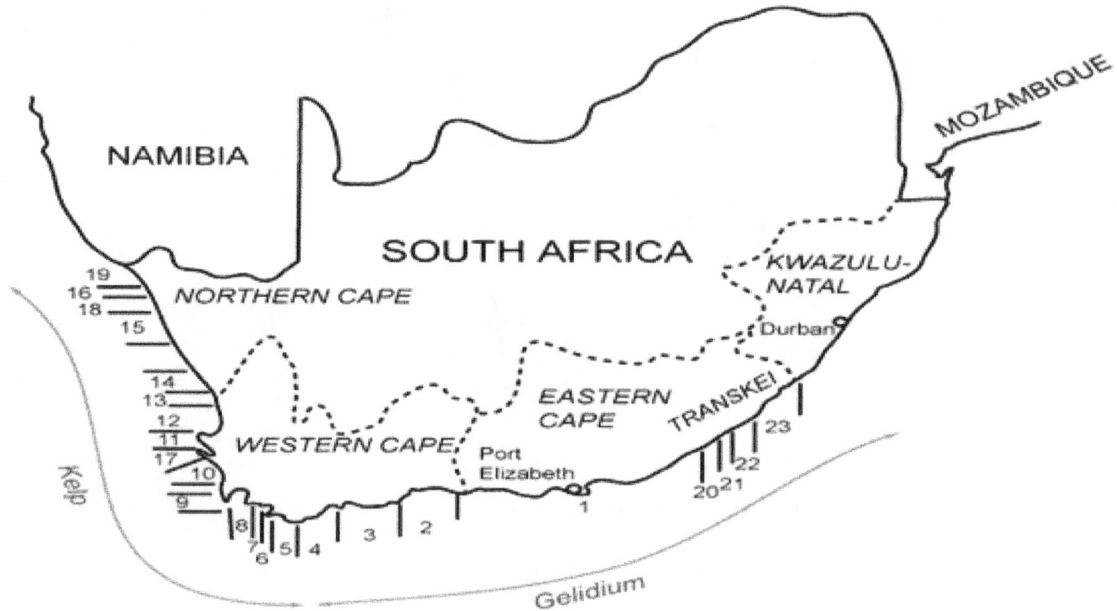

Figure 2. Map of South Africa showing area of potential commercial seaweed concession right areas (Anderson et al., 2003).

gelation, as stabilization of emulsions, in suspensions and foams, and in the control of crystal formation in dairy products and frozen foods. Seaweeds have been called the medical food of the 21st century (Khan and Satam, 2003). According to the World Health Organization (WHO) seaweed are among the healthiest foods on the planet as they contain vitamins, over 90 minerals and many antioxidants (FAO, 2003). Historic evidence shows that seaweeds have been eaten by coastal communities of many countries since ancient times (FAO, 2002) and according to research, such communities who have historically consumed large quantities of seaweed, on average lived longer and had a lower incidence of hypertension and arteriosclerosis (Tietze, 2004).

Seaweeds are also used in the manufacture of pharmaceuticals and cosmetic creams (Bhakuni and Rawat, 2005; Leonel, 2011; Lewis et al., 2011). For example, *Digenea* spp (Rhodophyta) produce an effective vermifuge (kainic acid) (Smit, 2004). *Laminaria* and *Sargassum* species have been used for the treatment of cancer (Khan and Satam, 2003). Anti-viral compounds discovered in *Undaria* spp have been used to inhibit the Herpes simplex virus (Barsanti and Gualtieri, 2006). Research is now being carried out into using seaweed extracts to treat breast cancer and HIV (Schaeffer and Krylov, 2000; Synytsya et al., 2010). Several calcareous species of *Corallina* have been used in bone-replacement therapy (Stein and Borden, 1984). *Asparagopsis taxiformis* and *Sarconema* spp. are used to control and cure goiter while heparin, a seaweed extract, is used in cardiovascular surgery (Khan and Satam, 2003).

GLOBAL SEAWEED TRADE

Globally the seaweed industry is estimated to have an annual value of some US$6 billion (McHugh, 2003) and the largest share of this is for food products. Currently there are 42 countries across the world with reports of commercial seaweed activity (Khan and Satam, 2003; Bixler and Porse, 2011). The primary wild-harvested genera include *Chondrus*, *Furcellaria*, *Gigartina*, *Sarcothalia*, *Mazzaella*, *Iridaea*, *Mastocarpus*, and *Tichocarpus* (Bixler and Porse, 2011). As already mentioned, seaweeds are an important food source, especially in Japan (FAO, 2003). Popular seaweed food stuffs include Wakame, Quandai-cai (*Undaria pinnatifida*), Nori (*Porphyra* spp), Kombu or Haidai (*Laminaria japonica*), Hiziki (*Hizikia fusiforme*), Mozuku (*Cladosiphon okamuranus*), Sea grapes or Green caviar (*Caulerpa lentillifera*), Dulse (*Palmaria palmata*), Irish moss or Carrageenan moss (*Chondrus crispus*), Winged kelp (*Alaria esculenta*), Ogo, Ogonori or Sea moss (*Gracilaria* spp), Carola (*Callophyllis variegata*), Leafy sea lettuce (*Ulva* spp), Arame (*Eisenia bicyclis*), and Kanten (agar-agar). Seaweed products for human consumption contribute about US$ 5 billion of which nori is worth US$ 2 billion per annum (FAO, 2003).

The production of seaweeds and other aquatic algae reached 19.9 million mts in 2010, of which aquaculture produced 19 million mt. Japanese kelp was the most cultivated seaweed species (5.1 million mt) in 2010 and most of it was grown in China (FAO, 2012). Major seaweed aquaculture production come from China, Indonesia, Philippines, North Korea, South Korea, Japan,

Malaysia, Chile, India, and Tanzania (Barsanti and Gualtieri, 2006; Bixler and Porse, 2011; FAO, 2010b, 2012). The most cultivated seaweed is the kelp *Saccharina japonica*, which accounts for over 60% of the total cultured seaweed; species from the genera *Porphyra*, *Kappaphycus*, *Undaria*, *Eucheuma*, and *Gracilaria* make up the majority of the remaining total (Barsanti and Gualtieri, 2006). The most valuable among the seaweed is the red alga (*Porphyra yezoensis*), an important constituent of sushi (FAO, 2010, 2012; Bixler and Porse, 2011).

High demand for carrageenan has similarly triggered the development of *Kappaphycus alvarezii* and *Eucheuma denticulatum* farming in several countries, the largest producers being the Philippines, Indonesia, Malaysia, Tanzania, Kiribati, Fiji, Kenya, and Madagascar (Bixler and Porse, 2011). World carrageenan production exceeded 50,000 mt in 2009, with a value of over US$527 million (Bixler and Porse, 2011). About 32,000 to 39,000 mt of alginic acid per annum is extracted worldwide from approximately 50,000 mt (wet weight) annual production of kelp (Barsanti and Gualtieri, 2006). Agar is relatively cheap, usually around US$18 per kg. In 2009, about 86,100 mt of hydrocolloids were traded comprising 58% of carrageen, approximately 31% alginates, and approximately 11% agar (10,000 mt with a value of $175 million); the major genera included *Ahnfeltiopsis*, *Gelidium*, *Gelidiella*, *Gracilaria*, *Pterocladiella* and *Pterocladia* (Bixler and Porse, 2011).

AFRICA REGIONAL SEAWEED AQUACULTURE DEVELOPMENT

The African continent comprises 29 coastal countries and five island nations, few of which are practicing some form of seaweed aquaculture (Machena and Moehl, 2001). However, the biogeographical features and shore characteristics in several of these countries suggest a high potential for seaweed resources exploitation, culture and utilization.

West Africa

Excluding Ghana (200 species), Senegal (241 species), and Sierra Leone (112 species), which have high seaweed diversities associated with upwelling events and rocky shores (Bolton et al., 2003), West Africa generally has a low seaweed diversity (John and Lawson, 1991). Nigeria (49 species), Benin (16 species), Togo (37 species) and Guinea Bissau (12 species) have coastlines characterized by sandy beaches and extensive mangroves, deltas, estuaries, and lagoons with correspondingly low algal diversity (John and Lawson, 1997). Recent research (Fakoya et al., 2011; Abowei and Tawari, 2011) has shown the potential of seaweed resources for exploitation, culture and utilization for Nigeria but as yet, no targeted commercial harvesting

and cultivation has commenced.

North Africa

North Africa (Morocco – 197 species, Libya – 178 species, Tunisia – 87 species, Western Sahara – 81 species, Sudan – 18 species) has a variable seaweed species richness. The Moroccan coast, however, has been most studied, due to its proximity to European countries (Gallardo et al., 1993) and this may explain the high species numbers. None the less, Morocco has a well-established seaweed industry based on the extraction of agar from wild *Gelidium* species. Steps are also being taken to identify suitable protected natural sites for seaweed cultivation, presumably with a view to cultivating *Gracilaria* to supplement the natural resources of *Gelidium* for agar production (FAO, 2003).

East Africa

The East African coastline is about 9500 km long and comprises the tropical coasts of Somalia (211 species), Kenya (403 species), Tanzania (428 species), Mozambique (243 species) and Madagascar (207 species). Seaweed aquaculture is a recent development in East Africa, occurring in all East African countries except Somalia. Tanzania's aquaculture production has increased steadily to become the largest producer of aquaculture products in Africa (FAO, 2012). *Eucheuma denticulatum* (previously *E. spinosum*) and *Kappa-phycus alvarezii* (previously *E. cottonii*) have been farmed in the region since 1989. These two species are found naturally in East Africa, and were previously collected from the wild for export to USA and Europe. Although the species are found locally, the farmed strains are mainly imported from the Philippines. Madagascar currently accounts for a very small proportion (about 4,000 mt of seaweed per year) of global seaweed production, despite the fact that much of its 5,000 km coastline provides perfect conditions for seaweed cultivation (FAO, 2012). With assistance from commercial sources, seaweed cultivation is proving to be promising in Mozambique. This will make Mozambique only the fourth seaweed producing nation in Africa (FAO, 2012). To support this industry and to promote aquaculture, the Mozambique government recently (2011) approved a decree establishing the marine aquaculture reserve. Approximately 10,600 ha have been set aside for seaweed aquaculture, potentially yielding 641,000 mt of seaweed (Nkutumula, 2011). The seaweeds of Kenya are well-studied relative to other East African/Indian Ocean countries (Bolton et al., 2003). However, Kenya does not present good prospects for a seaweed industry. None of the pilot studies carried out have produced any promising results that would encourage investors to venture into seaweed farming for Kenya.

Southern Africa

Namibia's proximity to South Africa greatly influenced the documentation of the former country's seaweed resources. Both countries have developed through technology sharing, but the seaweed aquaculture industry in Namibia is still not as developed as in South Africa. The 196 seaweed species of Namibia have been studied and documented (Engeldow, 1998; Rull Lluch, 1999, 2002; Engeldow and Bolton, 2003). As in South Africa, Namibian seaweed harvesting companies operate under a system of Concessions Areas. The industry provides employment opportunities for over 250 people in an area where job opportunities are severely lacking. Investment in polyculture of seaweeds and crustaceans has also been promoted in Namibia (Hasan and Chakrabarti, 2009). Of the 196 Namibian species of seaweeds, nine have shown potential use as animal feed supplements. Beach-cast *Gracilaria* is also collected and cultivation is being developed by a local company; the current market, however, is depressed.

The seaweeds of South Africa have been extensively detailed (Stegenga et al., 1997; De Clerck et al., 2005; Maneveldt et al., 2008). The known seaweed diversity of South Africa has increased from 547 species in 1984 to around 900 species in 2012, making the region one of the richest marine floras in the world, with a high level of endemism (Payne et al., 1989; Bolton, 1999; Bolton et al., 2003; Maneveldt et al., 2008, pers. obs.). South African seaweed aquaculture is focused on the abalone industry, particularly the abalone, *Haliotis midae* (Bolton et al., 2006; Troell et al., 2006). By far the most cultivated seaweed species is *Ulva* spp. The aquaculture of *Ulva* spp occurs on many abalone farms (DAFF, 2010) and here paddle-wheel raceways have proven to be the most suitable device for growing *Ulva* spp in large quantities (Chopin et al., 2008).

The South African abalone aquaculture industry has grown rapidly over the past few decades along the west coast of South Africa where suitable rocky habitat exists (Troell et al., 2006). On-land integrated culture units, which use shallow raceways, are the preferred method of production for the abalone industry (Bolton et al., 2006). There is growing evidence that suggests a mixed diet of kelp and other seaweeds can induce growth rates that meet or exceed those attained with artificial feed (Naidoo et al., 2006, Dlaza et al., 2008; Francis et al., 2008; Robertson-Andersson et al., 2011). Moreover, a natural diet can improve abalone quality and reduce parasite loads (Robertson-Andersson, 2003; Naidoo et al., 2006; Al-Hafedh et al., 2012).

PROBLEMS AND PROSPECTS OF SEAWEED AQUACULTURE IN AFRICA

Failures of some ill-conceived pilot projects (e.g. South-west Madagascar – De San, 2012) continue to remain a major constraint in convincing farmers and investors of the economic viability of seaweed aquaculture in most African coastal countries.

Several other constraints have prolonged the development of the industry in many African countries, and these can be summarized as: weak economies; *poor aquaculture development policies*; inappropriate technologies; weak extension services; weak impact of research institutions; inadequate information management systems; limited coordination between research and production sectors; scanty reliable production statistics and the high value/cost of coastal land; and the associated competition for this land from other coastal industries. In the countries (South Africa, Tanzania, Madagascar, Mozambique, Namibia, Burkina Faso, Central Africa Republic and Senegal) where thriving seaweed cultivation practices have been achieved, these industries provide a meaningful form of income for communities that might otherwise not be employable in the traditional sense.

LESSONS FOR OTHER COASTAL NATIONS

The general benefit of integrated multitrophic aquaculture[4] (IMTA) is the reduction of nutrient release to the environment (Neori et al., 2004, Bolton et al., 2009). This phenomenon is also true for integrated seaweed-abalone culture in South Africa. The technical and economic feasibility of IMTA using seaweeds as biofilters is already well established in South Africa (Nobre, 2010). Seaweeds grown in abalone effluent have an increased nitrogen content (sometime as much as 40% protein dry weight content), resulting in value-added seaweeds of excellent quality to feed abalone (Naidoo et al., 2006; Robertson-Andersson, 2007; Robertson-Andersson et al., 2011). Not only in South Africa but elsewhere, the increasing demand for abalone feed has seen the need for sustainable production of seaweed in IMTA aquaculture with aquatic animals (Brzeski and Newkirk, 1997, Troell et al., 1999, Buschmann et al., 2001), especially with abalone (Neori et al., 1991, 1996, 1998, 2004). To improve seaweed biomass estimations and to document the relative seaweed distributions, GIS mapping and diver-based sampling of the resource is regularly undertaken in South Africa as a government requirement. Monthly harvests of fresh kelp are routinely checked against the prescribed MSY as set in the annual permit conditions of all rights holders. Visual inspections by South African government officials, and reports received from right-holders, show that the kelp resource is stable and healthy (DAFF, 2011a).

Although, the South African seaweed sector is small in comparison to similar fisheries, it is currently worth

---

[4] Integrated Multitrophic Aquaculture (IMTA) is defined as an ecosystem based management approach that effectively mitigates the overabundance of nutrients introduced by fish farming.

US$3.7 million, generates approximately US$2 million[5] per year, but nevertheless employs up to 400 people, the majority of whom are women who earn an average annual salary of US$ 5 000 (Payne et al., 1989; DAFF, 2011a, b, 2012). More importantly, high proportions (92%) of the employees in the sector are classified as historically disadvantaged persons[6] (DAFF, 2011a). The South African aquaculture sector thus has an important local impact within previously disadvantaged coastal communities, where any increase in employment is valuable largely because such communities are generally characterized by high rates of unemployment (85.7%) and low skill levels (50%) (Nobre et al., 2010).

South Africa is currently spearheading a number of other research innovations. Research has shown that abalone farms incorporating an IMTA seaweed-abalone system can significantly reduce their green-house gas (GHG) emissions (Nobre et al., 2010; Troell et al., 2011). Due to their high carbohydrate contents, seaweeds can be fermented to CH_4 (biogas) and have subsequently been considered a potential CO_2-neutral and renewable energy supply (Bartsch et al., 2008; Roesijadi et al., 2008; Bruhn et al., 2011, Chung et al., 2011). Furthermore, recent research findings have shown that *Laminaria* and *Ulva* species are important prospects from an energy point of view (Bruton et al., 2009; Klaus et al., 2009; Abowei and Tawari, 2011; Bruhn et al., 2011, Chanakya et al., 2012). This important benchmarking knowledge could propel the commencement of research on seaweed as a substitute for liquefied petroleum gas (LPG). Aside from being a renewable resource and reducing CO_2 emissions (especially if seaweed cultivation is incorporated with a source of CO_2 production), seaweed cultivation could potentially have a major positive impact on global warming and ocean acidification. As a consequence of these findings, South Africa is currently investigating large-scale anaerobic digestion of methane gas from seaweed as well as the local species of seaweeds' potential for mitigation of ocean acidification.

CONCLUSION

Despite the fact that South Africa is currently not Africa's highest seaweed aquaculture producer, the country has the highest regional seaweed diversity and one of the richest in the world. As a third-world country with many first-world technologies, South Africa provides many important lessons for less developed coastal African nations. The South African seaweed aquaculture industry is well researched and has developed steadily due to the need for sustainable production of seaweed in IMTA.

Presently the South African seaweed aquaculture industry provides raw materials for other sectors of the economy, as well as the potential for bioremediation of both the atmospheric and aquatic environment.

ACKNOWLEDGEMENTS

We thank the Department of Biodiversity and Conservation Biology at the University of the Western Cape, and the Department of Biological Sciences at the University of Cape Town for providing funding and research equipment. The South African National Research Foundation (NRF) and Department of Agriculture, Forestry and Fisheries (DAFF) are thanked for research grants and the required permits to GWM, JJB and RJA.

REFERENCES

Abowei JFN, Tawari CC (2011). A Review of the Biology, Culture, Exploitation and Utilization Potentials Seaweed Resources: Case Study in Nigeria. Res. J. Appl. Sci. Eng. Technol. 3(4):290-303.

Al-Hafedh YS, Alam A, Buschmann AH, Fitzsimmons KM (2012). Experiments on an integrated aquaculture system (seaweeds and marine fish) on the Red Sea coast of Saudi Arabia: efficiency comparison of two local seaweed species for nutrient biofiltration and production. Rev. Aquacult. 4:21-31.

Amosu AO, Bashorun OW, Babalola OO, Olowu RA, Togunde KA (2012). Impact of climate change and anthropogenic activities on renewable coastal resources and biodiversity in Nigeria. J. Ecol. Nat. Environ. 4(8):201-211.

Anderson RJ, Rothman MD, Share A, Drummond H (2006). Harvesting of the kelp *Ecklonia maxima* in South Africa affects its three obligate, red algal epiphytes. J. Appl. Phycol. 18:343-349.

Anderson RJ, Bolton JJ, Molloy FJ, Rotmann KW (2003). Commercial seaweeds in southern Africa. In: Chapman ARO, Anderson RJ, Vreeland VJ, Davidson I (Eds) Proceedings of the 17th International Seaweed Symposium. Oxford University Press, Oxford, pp. 1-12.

Anderson RJ, Simons RH, Jarman NG (1989). Commercial seaweeds in southern Africa: A review of utilization and research. South Afr. J. Marine Sci. 8:277-299.

Barsanti L, Gualtieri P (2006). Algae Anatomy, Biochemistry, and Biotechnology Taylor and Francis Group LLC. Boca Raton London New York CRC Press.

Bartsch I, Wiencke C, Bischof K, Buchholz CM, Buck BH, Eggert A, Feuerpfeil P, Hanelt D, Jacobsen S, Karez R, Karsten U, Molis M, Roleda MY, Schumann R, Schubert H, Valentin K, Weinberger F, Wiese J (2008). The genus *Laminaria* sensu lato: recent insights and developments. Eur. J. Phycol. 43(1):1-86.

Bhakuni DS, Rawat DS (2005). Bioactive marine natural products. Anamaya Publishers, New Delhi, India.

Bixler HJ, Porse H (2011). A decade of change in seaweed hydrocolloids industry. J. Appl. Phycol. 23:321-335.

Bolton JJ (1999). Seaweed systematic and diversity in South Africa: An historical account. Trans. Royal Soc. South Afr. 54(1):167-177.

Bolton JJ, De Clerck O, John DM (2003). Seaweed diversity patterns in Sub-Saharan Africa. Proceedings of the Marine Biodiversity in Sub-Saharan Africa: The Known and the Unknown, 23-26 September, Cape Town, South Africa, pp. 229-241.

Bolton JJ, Robertson-Andersson DV, Shuuluka D, Kandjengo L (2009). Growing *Ulva* (Chlorophyta) in integrated systems as a commercial crop for abalone feed in South Africa: a SWOT analysis. J. Appl. Phycol. 21:575-583.

Bolton JJ, Robertson-Andersson DV, Troell M, Halling C (2006). Integrated systems incorporate seaweeds in South African abalone culture. Global Aquacult. Advocate 9:54-55.

---

[5] 1US$(US dollar) equals R9.55 (SA Rand) as at 22 May 2013.
[6] Historically disadvantaged persons are persons so classified as underdeveloped populations targeted by the SA Government for accelerated development (www.polity.org.za/html/govdocs/rd/rdp2.html. 2013).

Bruhn A, Dah J, Nielsen HB, Nikolaisen L, Rasmussen MB, Markager S, Olesen B, Arias C, Jensen PD (2011). Bioenergy potential of *Ulva lactuca*: Biomass yield, methane production and combustion. Bioresour. Technol. 102:2595-2604.

Bruton T, Lyons H, Yannick Lerat, Y, Rasmussen MB (2009). A Review of the Potential of Marine Algae as a Source of Biofuel in Ireland. Sustainable Energy Ireland, Glasnevin, Dublin 9. Ireland. www.sei.ie

Brzeski V, Newkirk G (1997). Integrated coastal food production systems — a review of current literature. Ocean Coast. Manage. 34:66–71.

Buschmann A, Troell M, Kautsky N (2001). Integrated algal farming: a review. Cah. Biol. Mar. 43:615–655.

Chanakya HN, Mahapatra DM, Ravi S, Chauhan SV, Abitha R (2012). Sustainability of large-sScale aAlgal bBiofuel pProduction in India. India Inst. Sci. 92(1):63-98.

Chopin T, Robinson SMC, Troell M, Neori A, Buschmann AH, Fang J (2008). Multitrophic integration for sustainable marine aquaculture. In: Jørgensen SE, Fath BD (eds) Encyclopedia of ecology, vol 3, Ecological engineering. Elsevier, Oxford. pp 2463–2475.

Chung IK, Beardall J, Mehta S, Sahoo D, Stojkovic S (2011). Using marine macroalgae for carbon sequestration: a critical appraisal. J. Appl. Phycol. 23:877–886.

Department of Agriculture, Forestry and Fisheries DAFF (2010). Marine Aquaculture Annual Farm Operation Report 2010.

DAFF (2011a). Status of the South African Marine Fishery Resources, 2010. Unpublished report, South African Department of Agriculture, Forestry and Fisheries.

DAFF (2011b). Aquaculture annual report 20011 South Africa. South African Department of Agriculture, Forestry and Fisheries.

DAFF (2012). Agriculture, Forestry and Fisheries- Integrated Growth and Development Plan 2012. . South African Department of Agriculture, Forestry and Fisheries.

De Clerck O, Bolton JJ, Anderson RA, Copperjans E (2005). Guide to the seaweeds of KwaZulu-Natal. Scripta Botanica Belgica 33:1–294.

De San M (2012). The farming of seaweeds. Implementation of a regional fisheries strategy for the eastern - southern Africa and India Ocean Region. Report: SF/2012/28.

Dlaza TS, Maneveldt GW, Viljoen C (2008). The growth of post-weaning abalone (*Haliotis midae* Linnaeus) fed commercially available formulated feeds supplemented with fresh wild seaweed. Afr. J. Mar. Sci. 30(1):199-203.

Engeldow HE (1998). The biogeography and biodiversity of the seaweed flora of the Namibian intertidal seaweed flora. PhD thesis, University of Cape Town, South Africa.

Engledow HE, Bolton JJ (2003). Factors affecting seaweed biogeographical and ecological trends along the Namibian coast. Proceedings of the 17th International Seaweed Symposium. Oxford University Press. pp. 285-291

ESS (2003). Mather D, Britz P J, Hecht T, Sauer, WH H (2003). An economic & sectoral study of the South African Fishing Industry. V .1. Economic & regulation principles, survey results, transformation and socio-economic impact. Report prepared for Marine and Coastal Management, Department of Environmental Affairs and Tourism by Rhodes University, Grahamstown, South Africa. P. 300.

Fakoya KA, Owodeinde FG, Akintola SL, Adewolu MA, Abass MA, Ndimele PE (2011). An Exposition on Potential Seaweed Resources for Exploitation, Culture and Utilization in West Africa: A Case Study of Nigeria. J. Fish. Aquat. Sci. 6:37-47.

FAO (2002). Prospects for seaweed production in developing countries, FAO Fisheries Circular No. 968 FIIU/C968 (En). FAO Rome.

FAO (2003). A guide to the seaweed industry, FAO fisheries technical paper 44, Rome.

FAO (2009). The state of world fisheries and aquaculture 2008. FAO fisheries and aquaculture department. Food and Agriculture Organization of United Nations, Rome. 2009.

FAO (2010a). The State of World Fisheries and Aquaculture. FAO Fisheries and Aquaculture Department. Food and Agricultural Organization of the United Nations: Rome.

FAO (2010b). FishStat fishery statistical collections: aquaculture production (1950–2008; released March 2010). Rome, Italy: Food and Agriculture Organization of the United Nations. http://www.fao.org/fishery/statistics/software/fishstat/en.

FAO (2011). FAO Fisheries Department, Fishery Information, Data and Statistics Unit. FishStatPlus. Universal Software for fishery statistical time series. Version 2.3 in 2000. Last database update in April 2011 http://www.fao.org/docrep/013/i1820e/i1820e00.htm, Cited 10 Oct 2011.

FAO (2012). The State of World Fisheries and Aquaculture. ISBN: 978-92-5-107225-7. ftp://ftp.fao.org/FI/brochure/SOFIA/2012/english_flyer.pdf.

Fleurence J (2004). Seaweed proteins, in: R.Y. Yada (Ed.), Proteins in Food Processing, Woodhead Publishing, pp. 197-210.

Francis TL, Maneveldt GW, Venter J (2008). Growth of market-size abalone (*Haliotis midae*) fed kelp (*Ecklonia maxima*) versus a low-protein commercial feed. Afr. J. Aquat. Sci. 33(3):279-282.

Gallardo T, Garreta GA, Ribera MA, Cormaci M, Furnari G, Giacconi G, Boudouresque CF (1993). Checklist of Mediterranean seaweeds. II Chlorophyceae. Bot. Marina 36:399-421.

GPR (2005). General reasons for the decisions on the allocation of rights in the seaweed sector. DEAT publication. Available at http: www.mcm-deat.gov.za.

Hasan MR, Chakrabarti R (2009). Use of algae and aquatic macrophytes as feed in small-scale aquaculture: a review. FAO Fisheries and Aquaculture Technical. Rome, FAO. 2009. 531:123.

John DM, Lawson GW (1991). Littoral ecosystems of tropical western Africa. In: Mathieson, A.C. & Nienhuis, P.H. (eds) Intertidal and Littoral Ecosystems. Ecosys. World 24:297-322.

John DM, Lawson GW (1997). Seaweed biodiversity in West Africa: a criterion for designating marine protected areas S.M. Evans, C.J. Vanderpuye, A.K. Armah (Eds.), The Coastal Zone of West Africa: Problems and Management, Penshaw Press, Sunderland UK (1997), pp. 111-123.

Khan SI, Satam SB (2003). Seaweed Mari culture. Scope and Potential in India. Aquaculture Asia, 8(4):26-29.

Khan W, Rayirath UP, Subramanian S, Jithesh NM, Rayorath P, Hodges DM, Critchley AT, Craigie JS, Norrie J, Prithiviraj B (2009). Seaweed Extracts as Biostimulants of Plant Growth and Development. J. Plant. Growth Regul 28:386-399.

Klaus JH, Uwe F, Rocio H, Anja E, Morchio R, Suzanne H, Boosya B (2009). Aquatic Biomass: Sustainable Bio-energy from Algae? - Issue Paper -,The workshop is co-funded by two German research projects "bio-global" (sponsored by the Federal Ministry for Environment, BMUthrough the Federal Environment Agency, UBA) and "conCISEnet"(sponsored by the Federal Ministry for Education and Research, BMBF) Bio-global / conCISEnet.www.oeko.de.

Leonel P (2011).A Review of the Nutrient Composition of Selected Edible Seaweeds. In: Seaweed (Ed) by Vitor H. Pomin. Nova Science Publishers, Inc.

Lewis J, Salam F, Slack N, Winton M, Hobson L (2011). 'Product options for the processing of marine macro-algae – Summary Report'.The Crown Estate, P. 44.

Lorentzen K, Amarasinghe US, Bartley DM, Bell JD, Bilio M, de Silva SS, Garaway CJ, Hartman WD, Kapetsky JM, Laleye P, Moreau J, Sugunan VV, Swar DB (2001). Strategic review of enhancement and culturebased fisheries. pp. 211-237 In Subasinghe, RP, Bueno, PB., Phillips, MJ. Hough C, McGladdery, SE, Arthur JR. (eds.). Aquaculture in the third millennium. Technical proceedings of the conference on aquaculture in the thirdmillennium, Bangkok, Thailand. 2025 February 2000. NACA, Bangkok and FAO, Rome.

Machena C, Moehl J (2001). Sub-Saharan African aquaculture: regional summary. In R.P. Subasinghe, P. Bueno, M.J. Phillips, C. Hough, S.E. McGladdery & J.R. Arthur, eds. Aquaculture in the Third Millennium. Technical Proceedings of the Conference on Aquaculture in the Third Millennium, Bangkok, Thailand, 20-25 February 2000. pp. 341-355. NACA, Bangkok and FAO, Rome.

Maneveldt GW, Chamberlain YM, Keats DW (2008). A catalogue with keys to the non-geniculate coralline algae (Corallinales, Rhodophyta) of South Africa. South African Journal of Botany 74: 555-566.

McHugh DJ (1987) (ed.), 1987. Production and utilization of products from commercial seaweeds. FAO Fish. Tech. Pap. 288:189.

McHugh, DJ (2001). Prospects for Seaweed Production in Developing Countries. Food and Agriculture Nations, Rome.Organization, Rome, Italy.

McHugh DJ (1987) (ed.). Production and utilization of products from commercial seaweeds. FAO Fish. Tech. Paper 288:189.

Mohammad RH, Chakrabarti R (2009).Use of algae and aquatic macrophytes as feed in small-scale aquaculture: A review. FAO, Fisheries and aquaculture technical P. 531. Rome.

Naidoo K, Maneveldt G, Ruck K, Bolton JJ (2006). A comparison of various seaweed-based diets and artificial feed on growth rate of abalone in land-based aquaculture systems. J. Appl. Phycol. 18:437-443.

Neori A, Chopin T, Troell M, Buschmann AH, Kraemer GP, Halling C, Shpigel M, Yarish C (2004). Integrated aquaculture: rationale, evolution and state of the art emphasizing seaweed biofiltration in modern aquaculture. Aquaculture 231:361–391.

Neori A, Krom M, Ellner S, Boyd C, Popper D, Rabinovitch R, Davison P, Dvir O, Zuber D, Ucko M, Angel D, Gordin H (1996). Seaweed biofilters as regulators of water quality in integrated fish-seaweed culture units. Aquaculture 141:183–199.

Neori A, Ragg NLC, Shpigel M (1998). The integrated culture of seaweed, abalone, fish and clams in modular intensive land-based systems: II. Performance and nitrogen partitioning within an abalone (Haliotis tuberculata) and macroalgae culture system. Aquacult. Eng. 17:215–239.

Neori A, Cohen I, Gordin H (1991). Ulva lactuca biofilters for marine fish pond effluents. II. Growth rate, yield and C: N ratio. Bot. Mar. 34:483–489.

Nkutumula A (2011). Mozambique: Government Promotes Aquaculture, Fisheries Ministry, National Aquaculture Development Institute (INAQUA), http://allafrica.com/stories/201109141399.html.

Nobre AM, Robertson-Andersson DV, Neori A, Sankar K (2010). Ecological–economic assessment of aquaculture options: comparison between abalone monoculture and integrated multi-trophic aquaculture of abalone and seaweeds. Aquaculture 306:116–126.

Paul NA, Tseng CK (2012). Seaweed and Microalgae .Aquaculture: Farming Aquatic Animals and Plant, Second edition, edited by John S Lucas. C South gate. Blackwell Publishing Ltd.

Payne AIL, Crawford RJM, Van Dalsen A (1989). Oceans of Life off Southern Africa. Vlaeberg Publishers, Cape Town.

Pia W, Ghosh D, Tapsell L (2009). Seaweed Culture in Integrated Multi-Trophic Aquaculture: Nutritional Benefits and Systems for Australia, Rural Industries Research and Development Corporation. Project No. PRJ-000162.

Robertson-Andersson DV, Maneveldt GW, Naidoo K (2011). Effects of wild and farm-grown macroalgae on the growth of juvenile South African abalone Haliotis midae (Linnaeus). Afr. J. Aquatic Sci. 36(3):331-337

Robertson-Andersson DV (2007). Biological and economical feasibility studies of using seaweeds Ulva lactuca (chlorophyta) in recirculation systems in abalone farming. PhD Dissertation, University of Cape Town, South Africa.

Robertson-Andersson DV, Leitao D, Bolton JJ, Anderson RJ, Njobeni A, Ruck K (2006). Can kelp extract (KELPAK™) be useful in seaweed mariculture. J. Appl. Phycol. 18:315–321.

Robertson-Andersson DV (2003). The cultivation of Ulva lactuca (Chlorophyta) in an integrated aquaculture system, for the production of abalone feed and the bioremediation of aquaculture effluent. MSc Dissertation, University of Cape Town, South Africa.

Roesijadi G, Copping AE, Huesemann MH, Forster J, Benemann JR (2008). Techno-economic feasibility analysis of offshore seaweed farming for bioenergy and biobased products. Independent research and development report IR # PNWD-3931, Battelle Pacific Northwest Division, P. 115.

Rull Lluch JR (1999). Algues benthõniques marines de Namibia. Ph.D. thesis, University of Barcelona, Spain.

Rull Lluch JR (2002). Marine benthic algae of Namibia. Sci. Marina 66(3):5-256.

Schaeffer DJ, Krylov VS (2000). Anti-HIV activity of extracts and compounds from algae and cyanobacteria. Ecotoxicol. Environmental Safe. 45:208-227.

Simons RH, Jarman NG (1981). Subcommercial harvesting of kelp on a South African shore. Proceedings of the Tenth International Seaweed Symposium. De Gruyter, Berlin. pp. 731–736.

Smit AJ (2004). Medicinal and pharmaceutical uses of seaweed natural products: A review. J. Appl. Phycol. 16:245–262.

Smith MD, Roheim CA, Crowder LB, Halpern BS, Turnipseed M, Anderson JL, Asche F, Bourillón L, Guttormsen AG, Kahn A, Liguori LA, McNevin A, O'Connor M, Squires D, Tyedemers P, Brownstein C, Carden K, Klinger DH, Sagarin R, Selkoe KA (2010) Sustainability and Global Seafood. Science 327:784-786.

Stegenga H, Bolton JJ, Anderson RJ (1997). Seaweeds of the South African west coast. Contributions from the Bolus Herbarium University of Cape Town, 8: 655.

Stein JR, Borden CA (1984). Causative and beneficial algae in human disease conditions: A review. Phycologia 23:485-501.

Synytsya A, Kim W, Kim S, Pohl R, Synytsya A (2010). Structure and antitumour activity of fucoidan isolated from sporophyll of Korean brown seaweed Undaria pinnatifida. Carbohydr. Polym. 81(1):41–48.

Tietze HW (2004). Spirulina - Micro Food Macro Blessing, Harald W. Tietze Publishing, Bermagui NSW 2546 Australia.

Troell MD, Robertson-Andersson DV, Anderson RJ, Bolton JJ, Maneveldt G, Halling C, Probyn T (2006). Abalone farming in South Africa: An overview with perspectives on kelp resources, abalone feed, potential for on-farm seaweed production and socio-economic importance. J. Appl. Phycol. 257(4):266-281.

Troell M, Ronnback P, Halling C, Kautsky N, Buschmann A (1999). Ecological engineering in aquaculture: use of seaweeds for removing nutrients form intense mariculture. J. Appl. Phycol. 11:89-97.

Troell M, Hecht T, Beveridge M, Stead S, Bryceson I, Kautsky N, Mmochi A, Ollevier F (eds.) (2011) Mariculture in the WIO region - Challenges and Prospects. WIOMSA. Book Series No. 11:8-59.

Zemke-White WL, Clements KD, Harris PJ (1999). Acid lysis of macroalgae by marine herbivorous fishes: myth or digestive mechanism. J. Exper. Mar. Biol. Ecol. 233:95–113.

Zhang G, Que H, Liu X, Xu H (2004). Abalone mariculture in China. J. Shellfish Res. 23:947-950.

24

Analysis of pumpkin (*Cucurbita pepo* Linn.) biomass yield and its components as affected by nitrogen, phosphorus and potassium (NPK) fertilizer rates

Oloyede F. M., Agbaje G. O. and Obisesan I. O.

Department of Crop Production and Protection, Obafemi Awolowo University, Ile-Ife, Nigeria.

Pumpkin (*Cucurbita pepo* L.) young leaves and vines are consumed as vegetables among the rural dwellers in Southwest Nigeria. It is a cheap source of protein, vitamins, fibres and antioxidants in their diet. Due to the intensive cultivation of available lands which affects the nutrient status of most soils and their productivity, the need to apply fertilizer on pumpkin became pertinent. Therefore, nitrogen, phosphorus and potassium (NPK) fertilizer influence on biomass accumulation in pumpkin was investigated for two consecutive seasons in 2010 at the Research Farm of Obafemi Awolowo University, Ile-Ife, Nigeria. NPK (15:15:15) compound fertilizer was applied at the rates of (0, 50, 100, 150, 200 and 250 kg ha^{-1}) to *C. pepo* at 2 weeks after planting (WAP). At 8 WAP, the following morpho-physiological traits: vine length, vine diameter, number of internodes, internodes length, vine weight, number of leaves, total fresh and dry biomass assessed were significantly influenced by season and fertilizer effects. The combined analysis of variance (ANOVA) results showed that the effects of season (S) and fertilizer (F) are significant effect on all the traits. Season effect showed that the early season was better (p = 0.05) while the application of fertilizer beyond 100 kg of NPK per hectare was not statistically beneficial for biomass yield and its components. However, the significant S X F interaction on the total dry matter fitted well into a quadratic equation with significant R^2 values. The response curves showed higher response to fertilizer in the early than late planting season and a maximum biomass yield was attained at the application of 205 kg and 244 kg ha^{-1} of NPK (15-15-15) for early and late season cultivations, respectively.

Key words: Biomass, nitrogen, phosphorus and potassium (NPK) fertilizer, pumpkin, dry matter, morpho-physiological trait, quadratic equation.

INTRODUCTION

Cucurbita pepo Linn. commonly known as pumpkin and locally called "Elegede" in Southwest Nigeria belongs to the Cucurbitaceae family. The family is among the most important plant families supplying humans with edible products and useful fibres (Smith, 1997). The palatable leafy vegetable deserts from this crop are relished in Southwestern Nigeria.

C. pepo L. produces a lot of biomass and its nutrient requirements are generally considered to be high

(Graifenberg et al., 1996; Colla and Saccardo, 2003). Nevertheless, excessive fertilization is common among farmers due to their poor knowledge of fertilizer types and nutrient requirement of crops (Martinetti and Paganini, 2006). Generally, tropical soils require additional fertilizer application for crops to yield optimally. The tropical soils are degraded and are of low soil fertility status due to excessive rainfall and intensive cultivation. Hence, most tropical soils are deficient in essential nutrients particularly nitrogen and phosphorus (Obalum et al., 2012). Optimal mineral nutrition is fundamental to the growth and productivity of plants (Liu et al., 2010). The optimum doses of nitrogen, phosphorus and potassium vary greatly with the length of growing season, fertility status of soil, soil type, cultivar, geographical location and the environmental factors. These factors will have marked effect on the growth and yield parameters of pumpkin (Manjunath Prasad et al., 2008). The macronutrient nitrogen (N), which is essential for amino acid, protein and enzyme biosynthesis, is quantitatively the most important element (Sinclair and Vadez, 2002). This study was designed to evaluate the influence of nitrogen, phosphorus and potassium (NPK) fertilizer application on the biomass and its components in pumpkin and determine the fertilizer rate required for maximum biomass yield.

MATERIALS AND METHODS

Location, soil characteristics and meteorological data

Field studies were conducted at the Teaching and Research Farm, Obafemi Awolowo University, Ile-Ife, Nigeria, for two seasons, the early season (May to August) and late season (August to November) of 2010 to determine the effect of NPK fertilizer on biomass yield and the fertilizer rate required for maximum biomass yield. Soil samples were taken randomly across the plots prior to ploughing of land. The soil samples were mixed to form a composite sample for physical and chemical composition analyses in the laboratory. Soil was ploughed twice and harrowed once before sowing. Two seeds per hole were sown and the seedlings were thinned to one plant per stand at 2 weeks after planting (WAP). The NPK fertilizer was added in two equal halves at 2 and 6 WAP. Insecticide (lambda-cyhalothrin) was applied fortnightly from 6 WAP. Post-emergence herbicide, glyphosate was applied at the rate of 200 ml in 15 L of H_2O at 4 and 7 WAP for weed control.

Climatological data regime

Rainfall pattern at Ile-Ife, Osun State, during the early season of the experiment from May to August, 2010 ranged from 7 to 16 mm. Maximum temperature ranged from 28 to 32°C while minimum temperature ranged from 21.0 to 23.0°C. The average rainfall, average maximum temperature and average minimum temperature during the period were 13.2 mm, 29.9°C and 22.5°C, respectively. During the late season of the experiment, August to November 2010, rainfall pattern ranged from 9 to 16 mm. Maximum temperature ranged from 28 to 31°C while minimum temperature ranged from 22.0 to 22.4°C. The average rainfall, average maximum temperature and average minimum temperature during the period were 15.4 mm, 30.1°C and 22.2°C, respectively (Table 2).

Experimental design, treatments and agronomic practices

The experimental treatments were laid in a randomized complete block design and replicated six times. Each plot size was 10 X 12 m and consisted of 7 rows. Alley was 3 m, while the plants were spaced 2 X 2 m. The treatments consisted of a local cultivar of C. pepo and 6 rates of NPK 15:15:15 fertilizer. Fertilizer was applied at the rate of 0, 50, 100, 150, 200, 250 kg ha^{-1} at 2 WAP of seeds. Plantings were done on the 15th of May and 1st of August,

2010, respectively.

Agronomic data collection

Data were collected on biomass yield at 8 WAP. Samples of the plants were obtained within the two last rows by destructive sampling. Five plants per plot were harvested above the ground level to collect data on fresh weight, vine length, vine diameter, number of internodes, internodes length, vine weight and number of leaves. Dry matter was obtained by drying the samples to constant weight in the oven at 70°C.

Statistical analysis

All data were subjected to combined analysis of variance (SAS, 2003). Means squares were significantly different and separated using Duncan multiple range test (DMRT) at 5% level of probability. Regression analysis was performed on the total dry matter using the quadratic equation to determine the fertilizer rate required for maximum biomass yield. The fertilizer rate for maximum biomass yield (N_y) was calculated using the formula: $N_y = -b/2c$ where b and c are the estimates of the regression coefficients in $Y = a + bN + cN^2$ (Gomez and Gomez, 1983).

RESULTS

Soil properties of the experimental site

The results of general chemical and physical properties of the soil in the experimental area before and after cropping for both early and late cropping seasons, respectively are presented in Table 1. The surface soil was slightly acidic with a pH of 6.4 and 6.0. The soil of the site was low in organic carbon (9.5 and 9.8 g kg^{-1}) and also moderate in total nitrogen (1.6 and 1.7 g kg^{-1}). Available P (Bray-P) was 5.4 and 5.80 mg kg^{-1}. Exchangeable Mg^{2+} was 0.38 and 0.43 cmol/kg. The values of exchangeable Ca^{2+} was 2.81 and 1.0 while for K^+ it was 0.40 and 0.28 cmol/kg. The soil of the site was classified as sandy loam. There was a reduction in the amount of N and organic carbon left in the soil after cropping in both seasons.

Biomass traits of pumpkin as influenced by NPK fertilizer at 8 WAP

The effects of season and fertilizer were significant on the biomass yield and its component traits. However, when the percentage contribution of the mean square to the variation in biomass traits were compared, only in the number of internodes and the number of leaves per plant were fertilizer effects more than that of season. Although, the season x fertilizer effect was significant on six of the assessed traits except in vine length, number of internodes and internodes length, its mean square contribution to traits variation was the least (Table 3).

The performance of the traits was better in early season than in late season. Table 4 showed that the vine growth traits from early season were significantly higher than in the late season. Fresh weight and dry matter was

Table 1. Pre-planting and post-planting soil chemical and physical properties at 0 to 15 cm depth in the experimental site for Season 1 (S1) and 2 (S2).

Chemical property	Pre-planting S1	Post-planting S1	Pre-planting S2	Post-planting S2
pH (H$_2$O) (1:2)	6.4	6.8	6.0	6.2
Organic carbon (g kg^{-1})	9.8	7.8	9.5	8.1
Total N (g kg^{-1})	1.7	0.85	1.6	0.66
Available P (mg kg^{-1})	5.80	6.50	5.4	6.1
Exchangeable cations (cmol/kg)				
K$^+$	0.4	0.32	0.28	0.26
Ca^{2+}	0.38	0.45	1.0	0.9
Mg^{2+}	2.81	2.85	1.03	0.54
Physical property				
Sand (g kg^{-1})	802	790	760	760
Silt (g kg^{-1})	97	99	140	120
Clay (g kg^{-1})	101	111	100	120

Table 2. Summary of weather data at Ile-Ife, Osun-State during the cropping year (2010).

Months	Total rainfall (mm)	Average temperature (Min) °C	Average temperature (Max) °C	Solar radiation (MJ/m²/day)
January	0	22.1	33.9	11.12
February	3.7	23.8	35.6	13.38
March	3.9	24.1	34.7	13.61
April	6.7	23.9	33.9	14.71
May	14.4	23.0	31.7	12.74
June	6.9	23.0	30.6	12.45
July	16	21.9	28.6	11.16
August	15.5	22.2	28.4	10.13
September	15.2	22.1	29.7	13.40
October	22	22.0	30.6	13.66
November	8.7	22.4	31.4	9.30
December	0	21.5	32.8	7.83

31 and 39% higher, respectively in the early season compared to the late season while vine length and vine diameter were 17 and 22% higher in the early season plants. Number of internodes and internodes length in early season plants was 15% each higher compared to the late season plants. Number of leaves and leaves weight reduced by 57 and 40%, respectively in late season plants while number of branch in early

Table 3. Combined analysis of variance showing means squares for biomass growth traits of pumpkin as influenced by season and NPK fertilizer at 8 WAP.

Source	DF	Biomass fresh weight (g/plant)	Vine length (cm)	Vine diameter (cm)	No of internodes (n)	Internodes length (cm)	No of leaves (n/plant)	Leaf weight (g/plant)	No of branch (n)	Total biomass dry weight (g/plant)
Season	1	9046131**	240471**	0.80**	480.5**	124.3*	188293**	6293561**	840.5**	3792601**
Rep within season	10	446844	7334	0.01	25.6	12.5	5667	40587	16.26	1413
Fertilizer	5	5838958**	205319**	0.28**	697.9**	72.2**	19805**	1949939**	276.72**	128182**
Season*Fertilizer	5	311701**	6592	0.02**	5.3	1.6	5824*	150651*	53.73**	5771**
Pooled error	50	80100	3067	0.01	14.0	3.5	2357	44689	3.85	859
CV (%)		14.9	13.2	7.3	11.8	11.4	38.2	17.9	17.0	9.7

* = significant at 0.05 level of probability, ** = significant at 0.01 level of probability.

Table 4. Biomass traits of pumpkin as influenced by season at 8 WAP.

Season	Fresh weight (g)	Dry matter (g)	Vine length (cm)	Vine diameter (cm)	No of internodes (n)	Internodes length (cm)	No of leaves (n)	Leaves weight (g)	No of branch (n)	Biomass growth rate (4-8 WAP) (g/day)
Early season	2255.3	375.52	477.36	1.10	34.39	17.62	178	1476.39	14.97	13.0
Late season	1546.4	230.36	361.78	0.89	29.22	14.99	76	885.08	8.14	8.0
LSD (0.05)	163.7	14.54	28.11	0.04	1.82	0.97	23.99	97.66	1.05	0.73

NS = not significant at 5% level of probability.

season plants was 47% higher than the late season plants.

The fresh and dry matter weights and values from other traits increased with the addition of fertilizer. The control traits had the lowest yields or values and these increased significantly when fertilizer rate applied was 100 kg ha⁻¹. However, there was no significant difference in fresh or dry matter weight or in the other traits when the applied fertilizer rate (NPK 15-15-15) increased above 100 kg ha⁻¹ (Table 5). The mean fresh and dry biomass yields at maximum fertilizer rates were depressed by about 50 and 37% when fertilizer rates of zero and 50 kg ha⁻¹ were applied, respectively. The addition of 50 kg ha⁻¹ increased vine length by 58% when compared to the control

treatment and this could be as high as 150% when fertilizer rates increased to 100 and 250 kg ha⁻¹. Vine diameter also increased by 21 and 50% with the addition of fertilizers at 50 kg ha⁻¹ and above, respectively.

The application of fertilizers above 100 kg ha⁻¹ increased the number of internodes by 100% while the internode length increased by 5 to 6 cm more when compared with the control treatment. The number of branches increased by over 200% and number of leaves increased by 170% and leaves weight by as much as 180% when fertilizer rates increased to 100 kg ha⁻¹ and above when compared to the control treatment.

The response of dry matter of biomass to fertilizer rates at 8 WAP fitted into quadratic

equation and the R^2 was significant at 0.95 and 0.91 for early and late seasons, respectively (Figure 1). The rate for dry matter production was higher in early season (b, 3.29 g dy⁻¹) than in late season (b, 1.94 g dy⁻¹) in response to fertilizer increase. Also, higher yield was obtained across fertilizer rates during the early than in the late season. The control had a biomass yield of 180 g plant⁻¹ while at the optimal fertilizer rate, 480 g plant⁻¹ was obtained. For late season, dry matter yield was 100 g plant⁻¹ in zero fertilizer but 300 g/plant at optimal fertilizer application of 200 kg ha⁻¹ NPK. From the equation on the fertilizer rates for required maximum yield were estimated to be 205 and 244 kg of NPK (15-15-15) ha⁻¹ for early and late season's production, respectively.

Table 5. Pumpkin biomass at 8 WAP as affected by NPK fertilizer.

NPK level (kg ha^{-1})	Fresh weight (g)	Dry matter (g)	Vine length (cm)	Vine diameter (cm)	No of internodes (n)	Internodes length (cm)	No of leaves (n)	Leaves weight (g)	No of branch (n)
Control	854.6[c]	134.9[c]	200.1[c]	0.73[c]	19.1[c]	12.2[c]	59[c]	532.7[c]	4.5[d]
50	1168.3[b]	211.6[b]	316.0[b]	0.88[b]	25.6[b]	14.5[b]	96[b]	820.0[b]	6.4[c]
100	2391.7[a]	370.2[a]	512.1[a]	1.11[a]	37.1[a]	18.2[a]	155[a]	1500.6[a]	14.3[ab]
150	2327.5[a]	364.1[a]	481.6[a]	1.06[a]	36.3[a]	17.4[a]	148[a]	1419.2[a]	13.8[b]
200	2268.4[a]	371.4[a]	499.2[a]	1.09[a]	36.6[a]	17.7[a]	145[a]	1376.8[a]	14.7[ab]
250	2394.4[a]	365.3[a]	508.5[a]	1.09[a]	36.2[a]	18.0[a]	160[a]	1435.3[a]	15.7[a]

Means with the same letter in each column are not significantly different at 5% level of probability using Duncan's multiple range test.

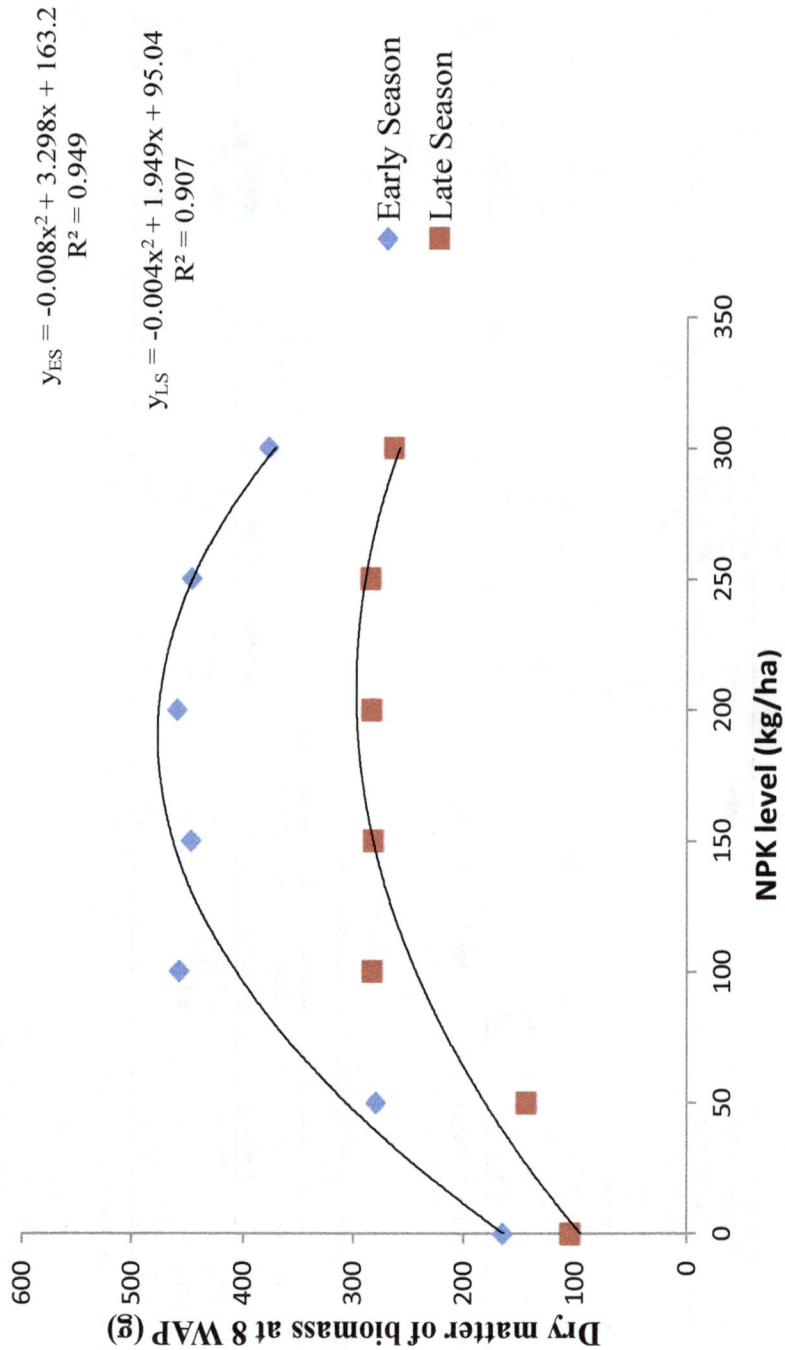

$$y_{ES} = -0.008x^2 + 3.298x + 163.2$$
$$R^2 = 0.949$$

$$y_{LS} = -0.004x^2 + 1.949x + 95.04$$
$$R^2 = 0.907$$

Figure 1. Biomass dry matter at 8 WAP as affected by season x fertilizer.

DISCUSSION

Available P (Bray-P) was considered inadequate when compared with the critical value of 10 to 16 mg kg^{-1} for Southwest Nigeria (Sobulo et al., 1975; Agbede and Aduayi, 1978). Exchangeable Mg^{2+} was considered adequate in Southwest Nigeria where the critical value is 0.2 to 0.4 cmol/kg. The values of exchangeable Ca^{2+} and K$^+$ were considered adequate considering the critical values of 2.5 and 0.16 to 0.25 cmol/kg, respectively for Ca^{2+} and K$^+$.

Biomass production in *C. pepo* was significantly influenced by the quantity of nutrient supplied in this experiment. Yield of pumpkin was found to have increased by 120% with the application of 200 to 280 kg of NPK fertilizer in temperate environment (Martinetti and Paganini, 2006). Although, the experimental site has moderate N status (0.17 to 0.18 g kg^{-1}), the plant biomass increased with fertilizer rates up to 200 kg NPK ha^{-1} but above this a reduction in biomass was observed. The result agreed with earlier reports that biomass yield increased with fertilizer application up to a certain optimal limit (Bradley et al., 1976; Shukla and Gupta, 1980; Almishaal et al., 1984; Alwan, 1986; Al-Mukhtar et al., 1987).

Vine length and its rate of extension have influence on plant biomass. Pumpkin vine length increased with NPK fertilizer rate in this study and could be as long as five meters from the point of sowing. This created a technical problem for maintenance and mechanization of pumpkin production. The extensive foliage covering though genetic was enhanced by fertilizer application due to the fundamental involvement of NPK in the large number of enzymatic reactions as well as other metabolic, energy transfer and biological processes which hasten cell division and growth in plants (Al-Mukhtar et al., 1987). Similar results were obtained in the application of NPK fertilizer to *Philodendron domesticum* L., *Cordyline terminalis* and *Codiaeum variegatum* (Zaghloul et al., 1996; Atta-Alla et al., 1996; Reem, 1997).

It is suggested that the development of plants with shorter vine length and more internodes or branches where more leaves can be harvested is preferred when pumpkin is to be cultivated for leafy vegetables. The use of growth regulators such as maleic hydrazide and ethephon has been found to be effective in reducing the length of the main stem and increasing the number of primary branches in cucumber (Mukesh et al., 2011). The inhibition of apical growth and the proliferation of ancillary branches by the growth regulators was responsible for profuse branching (Odejimi and Akpan, 2006). Similar findings were reported in long melon, gherkin, eggplant and pepper (Murthy et al., 2007; Arora et al., 1994; Miller et al., 1996). If this is applied on pumpkin, probably more shoots with shorter length will be produced per unit land area. Mukesh et al. (2011) reported that the application of growth hormone on cucumber is cost effective and profitable, considering the number of fruits produced per hectare and the additional income generated when compared with a control treatment.

In this study, biomass growth was higher in early season compared to the late season irrespective of the factors imposed. Also, early season require lower fertilizer for maximum yield than the late season. The excessive rainfall which is often accompanied by leaching could have contributed to the higher fertilizer requirement in the late season when compared to early season planting. Márton (2004) reported that excessive rainfall reduced the yield of Rye by 29%. Oloyede and Adebooye (2005) found out that early season's biomass of snake tomato (*Trichosanthes cucumerina* L.) significantly out yielded the late seasons. Even though biomass yield and hence profit returns to prospective Cucurbit leaf vegetable, farmers are better during early planting season, the calculation of economic fertilizer rate in this study from the cut marketable leaves would have captured the economic profitability of this new practice.

REFERENCES

Almishaal AJ, Buhairi AG, Gallum, AAA (1984). Effect of soaking seeds with some micronutrients on the flowering and fruit yield of squash cv. Eskandarany grown under plastic houses. Iraqi J. Agric. Sci. Zanco. 2:59-64.

Al-Mukhtar FA, Hummadi FM, Al-Sahaf FH (1987). Effect of different levels of NPK fertilizer on growth and yield of two summer squash cultivars. Acta. Hortic. 200:253-258.

Alwan OK (1986). Effect of nitrogen fertilization and yield of summer squash *Cucurbita pepo* L. M.Sc. Thesis. Horticultural Department University of Mosul, Iraq.

Arora SK, Pandita ML, Pratap PS, Batra BR (1994). Response of long melon (*Cucumis melo* var. *utilissimus*) to foliar application of plant growth substances. Indian J. Agric. Sci. 69(12):841-844.

Atta-Alla HK, Zaghloul MA, Waly AK, Khattab SH (1996). *In vitro* culture, establishment and effect of NPK fertilizer on *ex vitro* of *Cordyline terminals* cv. Atoom. Ann. Agric. Sci. Moshtohor 34:691-709.

Bradley GA, Barker EC, Motes DR (1976). Cultural and fertilizer studies on summer squash. Arkansas Farm Res. 25(5):11-12.

Colla G, Saccardo F (2003). Application of systematic variation method for optimizing mineral nutrition of soilless-grown zucchini squash. J. Plant Nutr. 9:1859-1872.

Gomez KA, Gomez AA (1983). Statistical procedures for agricultural research (2nd Edition). Publisher: John Wiley and sons, New York. pp. 323-327.

Graifenberg A, Botrini L, Giustiniani L, Lipucci Di Paola M (1996). Yield, growth and element content of zucchini squash grown under saline-sodic conditions. J. Hortic. Sci. 2:305-311.

Liu W, Zhu D, Liu D, Geng M, Zhou W, Mi W, Yang T, Hamilton D (2010). Influence of nitrogen on the primary and secondary metabolism and synthesis of flavonoids in *Chrysanthemum morifolium*. J. Plant Nutr. 33(2):240-254.

Manjunath Prasad CT, Ashok SS, Vyakaranahal BS, Nadaf HL, Hosamani RM (2008). Influence of Nutrition and Growth Regulators on Fruit, Seed Yield and Quality of Pumpkin cv. Arka Chandan. Karnataka J. Agric. Sci. 21(1):115-117.

Martinetti L, Paganini F (2006). Effect of organic and mineral fertilization on yield and quality of zucchini. Acta. Hortic. 700:125-128.

Miller CH, Lower RL, Mc Murray AL (1996). Some effects of etherel (2 chloroethyl phosphonic acid) on vegetable crops. Hortic. Sci. 4:248-249.

Mukesh T, Satesh K, Romisa R (2011). Influence of plant growth regulators on morphological, floral and yield traits of cucumber (*Cucumis sativus* L.). Kasetsart J. Nat. Sci. 45:177-188.

Murthy TC, Negegowda S, Basavaiah V (2007). Influence of growth regulators on growth, flowering and fruit yield of gherkin (*Cucumis anguria* L.). Asian J. Hortic. 2(1):44-46.

Obalum SE, Buri MM, Nwite JC, Hermansah L, Watanabe Y, Igwe CA, Wakatsuki T (2012). Soil Degradation-Induced Decline in Productivity of Sub-Saharan African Soils: The Prospects of Looking Downwards the Lowlands with the Sawah Ecotechnology, Applied and Environmental Soil Science,Volume 2012. Article ID 673926, 10pages doi:10.1155/2012/673926.

Odejimi RAO, Akpan GA (2006). Effect of mineral supplements (NPK) on sex expression in fluted pumpkin (*Telfairia occidentalis*) Hook F. Int. J. Nat. Appl. Sci. 1(1):56-58.

Oloyede FM, Adebooye OC (2005). Effect of season on growth, fruit yield and nutrient profile of two landraces of *Tricosanthes cucumerina* L. Afr. J. Biotechnol. 4(6):1040-1044.

Reem MSS (1997). Effect of some chemical fertilization on croton plant. M.Sc. Thesis, Faculty of Agriculture, Cairo University.

SAS (2003). Version 9.1. SAS Institute Inc., Cary, NC.

Shukla V, Gupta R (1980). Notes on the effect of levels of nitrogen, phosphorus fertilization on the growth and yield of squash. Indian J. Hortic. 37(2):160-161.

Sinclair TR, Vadez V (2002). Physiological traits for crop yield improvement in low N and P environments. Plant Soil 245:1-15.

Smith BD (1997). The initial domestication of *C. pepo* in the Americas 10,000 years ago. Science 276:932-934.

Zaghloul MA, Atta-Alla HK, Waly AK, Khattab SH (1996). *In vitro* culture, establishment and effect of potting mixture and NPK fertilizer on *ex vitro* of *Philodendron domesticum* L. Ann. Agric. Sci. Moshtohor. 34:711-725.

25

Quantitative and qualitative analyses of weed seed banks of different agroecosystems

Ademir de Oliveira Ferreira[1] , Cláudio Purissimo[2], Rafael Pivotto Bortolotto[1], Telmo Jorge Carneiro Amado[1], Klaus Reichardt[3], Larissa Bavoso[4] and Mariânne Graziele Kugler[4]

[1]Department of Soil Science, Federal University of Santa Maria (UFSM), Avenida Roraima, 100, 97105-900, Santa Maria, Rio Grande do Sul, Brazil.
[2]Departament of Crop Science and Plant Health, University of Ponta Grossa (UEPG), Avenida Carlos Cavalcanti, 4748, Campus Uvaranas, 84030-960, Ponta Grossa, Paraná, Brazil.
[3]Center for Nuclear Energy in Agriculture, University de São Paulo (CENA/USP), Avenida Centenario, 303, bairro São Dimas, caixa postal 96, 13400-970, Piracicaba, São Paulo, Brazil.
[4]University of Ponta Grossa (UEPG), Avenida Carlos Cavalcanti, 4748, Campus Uvaranas, 84030-960, Ponta Grossa, Paraná, Brazil.

One of the main survival mechanisms of weeds in constantly disturbed environments, specially the annual weeds, is their high production of seeds. In this study it was intended to evaluate the influence of different agroecosystems (vegetable garden, pasture, native field, soybean, dry bean and corn) on the strength of the seed bank, making quantitative and qualitative analyses. On each site, soil samples were collected, split to submit half to seed extraction by washing samples with water and counting the total number of seeds (quantitative analysis), and half to germination in trays placed in a greenhouse to evaluate weed emergence (qualitative analysis). The quantitative analysis of the agroecosystems showed that those cultivated with corn and vegetable garden presented best conditions for weed occurrence. The qualitative analysis resulted in the highest number of viable seeds for the vegetable garden (141,094,713 seeds, of which 74,965,862 were from monocotyledons plants and 66,128,851 dicotyledons). The weed seed concentration found for the vegetable garden is probably related to the management intensity in the area. The inverse is observed for the environments of less management intensity, as pasture and native field. Dry bean and soybean plots presented small seed bank and low emergence.

Key words: Vegetable garden, pasture, soybean, dry bean, corn, native field.

INTRODUCTION

The reserves of viable seeds in soil at the surface and in depth are known as seed bank (Gomes and Christoffoleti, 2008), other concepts or designations being also found. It is also known as seed reservoir, including the amount of non dormant seeds and other plant propagation structures present in the soil or in plant residues (Monquero and Christoffoleti, 2003). This reserve is the sum of all produced and introduced seeds along time that continue alive and dormant, with the seeds recently produced (Kuva et al., 2008). The variability and botanical density of a seed bank at a given time are the result of the balance between the input of new seeds

Table 1. Treatments and sampling collection sites.

Treatments	Soil sampling location
1	Vegetable Garden
2	Soybean Field at harvest[1]
3	Dry bean Field at harvest[1]
4	Corn Field at harvest[1]
5	Native Field
6	Pasture

[1]area of Grass-legume rotation under no tillage.

(by rain and dispersion) and losses by germination, deterioration, parasitism, predation and transport out of the area (Machado et al., 2013).

An accurate measure of weed emergency is a subsidy to farmers for a more efficient weed control without the inappropriate use of herbicides (Kuva et al., 2008). The quantitative and qualitative evaluation of seed banks can practically be made only by direct germination in soil samples and by physical and chemical seed extraction followed by viability essays (Luschei et al., 1998). The size and the composition of the weed seed bank are very important for the decision of integrated weed management strategies. The *in situ* observation of seedling emergence in the field may give a general indication of the size and composition of the vegetative population, and of the seed bank. However, this is not a precise method because several seeds can stay viable for a long period without germination, and some of the germinated seeds may not emerge due to unfavorable conditions or to deep positioning in the soil (Lacerda et al., 2005). The simplification of the environment that characterizes the modern agricultural systems, as for example mono-cropping, accelerates the ecological succession patterns (Gasparino et al., 2006), generating specialized "habitats" within ecosystems.

The cultivation system exerts an influence on the size of the seed bank. Carmona (1995) estimated the seed banks of four distinct agroecosystems: crop rotation (soybean, fallow, dry bean); lowland; citrus orchard and pasture of *Brachiaria brizantha*. The average quantities of seed per square meter were 22,313 for lowland; 6,768 for rotation; 3,595 for the orchard crowns; and 529 for pasture. He also found out that similarities of seed bank sizes among agroecosystems is greater for the most disturbed areas, as it is the case of crop rotation, lowland and orchards. In agricultural areas seed banks are comparatively greater than in non agricultural areas of low environmental disturbance, because weeds have a strategy of producing large numbers of seeds in much disturbed situations (Monquero and Christoffoletti, 2005). Environmental and management factors influence the seed consumption rate by predator organisms (Balbinot et al., 2002). The consumption of these seeds is made by a large number of species (animals, insects, fungi, etc) that are naturally present in the environment.

In agroecosystems the weed population is related to their seed bank, so that the knowledge of the seed bank size and of its species composition can be used to predict future infestations, to construct population models along time, and consequently for the definition of management programs that lead to a better rationalization of the use of herbicides (Gardarin et al., 2011; Soltani et al., 2013). In general, the decision-making of weed management strategies is based on visual evaluations of the needed weed control intensity without much technical criteria. It is therefore important, for emerging technologies like precision agriculture, to develop control strategies based on estimations of the potential of the weeds in the soil. They should be supported by research and be based on economical viability (Voll et al., 2003).

One of the important factors in studies of seed banks is related to the techniques used for their evaluation (Caetano et al., 2001). The quantification of a cultivated soil seed bank includes the problem of the minimum number of soil samples to be collected in order to have a precise estimation of the number of seeds per unit area (Voll et al., 2003). The understanding of the dynamics of a weed seed bank and the simulation of the emergency flux are among the most recent strategies used for weed control (Vivian et al., 2008). In this context, this experiment was carried out with the aim of evaluating the influence of different types of soil use, that is, different agroecosystems on their seed banks, making qualitative and quantitative analyses.

MATERIALS AND METHODS

Experiments were carried out in Ponta Grossa, PR, Brazil, where the climate is sub-tropical humid, mesotherimic, of Cfb type (Koeppen, 1931), with mild summers and frequent frosts in winter. Average temperature of the hottest month is less than 22°C and of the coldest month is less than 14°C. Average yearly rainfall is 1,545 mm, with no defined dry season (IAPAR, 2008). The soil is a red latosol, typical distrophic, according to EMBRAPA (2006), and an Oxisol according to the Soil Survey (2010). Natural vegetation is dominated by C_4 plants represented by some grasses like *Andropogon* sp., *Aristida* sp., *Paspalnm* sp., *Panicum* sp., and gallery forests are found along the natural drainage canals. The relief is softly ondulated with slopes between 2 and 7%. Data were collected during two different phases, first collecting soil samples for the qualitative analysis, and second manipulating tray soil samples for the quantitative analysis of the seed banks. The sites for sample collection, which represent the treatments, belong to different environments as shown in Table 1.

The fields of dry bean, soybean and corn where soil samples were collected, were cultivated by a long duration minimum tillage system with a rotation schedule shown in Table 2. The experimental field design was completely randomized with six treatments and 16 replicates (plots of 3 × 6 m), carried out as described above. Soil samples were composed of 20 sub-samples of each agroecosystem from the surface layer (0 to 0.5 m) in a randomized way. One composite soil sample of 3 to 4 kg was prepared after homogenizing the sub-samples, which were divided into parts A and B.

Sample A: This was used for the quantitative analysis of the seed bank through the number of seeds per ha and per 0.05 m of soil

Table 2. Rotation schemes where dry bean, soybean and corn soil samples were collected.

Season-year	Rotation scheme			
Winter - 2000	AP + E[1]	AP	T	P
Summer - 2000/2001	M[2]	S/F	F/S	S
Winter - 2001	AP[3]	T	P	AP+E
Summer -2001/2002	S/F[4]	F/S	S	M
Winter - 2002	T[5]	P	AP+E	AP
Summer - 2002/2003	F/S[6]	S	M	S/F
Winter - 2003	P[7]	AP+E	AP	T
Summer - 2003/2004	S[8]	M	S/F	F/S
Winter - 2004	AP/E	AP	T	P
Summer - 2004/2005	M	S/F	F/S	S

[1]Black oat + pea, [2]Corn, [3]Black oat, [4]Soybean/dry bean, [5]Wheat, [6]Dry bean/soybean, [7]Fallow, [8]Soybean.

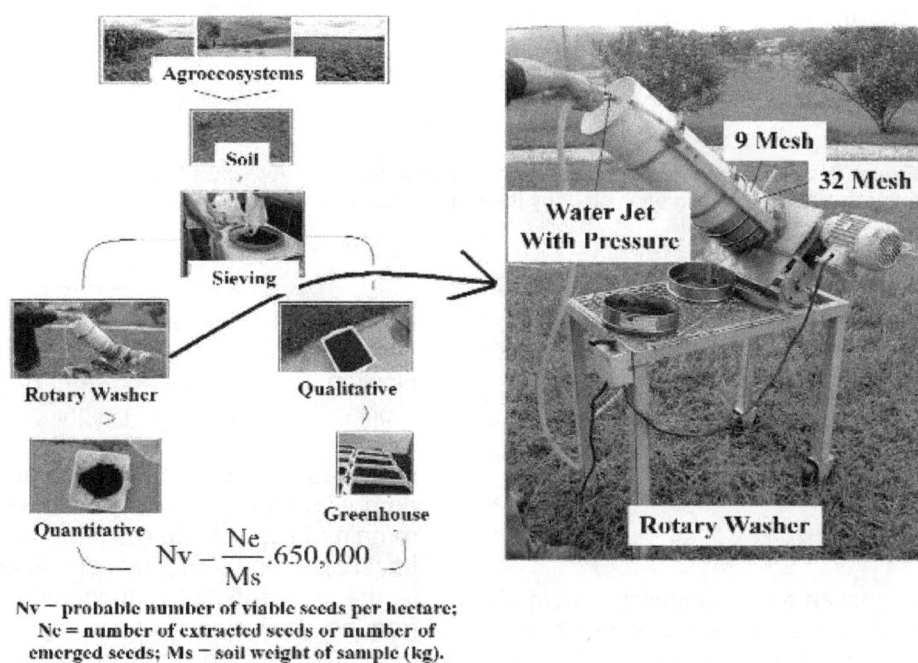

$$Nv = \frac{Ne}{Ms}.650,000$$

Nv − probable number of viable seeds per hectare;
Ne = number of extracted seeds or number of emerged seeds; Ms − soil weight of sample (kg).

Figure 1. Visual description of the used methodology.

depth. Samples were passed through a rotary washer (Figure 1) with two sieves, one with larger openings (9 mesh) to retain coarser materials like plant residues, and one with smaller openings (32 mesh) to retain seeds, coarser soil particles and aggregates, and plant material that was not retained by the previous sieve. After washing samples were air dried for 20 days and, with aid of a magnifying glass the inert material was separated and seeds counted.

Sample B: This was used for the qualitative analysis evaluating the number of non dormant viable seeds (germinated and emerging later) per ha and 0.05 per m depth. For the qualitative analysis field soil samples were displayed on plastic trays to form a 0.04 – 0.05 cm soil layer, so that seed depth would not be a limiting factor for their germination. Trays were displayed in a greenhouse and irrigated periodically to allow the germination of non dormant seeds. At fixed time intervals emerged seedling were counted, making the

distinction of the monocotyledons and dicotyledons. These seedlings were eliminated to give place to those germinating later. When the germination flux ended, soil was turned over to stimulate further germination, and counts continued for 53 days, a time admitted sufficient for this evaluation. The quantitative analysis was based on the estimative of the total number of probable seeds per hectare, per 0.05 m layer in depth, and the qualitative analysis on the estimative of the number of probable viable seeds in the same layer. For this layer of average soil bulk density of 1.3 g cm^{-3}, the soil mass totalized 650,000 kg of dry soil. The probable number of seeds per hectare in the 0.05 m soil layer was calculated through Equation (1) of Monqueiro and Christoffoleti (2003):

$$Nv = \frac{Ne}{Ms}.650,000 \tag{1}$$

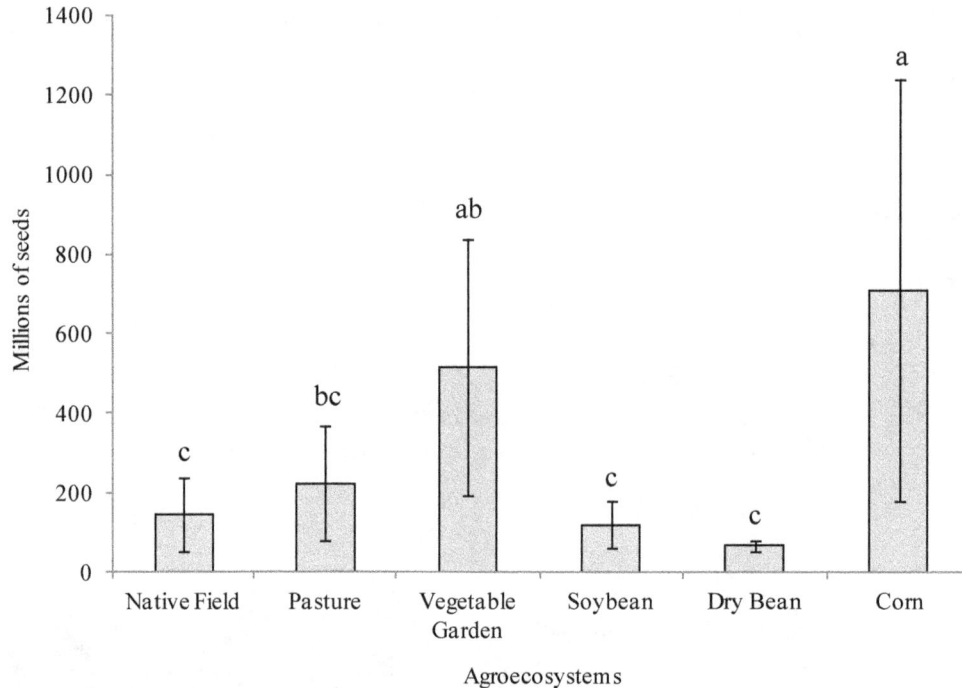

Figure 2. Seed bank size per hectare – quantitative analysis. Averages followed by the same letter do no differ among each other by the Duncan test at 5% significance level.

Where: Nv = probable number of viable seeds per hectare; Ne = number of extracted seeds or number of emerged seeds; Ms = soil weight of sample (kg).

Results were submitted to analysis of variance and means were compared by the Duncan test (p< 0.05), using the SASM - Agric (2001) program.

RESULTS AND DISCUSSION

In a general way, data show a great variability in both, the seed banks as well as the total germination of these banks in the different ecosystems. According to Carmona (1995), such variability is normal in these types of studies. The largest weed seed bank was found in the agroecosystem of corn (Figure 2), however not statistically different from the vegetable garden. The agroecosystem pasture presented the third largest seed bank, also not differing from the other, followed by native field, soybean and dry bean.

The history of the experimental fields (Table 2) points to a winter fallow interval that may have favored a renovation of the seed bank since weeds could complete their reproduction cycles. However, there was no difference between the agroecosystems corn and vegetable garden. This happened due to a better relation between the environment (soil and climate) and the weed species present in this environment that had a better ability to contribute to the establishment of a seed bank. In relation to the vegetable garden agroecosystem, soil

revolvement stimulates an increase of seed viability. Practices that promote the inversion of soil layers as plowing, foment a better seed distribution within the soil profile, and also bury a significant amount of seeds so that the regeneration capacity of part of certain seed populations is derailed. On the other hand, practices that do not invert soil layers allow the majority of the seeds to remain at soil surface (Lacerda et al., 2005). According to Lacerda et al. (2005), higher values of weed viable seeds in the conventional management system is due to the frequent soil perturbations by mechanical implements during a corn field establishment in summer.

For the pasture agroecosystem, the lack of soil perturbation added to the low fertility, promoted a more stable environment that is propitious only for few species with less individuals, reducing the strength of the seed bank (Carmona, 1995). Marquezan et al. (2003) analyzing the dynamic of a red rice seed bank, concluded that during the fallow period rice seed was reduced on average by 85% per year because the soil surface seeds lost their viability more rapidly in relation to deeper seeds. Another explanation is in the way data obtained in small samples are transformed to hectares through the average relation of soil mass per unit volume. Ideally, soil bulk density should be measured along sampling points in order to have more representative data. Observing again the history of the area (Table 2), now in relation to the agroecosystems soybean and dry bean, we can see a much lower seed number in comparison to the vegetable

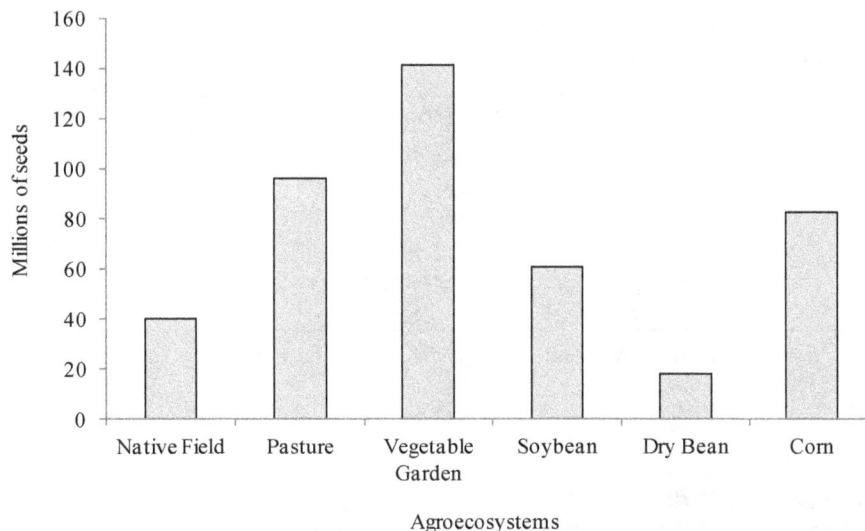

Figure 3. Probable total number of viable seeds (mono and dicotiledons) per hectare – qualitative analysis.

Table 3. Percentage of dormant seeds and emerged seeds, in relation to the total number of the seed bank.

Agroecosystem	Dormant seeds (%)	Emerged seeds (%)
Native field	72.19	27.81
Pasture	57.08	42.92
Vegetable garden	72.63	27.37
Soybean	48.54	51.46
Dry bean	73.40	26.60
Corn	88.30	11.70

garden and corn (Figure 3). This is probably due to the rapid and good covering of the soil surface by the crop and use of herbicides, which do not favor weed germination. This is the case of pastures and native fields because according to Caetano et al. (2001) the application of herbicides for weed control influences the distribution of the seed bank in the soil profile.

The variability of the data obtained in this type of study is normally high due to the relatively great non uniformity of seed distribution in the soil (Carmona, 1995). Observing the number of viable seeds (Figure 3) obtained from seeds germinated in soil trays, we can see little differences between treatments. However, observing the vegetable garden environment in relation to the other, it can be noted that the viable seed bank of this environment is much larger. This is mainly due to the fact that this environment involves manual control of weeds, little use of herbicides, constant soil revolvement, in this way favoring the renovation of the seed bank. This is confirmed when looking at the dry bean environment that due to soil surface shading and use of herbicides,

diminishes drastically the renovation of the soil seed bank.

Despite the obtained data, it can be seen in Table 3 that most of the seeds remain dormant for variable periods. To analyze a dormant seed bank there is need of longer evaluations. However, observing the total seed number together with the number of viable seeds, it can be noted that the viability potential of the seeds is only manifested when stimulated and submitted to ideal conditions of development. Dormancy and its seasonal changes are related to the persistence of seeds in the soil and, consequently, to the problems faced during the infestation of the crops. Weed seeds pass through annual cycles of more or less intensity of dormancy. These changes are attributed to variations in temperature, light, rainfall, agricultural practices and seed depth (Vivian et al., 2008). For the vegetable garden environment (Table 3), 72.63% of the seeds are dormant. This is due to soil mixing that lead to a more uniform seed distribution in the profile, and in a burying of a greater amount of seeds making them unviable (Lacerda et al., 2005). For the

Table 4. Qualitative analysis of mono and dicotyledon weeds in the surface layer (0.05 m of deep).

Agroecosystems	Probable number of viable seeds	
	Monocotyledons	dicotyledons
Native field	22360435[ab]*	17903565[b]*
Pasture	75402000[a]	20353111[b]
Vegetable Garden	74965862[a]	66128851[a]
Soybean	40002667[ab]	20958000[b]
Dry bean	11017804[b]	6859021[b]
Corn	58793967[ab]	24020134[b]

*Averages followed by the same letter do no differ among each other by the Duncan test at 5% significance level.

other agroecosystems the percent of dormant seeds was also high since seeds can remain viable in the soil for long periods without germination (Caetano et al., 2001). In this respect, Lacerda et al. (2005) states that in fallow areas the number of species and viable seeds in the soil are smaller. For all agroecosystems the majority of the weeds were monocotyledons (Table 4). For the corn environment the development of monocotyledons was favored because corn itself is a monocotyledon and herbicides used in this situation did not control weeds from the family Poaceae. For the agroecosystem vegetable garden, a large number of monocotyledons were observed, which can be explained by the intense cultivation of this area, with several stimulations of the seed bank. Blanco and Blanco (1991) observed that the weed management through soil movement with rotary hoes stimulated weed seed emergence.

For the dry bean environment the development of monocotyledons was also favored, despite being a dicotyledon plant. In this environment, however, there was an equilibrium of viable mono and dicotyledons seeds because besides soil chemical and physical effects, there were biological effects due to the interference of plant residues on the survival of seeds from the bank (Gomes and Christoffoleti, 2008) taking into account that the crop was managed under minimum tillage. For the natural ecosystem of native fields, an even greater equilibrium between viable mono- and dicotyledon seeds would be expected. This is however explained by the fact that the monocotyledon seed population was larger than that of the dicotyledons.

Cultivation systems favored the renovation of seed banks, with different intensities, and promoted a better development of monocotyledons in relation to dicotyledons. The size and composition of the soil seed banks are extremely variable among different habitats (Kuva et al., 2008). This is a response of the strategy of invasive plants producing a large number of seeds and having good dissemination mechanisms, longevity and dormancy to survive in hostile environments. The evaluation of the need for control of weeds is a function of the emergence rate of the species present in the soil seed bank and has to be established for each management system of the implanted crop (Voll et al., 2003).

In general, seed banks are composed of many species, few of them dominant, corresponding to 70 to 90% of the total seed number in the soil. These species are considered harmful because they resist control measures and are more adaptable to different climatic conditions.

Conclusions

The emergence of monocotyledon plant seeds was greater in all agroecosystems, especially in corn and pasture. The agroecosystem vegetable garden favored the increase of the soil seed bank because of its more intense soil revolvement and less use of chemical weed control. The agroecosystems dry bean and soybean presented low emergency and a smaller seed bank.

REFERENCES

Balbinot Junior AA, Fleck NG, Agostinetto D, Rizzardi, MA (2002) Weed seeds predation in cultivaded fields. Ciência Rural, 32:707-714.

Blanco HG, Blanco FMG (1991) Effects of soil management of emergence of annual weeds. Pesquisa Agropecuária Brasileira, 26:215-220.

Caetano RSX, Christoffoleti PJ, Victoria Filho R (2001) Weed seed bank of a 'pera' citrus orchard. Scientia Agricola, 58:509-517.

Carmona R (1995) Seed bank in the soil and the establishment of weeds in agro-ecosystems. Planta Daninha, 13:3-9.

Canteri MG, Althaus RA, Virgens Filho JS, Giglioti EA, Godoy CV (2001) SASM-Agri – System for analysis and mean separation in agricultural assays using Scott Knott, Tukey and Duncan methods. Revista Brasileira de Agrocomputação, 1:18-24.

EMBRAPA - Brazilian Agricultural Research Corporation (2006) Brazilian System of Soil Classification. 2ᵃ ed. Rio de Janeiro, National Research Center for Soil. 306p.

Gardarin A, Dürrb C, Colbacha N (2011) Prediction of germination rates of weed species: Relationships between germination speed parameters and species traits. Ecological Modelling, 222:626-636.

Gasparino D, Malavasi UC, Malavasi MM, Souza I (2006) Evaluation of seed bank under different soil uses. Revista Árvore, 30:1-9.

Gomes Jr FG, Christoffoleti PJ (2008) Weed biology and management in no-tillage areas. Planta Daninha, 26:789-798.

IAPAR - Agronomic Institute of Paraná (2008) Climatological data of

Ponta Grossa - 2008. Ponta Grossa, IAPAR. P. 22.

Köppen W (1931) Grundriss der Klimakunde. Berlin, Leipzig: Walter de Gruyter. P. 388.

Kuva MA, Pitelli RA, Alves PLCA, Salgado TP, Pavani MCDM (2008) Weed seedbank and its correlation with the established flora in no-burn sugarcane area. Planta Daninha, 26:735-744.

Lacerda ALS, Victoria Filho R, Mendonça CG (2005) Seeds bank assessment under two systems of management of soil irrigated by central pivot. Planta Daninha, 23:1-7.

Luschei EC, Buhler DD, Dekker JH (1998) Effect of separating giant foxtail (Setaria faberi) seeds from soil using potassium carbonate and centrifugation on viability and germination. Weed Science, 46:545-548.

Machado VM, Santos JB, Pereira IM, Lara RO, Cabral, CM e Amaral CS (2013) Evaluation of the seed bank in a campestre cerrado area under recovery. Planta Daninha, 31:303-312.

Marquezan E, Oliveira APBB, Avila LA, Bundt ALP (2003) Dinâmica do banco de sementes de arroz-vermelho afetado pelo pisoteio bovino e tempo de pousio da área. Planta Daninha, 21:55-62.

Monquero PA, Christoffoleti PJ (2003) Seed bank dynamics in areas with frequent glyphosate application. Planta Daninha, 21:63-69.

Monquero PA, Christoffoleti PJ (2005) Weed seed bank and herbicides as selection factor. Bragantia, 64:203-209.

Soil Survey Staff (2010) Keys to Soil Taxonomy. 11[a] ed. Washington DC, USDA-Natural Resources Conservation Service. P. 338.

Soltani E, Soltani A, Galeshi S, Ghaderi-Far F, & Zeinali E (2013) Seed bank modelling of volunteer oil seed rape: from seeds fate in the soil to seedling emergence, 31:267-279.

Vivian R, Silva AA, Gimenes Junior M, Fagan EB, Ruiz ST & Labonia, V (2008) Weed seed dormancy as a survival mechanism – brief review. Planta Daninha, 26:695-706.

Voll E, Adegas FS, Gazziero DLP, Brigheti AM & Oliveira MCN (2003) Sampling of weed seedbank and emerged populations. Pesquisa Agropecuaria Brasileira, 38:211-218.

26

Effect of weed control methods on yield and yield attributes of soybean

F. A. Peer, Badrul Hassan, B. A. Lone, Sameera Qayoom, Latief Ahmad, B. A. Khanday, Purshotam S singh and Gurdeep Singh

Division of Agronomy, S. K. University of Agricultural Sciences and Technology of Kashmir, India.

Two field experiments were conducted at the experimental farm of Agronomy, Sher-e-Kashmir University of Agricultural Sciences and Technology of Kashmir, Shalimar Campus in the kharif seasons of 2004 and 2005 to study the effect of various weed control methods on yield and yield attributes of soybean. All weed control measures registered significantly higher seed yields of soybean than weedy check. However, weed free treatments, hand weeding twice and both fluchloralin and pendimethalin integrated with hand weeding recorded far superior yields of soybean seed. Integrated use of herbicides gave better seed yield than their individual application. Similarly, higher doses of both herbicides gave more yield than their lower doses. Pendimethalin 1.5 kg ha^{-1} and hand weeding once recorded comparable yields of soybean. Seed protein content was significantly greater under all weed control measures. Weed free treatment, hand weeding twice and pendimethalin integrated with hand weeding recorded comparable percentage of both these parameters; while, in case of oil content, hand weeding twice, higher dose of pendimethalin and both the herbicides at low rates, integrated with hand weeding recorded comparable oil content in soybean seed. Fluchloralin and pendimethalin at either of the two rates namely 1.0 and 1.5 kg ha^{-1} gave statistically similar values of oil content as that recorded by hand weeding once. Lowest oil percentage was seen in the weedy check plots.

Key words: treatments, fluchloralin, pendimethalin, weed, soybean.

INTRODUCTION

Soybean [*Glycine max* (L.) Merill] popular as golden bean has become the miracle crop of 21st century. It serves the dual purpose for being grown both as an oilseed crop and pulse crop as well (Thakare et al., 2006). It is an excellent health food containing 40 to 44% good quality protein, 20% cholesterol free oil, 20% carbohydrates and 0.69% phosphorus. It also fixes atmospheric nitrogen (45 to 60 kg ha^{-1}) through root nodules and adds about 0.5 to 1.5 ton organic matter per hectare through leaf fall (Kanase et al., 2006). Reduction in soybean yield due to weed infestation varies from 27 to 77% (Gogoi et al., 1991), depending on type of weed, soil, seasons and weed infestation intensities. Some have reported the yield decline as high as 84% (Kachroo et al., 2003).

Weed infestation removed 21.4 kg N and 3.4 kg P ha^{-1} in soybean (Pandya et al., 2005). Two hand hoeings are recommended for effective weed control in soybean (Jain, 2000; Rakesh and Shirvastava, 2002; Galal, 2003; Singh and Jolly, 2004). Ahmed et al. (2001) reported that application of two hand hoeings is more effective in suppressing weeds and increasing soybean seed yield. Today, there is a great manual labor shortage and a rise in wage scale. Thus, chemical weed control is necessary to decrease cost and to increase soybean productivity. This crop is a large herbicide consumer, and almost 90% of the planted area in India is herbicide-treated. The advantages of herbicide use are high efficiency in weed control, the presence of selective products soybean at

Table 1. Weed flora of the experimental field.

Scientific name	Common name	Kashmiri name
Broad leaf weeds		
Amaranthus spp.	Pig weed	Lisa
Chenopodium album L.	Common lambsquarters	Kon'e/von palak
Convolvulus arvensis L.	Bind weed	Thrir
Portulaca oleracea L.	Purselane	Nuner
Capsella bursa-pastoris L.	Shepherd's purse	Kralmund
Solanum nigram L.	Black night shade	Kambal
Grassy weeds		
Cynodon dactylon	Bermuda grass	Dramun
Echinochloa colonum	Jungli rice	Hama
Poa annua	Blue grass	Mahigase
Sedges		
Cyperus rotundus	Nut sedge	Mothe

the lowest cost, compared to other available weed control methods. Despite the satisfactory weed control results, many questions remain on the effect of herbicides on the N_2 fixation process, since the soybean crop is dependent on symbiosis with bradyrhizobium (Zawoznik et al., 1995).

Pre-emergence herbicide application can help control weeds, to some extent, during the early crop growth stage. Soybean undergoes heavy weed competition especially in the early growth stages. Crop-weed competition is minimized by pre-emergence herbicide spray, resulting in decreasing weed dry matter and increasing crop yield (Jeyabal et al., 2001; Mohamed, 2004; Sha, 2004). Regarding chemical weed control, selective herbicides may be effective against annual weeds and achieve high soybean and legume yield such as butralin (Hassanein, 2000; El-Metwally and Saad El-Din, 2003), prometryn (Sha, 2004; Abd El-Razik, 2006) and oxadiargyl (Dobrzanski et al., 2001). Hence, two field experiments were conducted to examine the effects of different herbicides, applied at pre-emergence on weed infestation, yield and yield attributes of soybean plants.

MATERIALS AND METHODS

Two field experiments were conducted at the experimental farm of Agronomy, Sher-e-Kashmir University of Agricultural Sciences and Technology of Kashmir, Shalimar Campus in the kharif seasons of 2004 and 2005. The soil of the experimental field was silty clay loam with normal pH and EC, high in organic carbon, medium in nitrogen and potash and low in phosphorous. The experiment was laid out in a split pot design with 10 treatments and three replications. The treatments consisted of weedy check, eight weed control methods namely; 1 hand weeding (HW) 25DAS, 2 HW 25 and 45 DAS, pre-plant incorporation of fluchloralin 1.0 and 1.5 kg a.i ha^{-1} pre-plant incorporation of flucholarin 1.0 kg a.i + 1 HW 35 DAS, pre-emergence application of pendimethalin 1.0 kg a.i ha^{-1} + 1 HW 35 DAS and weed free treatment obtained by continuous hand weeding. After application of the pre-emergence herbicides, all the experimental plots were irrigated. When soil moisture became

adequate (3 to 4 days later), the seeds of soybean (*Glycine max*) cv. 'PS-1092' were sown on hill 20 cm apart in both sides of the ridge. The crop was sown on 23rd May during both 2004 and 2005. After complete germination, soybean seedlings were thinned to secure two plants per hill. The first irrigation was carried out 40 days after sowing. Fertilizers N, P and K were applied during soil preparation and before sowing. All recommended agricultural practices were adopted throughout the two seasons.

After maturity, soybean plants were harvested to estimate number of pods per plant, pod weight per plant (g), number of seeds per plant, seed yield per plant (g), seed yield (kg ha^{-1}), biological yield per plant (g) and 100-seed weight (g). The seeds were ground to pass a 0.5 mm sieve to estimate N and oil contents. Total nitrogen content of the seeds was determined according to AOAC (1980). N values were multiplied by 6.25 to calculate total crude protein (TCP). The oil content was determined with the help of nuclear magnetic resonance (NMR) spectroscopy technique (Alexander et al., 1967) for each representative sample. The oil content was worked out as the following and expressed as percent. The data recorded for different parameters were subjected to statistical analysis as per the method of analysis of variance as suggested by Gomez and Gomez (1984). Wherever, the 'F' test was found significant at 5% probability, the critical difference value was used to compare the treatment means and their interaction effects wherever required data was subjected to square root $\sqrt{x + 0.5}$ transformation. The software used for this analysis was $CPCS_1$.

RESULTS AND DISCUSSION

Weed species

Major weed floras found in the experimental field were grouped into broad leaf weeds, grassy weeds and sedges (Table 1 and Plate 1). The physiological and yield responses of soybean to an herbicide may vary, and may also depend on geographical location, environmental conditions, soil types, sensitivity of native populations of *Bradyrhizobium japonicum* etc. (Zablotowicz and Reddy, 2007). Significant differences were observed in function of weed management practices in yield and its attributes

Convolvulus arvensis

Portulaca oleraceae

Solanum nigrum

Cyperus rotundus

Capsella bursa - pastoris

Amaranthus spp.

Plate 1. Soybean yield and yield components.

(Tables 2 and 3). Greater weed competition in weedy check resulted in reduced number of branches per plant under this treatment at harvest. Consequently, weed control measures offered a better environment for enhanced branching by crop. Significantly, highest number of branches was recorded in weed free plots comparable with twice hand weeding treatments. Herbicides proved more effective at higher rates when applied alone. However, when combined with one hand weeding, they were more effective. Increased number of branches as a result of chemical and hand weeding methods has also been reported by Kushwah and Vyas (2005). Various yield components were markedly influenced by different weed control measures. Maximum number of pods was produced by weed free treatments (W_{10}) which was at par with hand weeding twice (W_3). Other weed control treatments also affected significantly higher number of pods as compared to un-weeded control (W_1) which gave the lowest number of pods per plant.

Severe weed competition in the weedy check might have reduced the number of pods per plant. Weed free treatment produced 60.08 and 56.67% extra pods than control. Jain (2000) also got highest pods in weed free treatment. Fluchloralin and pendimethalin at lower rates (1 kg ha^{-1} each) when integrated with hand weeding resulted in greater values of pods per plant than when applied alone recording at par influence with weed free in 2005. This is clearly indicative of more pronounced affect of their integrated use because of the fact that initial achievement of limiting weed growth by the herbicides is maintained as hand weeding eliminates the fresh flush of weeds that may regenerate due to loss of persistence of the applied herbicides as in the case of herbicides applied alone. A number of researchers like Veeramani et al. (2001) held similar views and reported more pods with integrated use of herbicides with hand weeding. Herbicides applied alone recorded pods at par with hand weeding once at 25 DAS. Number of pods per unit area basis was significantly influenced by different weed control measures. Weed free treatment (W_{10}) and hand weeding twice (W_3) affected number of pods per square metre that were at par with each other. Herbicides applied individually and in integration with one hand weeding at 35 DAS also caused significant enhancement in the number of pods per square metre as compared to un-weeded control. Both the number of seeds per pod and 100-seed weight were benefited by various weed control measures.

Weed free treatment (W_{10}) and hand weeding twice (W_3) were at par with each other in producing significantly highest number of seed per pod and also affecting highest 100-seed weight. Un-checked growth of weeds in weedy check caused lowest number of seeds per pod and 100-seed weight. Hand weeding twice (W_3) was found statistically at par with fluchloralin 1 kg integrated with hand weeding (W_8) and pendimethalin 1 kg integrated with hand weeding (W_9) with respect to the number of seeds per pods and 100-seed weight. Herbicides applied alone too had a significant promising influence on test weight giving higher values than the weedy check. Reduced weed competition as a

Table 2. Yield attributes of soybean as influenced by weed control methods.

Treatments	Branches per plant		Pods/plant (no.)		Pods/square metre (no.)		Seeds/pod (no.)		100-seed weight (g)		Biological yield (g)		Harvest index (%)	
	2004	2005	2004	2005	2004	2005	2004	2005	2004	2005	2004	2005	2004	2005
W_1	6.57	7.74	42.49	44.70	756.33	793.66	1.80	1.82	10.015	10.100	47.25	46.96	34.52	34.25
W_2	8.60	8.84	50.80	53.59	1086.08	1053.55	2.10	2.15	10.797	10.930	58.28	59.43	35.08	35.00
W_3	9.30	10.07	58.07	61.87	1161.55	1182.66	2.32	2.40	11.224	11.617	67.07	67.75	36.01	36.07
W_4	7.80	8.23	51.43	50.14	965.00	999.33	2.03	1.98	10.745	10.910	54.46	56.55	34.12	33.60
W_5	8.00	7.69	45.75	49.37	1002.11	952.88	1.91	1.99	10.822	11.020	56.75	58.33	34.88	34.48
W_6	8.30	8.75	46.81	49.72	926.88	958.41	2.04	1.87	10.693	10.961	55.43	55.74	35.24	35.88
W_7	8.73	8.32	50.71	52.86	948.33	1013.84	2.06	2.00	10.953	10.980	57.74	59.46	35.89	35.45
W_8	8.70	8.90	56.75	63.68	1052.12	1139.56	2.18	2.26	11.050	11.313	62.72	61.86	35.22	35.30
W_9	9.07	9.36	59.90	64.64	1085.34	1129.17	2.29	2.18	11.174	11.200	65.07	66.56	36.03	36.11
W_{10}	10.02	10.80	68.02	70.03	1174.21	1212.89	2.52	2.80	11.08	12.05	69.29	70.60	36.19	36.26
SE m±	0.27	0.35	2.27	3.06	20.87	23.69	0.07	0.08	0.242	0.259	0.33	0.33	0.31	0.34
CD (P = 0.05)	0.77	0.99	6.41	8.65	60.11	68.24	0.20	0.22	0.678	0.737	0.93	0.94	0.90	0.95

Where W_1 (weedy check), W_2 (1 HW 25 DAS), W_3 (2 HW 25 and 45 DAS), W_4 (fluchloralin 1 kg ha^{-1}), W_5 (fluchloralin 1.5 kg ha^{-1}), W_6 (pendimethalin 1 kg ha^{-1}), W_7 (pendimethalin 1.5 kg ha^{-1}), W_8 (fluchloralin 1 kg ha^{-1} + 1 HW 35 DAS), W_9 (pendimethalin 1 kg ha^{-1} + 1 HW 35 DAS), W_{10} (weed free).

consequence of weed control measures enabled to affect improved 100-seed weight in soybean possibly due to enhanced availability of nutrients etc. The results are akin to those reported by Vyas and Jain (2003). Severe weed competition due to unchecked weed growth and consequent reduction in seeds per pod and test weight was also observed by Rathman and Miller (1981). Seed yield is the most important criterion and the ultimate test to estimate and compare the efficiency of a particular treatment. As such, this parameter needs a thorough and comprehensive discussion here.

All weed control measures gave significantly higher seed yields than weedy check (Table 3). However, weed free treatment (W_{10}), hand weeding twice (W_3) and fluchloralin and pendimethalin 1 kg ha^{-1} integrated with hand weeding once (W_8 and W_9) procured far superior seed yields of soybean. The increase in seed yield due to these treatments on pooled basis was to the tune of 59.81, 53.25, 38.42 and 49.78%, respectively. Pendimethalin when applied alone or integrated with hand weeding was more effective than similar application of fluchloralin. Integrated use of both fluchloralin and pendimethalin with hand weeding yielded 4.72 and 8.21%, respectively more than their individual application. Higher doses (1.5 kg ha^{-1}) of herbicides proved more effective and produced superior seed yields than their lower doses (1.0 kg ha^{-1}), the increase being 6.34 and 5.98% for fluchloralin and pendimethalin, respectively. The yield given by pendimethalin 1.5 kg ha^{-1} (W_7) was comparable to that produced by hand weeding once (W_2) in both the years. The enhancement in the seed yield due to various weed control measures was because of the fact that they helped to keep the field comparatively free from weeds, thus resulting in better utilization of resources namely, nutrients, moisture, solar light etc. This consequently led to the production of more vigorous and healthy plants having more pod bearing capacity, more seed per pod and 100-seed weight. The cumulative effect of all these resulted in higher seed yields, making it amply clear that these weed control measures exerted a profound influence in curtailing the weed population and thereby reducing the weed biomass at important growth stages of the crop.

The results corroborate the findings of Vyas et al. (2000) and Pandya et al. (2005) and many others who reported enhanced soybean yield due to various weed control treatments. Weedy check produced lowest yield of soybean which was significantly inferior to different weed control treatments. Drastic yield reduction in weedy check

Table 3. Seed yield and straw yield of soybean as influenced by weed control methods.

Treatment	Seed yield (q/ha)		Pooled	Straw yield	
	2004	2005		2004	2005
W_1 (Weedy check)	15.97	15.74	15.85	30.28	30.20
W_2 (1 HW 25 DAS)	20.91	21.20	21.05	37.37	38.23
W_3 (2 HW 25 and 45 DAS)	24.10	24.44	24.29	42.42	43.31
W_4 (fluchloralin 1 kg ha^{-1})	18.58	19.00	18.76	35.88	37.35
W_5 (fluchloralin 1.5 kg ha^{-1})	19.80	20.11	19.95	36.95	38.22
W_6 (pendimethalin 1 kg ha^{-1})	19.52	20.00	19.72	35.86	35.73
W_7 (pendimethalin 1.5 kg ha^{-1})	20.73	21.07	20.90	37.00	38.09
W_8 (fluchloralin 1 kg ha^{-1} + 1 HW 35 DAS)	22.10	21.79	21.94	40.62	40.08
W_9 (pendimethalin 1 kg ha^{-1} + 1 HW 35 DAS)	23.45	24.04	23.74	41.57	42.49
W_{10} (weed free)	25.08	25.60	25.33	44.21	45.00
SE m±	0.23	0.29	0.25	0.25	0.26
CD (P = 0.05)	0.66	0.81	0.70	0.71	0.75

was due to heavy infestation of weeds, especially broad leaved weeds which grow faster and suppressed the crop growth, thus causing reduced yields. The broad leaved weeds on an average contributed 62.65% of total weed population. Howe and Oliver (1987) also reported reduced yield in weedy check due to higher density of weeds especially broad leaved weeds. The straw yield depicted a trend similar to seed yield. Significantly, superior straw yield was seen in different weed control treatment especially weed free treatment (W_{10}), hand weeding twice (W_3) and fluchloralin and pendimethalin (each 1 kg ha^{-1}) integrated with hand weeding (W_8 and W_9). Biological yield was favourably influenced by various weed control treatments. Weed free plots (W_{10}), hand weeding twice (W_3), fluchloralin and pendimethalin 1 kg/ha integrated with hand weeding (W_8 and W_9) far excelled in their influence in recording higher biological yield over weedy check and produced 51.76, 46.20, 35.12 and 43.06% more biological yield than un-weed control. Herbicides applied alone under different concentrations (W_4, W_5, W_6, W_7) too were efficient in producing higher biological yields, however, the treatments lagged behind hand weeding once (W_2) except pendimethalin 1.5 kg ha^{-1} with which it was comparable. Harvest index of soybean exhibited pronounced influence of various weed control treatments. Weed free treatment (W_{10}), hand weeding twice (W_3) and pendimethalin 1 kg ha^{-1} in integration with hand weeding once (W_9) produced statistically similar harvest index. Weedy check (W_1) affected significantly least harvest index compared to all the weed control treatments. The higher doses of both fluchloralin and pendimethalin (W_5 and W_7) proved significantly more effective than their corresponding low doses (W_4 and W_6). This was possibly due to persistence of these herbicides for longer duration at the higher concentration compared to their lower ones. Bhandiwaddar and Itnal (1998) also reported superiority of various weed control methods with respect to harvest

index of soybean over unweeded control.

Seed composition

Seed protein content of soybean was favourably and significantly influenced by different weed control treatments (Table 4). Weed free (W_{10}) and hand weeding twice (W_3) exhibited statistically similar protein content. Hand weeding twice (W_3) was also comparable to pendimethalin 1 kg/ha integrated with hand weeding once (W_9) with respect to protein content but superior to rest of the treatments besides the weedy check. Fluchloralin and pendimethalin (1 kg/ha each) supplemented with hand weeding (W_8 and W_9) were at par with each other for protein content. On an average, 14.04, 12.54 and 11.37% more protein content in seed was affected by weed free treatment, hand weeding twice and pendimethalin 1 kg ha^{-1} integrated with hand weeding once over that given by weedy check. The better protein content in soybean crop as a result of weed control measures could be attributed to better nitrogen content under these treatments favoured by effective elimination of weeds. Presence of weeds throughout the growing season in weedy check plots was instrumental in reduced protein content in these plots. The results corroborate the findings of Mohamed (2004) and EL-Metwally and Shalby (2007). So for as the oil content is concerned, hand weeding twice (W_3), higher dose of pendimethalin (1.5 kg/ha) (W_7) and both fluchloralin and pendimethalin 1 kg/ha supplemented with hand weeding (W_8 and W_9) recorded comparable oil content in soybean seed in both years. Both fluchloralin and pendimethalin at either of two rates namely, 1.0 and 1.5 kg ha^{-1} gave oil content that was at par with hand weeding once. Although, hand weeding twice was statistically similar to hand weeding once in 2004 with respect to oil content, it produced a significant improvement in this important quality

Table 4. Oil content (%) and crude protein content (%) of soybean as influenced by weed control.

Treatment	Oil content (%)		Crude protein content (%)	
	2004	2005	2004	2005
W_1 (Weedy check)	17.81	18.05	36.57	35.68
W_2 (1 HW 25 DAS)	19.07	19.00	38.19	38.66
W_3 (2 HW 25 and 45 DAS)	19.97	19.88	39.94	41.36
W_4 (fluchloralin 1 kg ha^{-1})	18.67	19.00	38.14	37.46
W_5 (fluchloralin 1.5 kg ha^{-1})	18.93	19.03	38.30	37.50
W_6 (pendimethalin 1 kg ha^{-1})	18.98	18.84	38.24	38.12
W_7 (pendimethalin 1.5 kg ha^{-1})	19.09	19.28	38.55	39.12
W_8 (fluchloralin 1 kg ha^{-1} + 1 HW 35 DAS)	19.28	19.48	39.09	40.00
W_9 (pendimethalin 1 kg ha^{-1} + 1 HW 35 DAS)	19.35	19.73	39.78	40.68
W_{10} (weed free)	19.69	20.19	40.62	41.76
SE m±	0.45	0.36	0.42	0.60
CD (P = 0.05)	0.92	0.72	0.84	1.22

parameter in 2005.

Weedy check proved very poor exhibiting significantly inferior values of oil content in soybean seed which was 10.82 and 9.12% deficit as compared to hand weeding twice in 2004 and 2005, respectively. Enhancement in the oil content of soybean as affected by various weed control measures may be attributed to better nutrition of the soybean which play a vital role in improving oil value of soybean. Increased oil content in soybean under weed control treatments has also been reported by Mohamed (2004) and EL-Metwally and Shalby (2007).

Conclusion

Pendimethalin 1.0 kg ha^{-1} integrated with one hand weeding at 35 DAS (critical period of weed removal) is the most appropriate method for effective weed management and profitable cultivation of soybean. Other methods are either less profit earners or are labour expensive.

REFERENCES

Abd El-Razik MA (2006). Effect of some weed control treatments on growth, yield and yield components and some seed technological characters and associated weeds of faba bean plants. J. Agric. Sci. 31(10):6283-6292.

Ahmed SA, Saad El-Din SA, El- Metwally IM (2001). Influence of some micro- elements and some weed control treatments on growth, yield and its components of soybean plants. Annals Agric. Sci. Moshtohor, 39(2):805-823.

BANDIWADDAR TT, ITNAL CJ (1998). Weed management in soybean on black soils of Northern Transitinal Track of Karnataka. Karnataka, Journal Agricultural Sciences.11:599-602. Dobrzanski A, Paczynski J, Anyszka Z (2001). The response of onion and weeds to oxadiargyl (Raft 400 Sc.) Progr. Plant Protect. 41(2):901-903.

El-Metwally IM, Saad El-Din SA (2003). Response of pea (Pisum sativum L.) plants to some weed control treatments. J. Agric. Sci.

28(2):947-969.

El-Metwally IM, Shalby EM (2007). Bio-Remediation of fluazifop-p-butyl herbicidecontaminated soil with special reference to efficacy of some weed control treatments in faba bean plants. Res. J. Agric. Bio. Sci. 3(3):157-165.

Galal AH (2003). Effect of weed control treatments and hill spacing on soybean and associated weeds. Assiut J. Agric. Sci. 34(1):15-32.

Gogoi AK, Kalita H, Pathak AK (1991). Integrated weed management in soybean [Glycine max. (L.) Merrill]. Indian J. Agron. 36(3):453-454.

Hassanein EE (2000). Effect of some weed control treatments on soybean and associated weeds. Egypt. J. Agric Res. 78(5):1979-1993.

Howe OWI, Oliver LR (1987). Influence of soybean (Glycine max) row spacing on pitted morning glory (Ipomea lacunose) interferences. Weed Sci. 35:185-193.

Jain VK (2000). Chemical weed control in soybean (Glycine max). Indian J. Agron. 45(1):153-157.

Jeyabal A, Palaniapan SP, Chelliah S (2001). Efficacy of metribuzin and trifulralin on weed management in soybean (Glycine max). Indian J. Agron. 46(2): 339-342.

Kachroo D, Dixit AK, Bali AS (2003). Weed management in oilseed crop - a review. J. Res. SKUAST – Jammu, 2(1):1-12.

Kanase AA, Mendhe SN, Khawale VS, Jarande NN, Mendhe JT (2006). Effect of integrated nutrient management and weed biomass addition on growth and yield of soybean. J. Soils Crops. 16(1):236-239.

Kushwah SS, Vyas MD (2005). Herbicides weed control in soybean (Glycine max). Indian J. Agron. 50(3):225-227.

Mohamed SA (2004). Effect of basagran herbicide and indole acetic acid (IAA) on growth, yield, chemical composition and associated weeds of soybean plants. Egypt. J. Appl. Sci. 19(10):79-91.

Pandya N, Chouhan GS, Nepalia V (2005). Effect of varieties, crop geometries and weed management on nutrient up take by soybean (Glycine max) and associated weeds. Indian J. Agron. 50(3):218-220.

Rakesh KS, Shirvastava UK (2002). Weed control in soybean (Glycine max). Indian J. Agron. 47(2):269-272.

Rathman DP, Miller SD (1981). Wild oat (Avena fatua) competition in soyben (Glycine max). Weed Sci. 29:410-414.

Sha HZ (2004). Test on the efficacy of 40% emulsifiable concentrate of prometryn and acetochlor against soybean weeds. J. Jilin Agric. Univ. 26(4):452-454.

Singh G, Jolly RS (2004). Effect of herbicides on the weed infestation and grain yield of soybean (Glycine max). Acta Agron., Hungarica 52(2):199-203.

Thakare KG, Chore CN, Deotale RD, Kamble PS, Sujata BP, Shradha RL (2006). Influence of nutrients and hormones on biochemical and yield and yield contributing parameters of soybean. J. Soils Crops 16(1):210-216.

Veeramani A, Palchamy A, Ramasamy S, Rangaraju G (2001). Integrated weed management in soybean [*Glycine max*. (L.) Merrill] under various plant densities. Madras Agric. J. 88(7-9):451-456.

Vyas MD, Jain AK (2003). Effect of pre-and post-emergence herbicides on weed control and productivity of soybean (*Glycine max*). Indian J. Agron. 48(4):309-311.

Vyas MD, Singh SS, Singh PP (2000). Weed management in soybean (*Glycine max*) Merill. Ann. Plant Protec. Sci. 8(1):76-78.

Zablotowicz RM, Reddy KN (2007). Nitrogenase activity, nitrogen content, and yield responses to glyphosate in glyphosate-resistant soybean. Crop Protec. 26(1):37.

Zawoznik MS, Benavides MP, Tomaro ML (1995). Effect of herbicide diuron on growth and symbiotic behavior of Rhizobium and Bradyrhizobium species. Eur. J. Soil Biol. 31:183-188.

Permissions

All chapters in this book were first published in AJAR, by Academic Journals; hereby published with permission under the Creative Commons Attribution License or equivalent. Every chapter published in this book has been scrutinized by our experts. Their significance has been extensively debated. The topics covered herein carry significant findings which will fuel the growth of the discipline. They may even be implemented as practical applications or may be referred to as a beginning point for another development.

The contributors of this book come from diverse backgrounds, making this book a truly international effort. This book will bring forth new frontiers with its revolutionizing research information and detailed analysis of the nascent developments around the world.

We would like to thank all the contributing authors for lending their expertise to make the book truly unique. They have played a crucial role in the development of this book. Without their invaluable contributions this book wouldn't have been possible. They have made vital efforts to compile up to date information on the varied aspects of this subject to make this book a valuable addition to the collection of many professionals and students.

This book was conceptualized with the vision of imparting up-to-date information and advanced data in this field. To ensure the same, a matchless editorial board was set up. Every individual on the board went through rigorous rounds of assessment to prove their worth. After which they invested a large part of their time researching and compiling the most relevant data for our readers.

The editorial board has been involved in producing this book since its inception. They have spent rigorous hours researching and exploring the diverse topics which have resulted in the successful publishing of this book. They have passed on their knowledge of decades through this book. To expedite this challenging task, the publisher supported the team at every step. A small team of assistant editors was also appointed to further simplify the editing procedure and attain best results for the readers.

Apart from the editorial board, the designing team has also invested a significant amount of their time in understanding the subject and creating the most relevant covers. They scrutinized every image to scout for the most suitable representation of the subject and create an appropriate cover for the book.

The publishing team has been an ardent support to the editorial, designing and production team. Their endless efforts to recruit the best for this project, has resulted in the accomplishment of this book. They are a veteran in the field of academics and their pool of knowledge is as vast as their experience in printing. Their expertise and guidance has proved useful at every step. Their uncompromising quality standards have made this book an exceptional effort. Their encouragement from time to time has been an inspiration for everyone.

The publisher and the editorial board hope that this book will prove to be a valuable piece of knowledge for researchers, students, practitioners and scholars across the globe.

List of Contributors

Carita Liberato do Amaral
Programa de Pós-Graduação em Agronomia (Produção Vegetal) – Departamento de Biologia Aplicada à Agropecuária, FCAV-UNESP, Jaboticabal-SP, Brazil

Marcelo Claro de Souza
Programa de Pós-Graduação em Biologia Vegetal – Departamento de Botânica, IB-UNESP, Rio Claro-SP, Brazil

Guilherme Bacarim Pavan
Bolsista de Iniciação Científica - Departamento de Biologia Aplicada à Agropecuária, FCAV-UNESP, Jaboticabal-SP, Brazil

Marina Alves Gavassi
Bolsista de Iniciação Científica - Departamento de Biologia Aplicada à Agropecuária, FCAV-UNESP, Jaboticabal-SP, Brazil

Pedro Luis da Costa Aguiar Alves
FCAV-UNESP - Departamento de Biologia Aplicada à Agropecuária, Jaboticabal-SP, Brazil

Daniel Valadão Silva
Universidade Federal dos Vales do Jequitinhonha e Mucuri – UFVJM, Bahia, Brazil

José Barbosa Santos
Programa em Pós Graduação em Produção Vegetal – UFVJM, Bahia, Brazil

João Pedro Cury
Universidade Federal dos Vales do Jequitinhonha e Mucuri – UFVJM, Bahia, Brazil

Felipe Paolinelli Carvalho
Universidade Federal dos Vales do Jequitinhonha e Mucuri – UFVJM, Bahia, Brazil

Enilson Barros Silva
Programa em Pós Graduação em Produção Vegetal – UFVJM, Bahia, Brazil

José Sebastião Cunha Fernandes
Programa em Pós Graduação em Produção Vegetal – UFVJM, Bahia, Brazil

Evander Alves Ferreira
Programa em Pós Graduação em Produção Vegetal – UFVJM, Bahia, Brazil

Germani Concenço
Empresa Brasileira de Pesquisa Agropecuária-Agropecuária Oeste, Brazil

Asma Chbani
Doctoral School for Sciences and Technology, Azm Centre for Research in Biotechnology and its application, Lebanese University, El Miten Street, Tripoli, Lebanon

Hiba Mawlawi
Doctoral School for Sciences and Technology, Azm Centre for Research in Biotechnology and its application, Lebanese University, El Miten Street, Tripoli, Lebanon

Laurence Zaouk
Doctoral School for Sciences and Technology, Azm Centre for Research in Biotechnology and its application, Lebanese University, El Miten Street, Tripoli, Lebanon

A. O. Eni
Department of Biological Sciences, College of Science and Technology, Covenant University, KM 10 Idiroko Road, Canaanland, P. M. B. 1023 Ota, Ogun State, Nigeria

D. K. Fasasi

Department of Biological Sciences, College of Science and Technology, Covenant University, KM 10 Idiroko Road, Canaanland, P. M. B. 1023 Ota, Ogun State, Nigeria

J. A. Orluchukwu
Department of Crop and Soil Science, University of Port Harcourt, P. M. B 5323 Port Harcourt, Nigeria

Udensi E. Udensi
Department of Crop and Soil Science, University of Port Harcourt, P. M. B 5323 Port Harcourt, Nigeria
International Institute of Tropical Agriculture (IITA) Cassava Project, South-South and South Eastern Nigeria Zones, Nigeria

Eda Aksoy
Biological Control Research Station, Adana-Turkey

Z. Filiz Arslan
Biological Control Research Station, Adana-Turkey

Naim Öztürk
Biological Control Research Station, Adana-Turkey

Biswajit Pramanick
Department of Agronomy, Bidhan Chandra Krishi Viswavidyalaya, Mohanpur, Nadia, West Bengal, 741252, India

Koushik Brahmachari
Department of Agronomy, Bidhan Chandra Krishi Viswavidyalaya, Mohanpur, Nadia, West Bengal, 741252, India

Arup Ghosh
Central Salt and Marine Chemicals Research Institution (Council of Scientific and Industrial Research), G. B. Marg, Bhabnagar, Gujrat, 364002, India

José M. G. CALADO
Department of Crop Science, University of Évora, Apartado 94, 7002-554 Évora, Portugal
Institute of Mediterranean Agricultural and Environmental Sciences (ICAAM), Apartado 94, 7002-554 Évora, Portugal

Gottlieb BASCH
Department of Crop Science, University of Évora, Apartado 94, 7002-554 Évora, Portugal
Institute of Mediterranean Agricultural and Environmental Sciences (ICAAM), Apartado 94, 7002-554 Évora, Portugal

José F. C. BARROS
Department of Crop Science, University of Évora, Apartado 94, 7002-554 Évora, Portugal
Institute of Mediterranean Agricultural and Environmental Sciences (ICAAM), Apartado 94, 7002-554 Évora, Portugal

Mário de CARVALHO
Department of Crop Science, University of Évora, Apartado 94, 7002-554 Évora, Portugal
Institute of Mediterranean Agricultural and Environmental Sciences (ICAAM), Apartado 94, 7002-554 Évora, Portugal

Zelalem Bekeko
Department of Plant Sciences, Haramaya University Chiro Campus, P. O. Box 335, Chiro, Ethiopia

D. Singh
Centre of Advanced Studies in Plant Pathology, G.B. Pant University of Agriculture and Technology, Pantnagar-263145, Udham Singh Nagar (Uttarakhand), India

A. Kumar
Department of Plant Pathology, IARI, Regional Station, Pusa-848125 Samastipur (Bihar), India

A. K. Singh
Department of Agronomy, ICAR Research Complex for Eastern Region Patna-800 014, Bihar, India

C. R. Prajapati
Department of Plant Protection, Krishi Vigyan Kendra, Baghpat, S.V.P.University of Agriculture and Technology, Meerut- 250 609, UP, India

H. S. Tripathi
Centre of Advanced Studies in Plant Pathology, G.B. Pant University of Agriculture and Technology, Pantnagar-263145, Udham Singh Nagar (Uttarakhand), India

Vijaymahantesh
Department of Agronomy, University of Agricultural Sciences, Bangalore, Karnataka, India

H. V. Nanjappa
Department of Agronomy, University of Agricultural Sciences, Bangalore, Karnataka, India

B. K. Ramachandrappa
Department of Agronomy, University of Agricultural Sciences, Bangalore, Karnataka, India

M. V. N. L. Chaitanya
Department of Pharmacognosy and Phytopharmacy, JSS College of Pharmacy (A constituent College of JSS University, Mysore), Rock lands, Ootacamund-643001, Tamilnadu, India

S. P. Dhanabal
Department of Pharmacognosy and Phytopharmacy, JSS College of Pharmacy (A constituent College of JSS University, Mysore), Rock lands, Ootacamund-643001, Tamilnadu, India

Rajendran
Department of Pharmacognosy and Phytopharmacy, JSS College of Pharmacy (A constituent College of JSS University, Mysore), Rock lands, Ootacamund-643001, Tamilnadu, India

S. Rajan
Survey of Medicinal plants and collection Unit, Indira Nagar, Emerald – 643209, India

Regine Cristina Urcoviche
Biotechnology Applied to Agriculture, Paranaense University – UNIPAR, Umuarama, Paraná, Brazil

Murilo Castelli
Biotechnology Applied to Agriculture, Paranaense University – UNIPAR, Umuarama, Paraná, Brazil

Régio Márcio Toesca Gimenes
Paranaense University – UNIPAR, Umuarama, Paraná, Brazil

Odair Alberton
Biotechnology Applied to Agriculture, Paranaense University – UNIPAR, Umuarama, Paraná, Brazil
Paranaense University – UNIPAR, Umuarama, Paraná, Brazil

Amadou Touré
Africa Rice Center (AfricaRice), 01 BP 2031 Cotonou, Benin

Jean Mianikpo Sogbedji
University of Lomé, BP 1515, Lomé, Togo

Yawovi Mawuena Dieudonné Gumedzoé
University of Lomé, BP 1515, Lomé, Togo

Fateme Nuraky
Shoushtar Branch, Islamic Azad university, Shoushtar, Iran

Hassan Rahmany
Payame Noor University of Khuzestan, Iran

Talemos Seta
Department of Biology, Dilla University, Ethiopia

Abreham Assefa
Department of Biology, Dilla University, Ethiopia

Fisseha Mesfin
Department of Biology, Dilla University, Ethiopia

Alemayehu Balcha
Department of Biology, Dilla University, Ethiopia

M. F. Baqual
Sher-e-Kashmir University of Agricultural, Sciences and Technology of Kashmir (K), Division of Sericulture, Mirgund, P. Box 674, G.P.O., Srinagar – 190001 (J & K), India

Pedro José da Silva Júnior
Instituto Federal de Educação, Ciência e Tecnologia de Pernambuco (IFPE), Campus Belo Jardim, PE, Brasil

Francisco de Assis Cardoso Almeida
Departamento de Engenharia Agrícola, Universidade Federal de Campina Grande (UFCG), Campus I, Campina Grande, PB, Brasil

Francisco Braga da Paz Júnior
Instituto Federal de Educação, Ciência e Tecnologia de Pernambuco (IFPE), Campus Recife, PE, Brasil

Mucksood Ahmad Ganaie
Section of Plant Pathology and Nematology, Department of Botany, Aligarh Muslim University, Aligarh, India

Tabreiz Ahmad Khan
Section of Plant Pathology and Nematology, Department of Botany, Aligarh Muslim University, Aligarh, India

D. Martins
São Paulo State University, Botucatu/SP, Brazil

S. R. Marchi
Federal University of Mato Grosso, Barra do Garça/MT, Brazil

N. V. Costa
University of Paraná Weast, Marechal Cândido Rondon/PR, Brazil

Sivagamy Kannan
Department of Agronomy, Tamil Nadu Agricultural University, Coimbatore, India

Chinnusamy Chinnagounder
Department of Agronomy, Tamil Nadu Agricultural University, Coimbatore, India

Marek Gugała
Department of Plant Cultivation, Faculty of Life Sciences, University of Natural Sciences and Humanities in Siedlce, Siedlce, Poland

Krystyna Zarzecka
Department of Plant Cultivation, Faculty of Life Sciences, University of Natural Sciences and Humanities in Siedlce, Siedlce, Poland

Albert O. Amosu
Department of Biodiversity and Conservation Biology, Faculty of Natural Sciences, University of the Western Cape, Private Bag X17, Bellville 7535, South Africa

Deborah V. Robertson-Andersson
Department of Biodiversity and Conservation Biology, Faculty of Natural Sciences, University of the Western Cape, Private Bag X17, Bellville 7535, South Africa

Gavin W. Maneveldt
Department of Biodiversity and Conservation Biology, Faculty of Natural Sciences, University of the Western Cape, Private Bag X17, Bellville 7535, South Africa

Robert J. Anderson
Fisheries Research, Department of Agriculture, Forestry and Fisheries, Private Bag X2, Roggebaai 8012, South Africa

John J. Bolton
Department of Biological Sciences, Faculty of Science, University of Cape Town, Cape Town. Private Bag X3, Rondebosch 7701, South Africa

F. M. Oloyede
Department of Crop Production and Protection, Obafemi Awolowo University, Ile-Ife, Nigeria

G. O. Agbaje
Department of Crop Production and Protection, Obafemi Awolowo University, Ile-Ife, Nigeria

I. O. Obisesan
Department of Crop Production and Protection, Obafemi Awolowo University, Ile-Ife, Nigeria

Ademir de Oliveira Ferreira
Department of Soil Science, Federal University of Santa Maria (UFSM), Avenida Roraima, 100, 97105-900, Santa Maria, Rio Grande do Sul, Brazil

Claúdio Purissimo
Departament of Crop Science and Plant Health, University of Ponta Grossa (UEPG), Avenida Carlos Cavalcanti, 4748, Campus Uvaranas, 84030-960, Ponta Grossa, Paraná, Brazil

Rafael Pivotto Bortolotto
Department of Soil Science, Federal University of Santa Maria (UFSM), Avenida Roraima, 100, 97105-900, Santa Maria, Rio Grande do Sul, Brazil

Telmo Jorge Carneiro Amado
Department of Soil Science, Federal University of Santa Maria (UFSM), Avenida Roraima, 100, 97105-900, Santa Maria, Rio Grande do Sul, Brazil

Klaus Reichardt
Center for Nuclear Energy in Agriculture, University de São Paulo (CENA/USP), Avenida Centenario, 303, bairro São Dimas, caixa postal 96, 13400-970, Piracicaba, São Paulo, Brazil

Larissa Bavoso
University of Ponta Grossa (UEPG), Avenida Carlos Cavalcanti, 4748, Campus Uvaranas, 84030-960, Ponta Grossa, Paraná, Brazil

Mariânne Graziele Kugler
University of Ponta Grossa (UEPG), Avenida Carlos Cavalcanti, 4748, Campus Uvaranas, 84030-960, Ponta Grossa, Paraná, Brazil

F. A. Peer
Division of Agronomy, S. K. University of Agricultural Sciences and Technology of Kashmir, India

Badrul Hassan
Division of Agronomy, S. K. University of Agricultural Sciences and Technology of Kashmir, India

B. A. Lone
Division of Agronomy, S. K. University of Agricultural Sciences and Technology of Kashmir, India

Sameera Qayoom
Division of Agronomy, S. K. University of Agricultural Sciences and Technology of Kashmir, India

Latief Ahmad
Division of Agronomy, S. K. University of Agricultural Sciences and Technology of Kashmir, India

B. A. Khanday
Division of Agronomy, S. K. University of Agricultural Sciences and Technology of Kashmir, India

Purshotam S singh
Division of Agronomy, S. K. University of Agricultural Sciences and Technology of Kashmir, India

Gurdeep Singh
Division of Agronomy, S. K. University of Agricultural Sciences and Technology of Kashmir, India